Cause and Correlation in Biology

A User's Guide to Path Analysis, Structural Equations
and Causal Inference with R

Second Edition

Many problems in biology require an understanding of the relationships among variables in a multivariate causal context. Exploring such cause–effect relationships through a series of statistical methods, this book explains how to test causal hypotheses when randomised experiments cannot be performed.

This completely revised and updated edition features detailed explanations for carrying out statistical methods using the popular, and freely available, R statistical language. Sections on d-sep tests, latent constructs that are common in biology, missing values, phylogenetic constraints and multilevel models are also an important feature of this new edition.

Written for biologists and using a minimum of statistical jargon, the concept of testing multivariate causal hypotheses using structural equations and path analysis is demystified. Assuming only a basic understanding of statistical analysis, this new edition is a valuable resource for students and practising biologists alike.

Bill Shipley is a Professor in the Department of Biology at Université de Sherbrooke, Canada. His research interests centre upon plant ecophysiology, functional and community ecology and statistical modelling. He is the author of *From Plant Traits to Vegetation Structure: Chance and Selection in the Assembly of Ecological Communities*, published by Cambridge University Press.

Cause and Correlation in Biology

A User's Guide to Path Analysis,
Structural Equations and
Causal Inference with R

Second Edition

BILL SHIPLEY
Université de Sherbrooke, Canada

CAMBRIDGE
UNIVERSITY PRESS

CAMBRIDGE
UNIVERSITY PRESS

University Printing House, Cambridge CB2 8BS, United Kingdom

One Liberty Plaza, 20th Floor, New York, NY 10006, USA

477 Williamstown Road, Port Melbourne, VIC 3207, Australia

314-321, 3rd Floor, Plot 3, Splendor Forum, Jasola District Centre, New Delhi - 110025, India

79 Anson Road, #06-04/06, Singapore 079906

Cambridge University Press is part of the University of Cambridge.

It furthers the University's mission by disseminating knowledge in the pursuit of education, learning and research at the highest international levels of excellence.

www.cambridge.org
Information on this title: www.cambridge.org/9781107442597

© Cambridge University Press 2016

First published 2000
Second edition 2016

A catalogue record for this publication is available from the British Library

ISBN 978-1-107-44259-7 Paperback

À ma petite Rhinanthe toujours aussi belle, à David et à Élyse.

Contents

Preface

This book describes a series of statistical methods for testing causal hypotheses using observational data – but it is not a statistics book. It describes a series of algorithms, derived from research in artificial intelligence (AI), that can discover causal relationships from observational data – but it is not a book about artificial intelligence. It describes the logical and philosophical relationships between causality and probability distributions – but it is certainly not a book about the philosophy of statistics. Rather, it is a *user's guide*, written for biologists, whose purpose is to allow the practising biologist to make use of these important new developments when causal questions cannot be answered with randomised experiments.

I have written the book assuming that you have no previous training in these methods. If you have taken an introductory statistics course – even if it was longer ago than you want to acknowledge – and have managed to hold on to some of the basic notions of sampling and hypothesis testing using statistics then you should be able to understand the material in this book. I recommend that you read each chapter through in its entirety even if you do not feel that you have mastered all the notions. This will at least give you a general feeling for the goals and vocabulary of each chapter. You can then go back and pay closer attention to the details.

The book is addressed to biologists, mostly because I am a practising biologist myself, but I hope that it will also be of interest to statisticians, scientists in other fields and even philosophers of science. I have not written the book as a textbook simply because the discipline to which the material in this book naturally belongs does not yet exist. Whatever the name eventually given to this new discipline, I firmly believe that it will exist, and be generally recognised as a distinct discipline, in the future. The questions that this new discipline addresses, and the elegance of its results, are too important for this not to be the case. Nonetheless, the chapters follow a logical progression that would be well suited to an upper-level undergraduate, or graduate, course. I have used the manuscript of this book for such a purpose, and every one of my students is still alive.

It is a pleasure and an honour to acknowledge the many people who have contributed to this project. First, Jim and Marg Shipley started everything. Robert van Hulst supplied much of the initial impulse through our conversations about science and causality while I was still an undergraduate. He has also read every one of the manuscript chapters and suggested many useful changes. Paul Keddy kept my interest burning during my PhD studies and also commented on the first two chapters. As usual, his comments went to the heart of the matter.

The late Robert Peters had a large impact on my thoughts about causality and even convinced me, for a number of years, that ecologists are best to give up on the concept – not because he viewed the notion of causality as meaningless (he never believed this, despite his empiricist reputation) but because it was simply too slippery a notion to demonstrate without randomised experiments. His constant prodding must have caused me to stop while wandering through the library one day when, almost subconsciously, I saw a book with the following provocative title: *Discovering Causal Structure: Artificial Intelligence, Philosophy of Science, and Statistical Modeling* (Glymour et al. 1987). That book was my introduction to a more sophisticated understanding of causality. Rob Peters was much too young when he passed away, and I am sorry that he never got to read the book that you are about to begin. I am not sure that he would have approved of everything in it but I know that he would have appreciated the effort.

Martin Lechowicz introduced me to the notion of path analysis at a time when this method had been mostly forgotten by biologists. He and I have collaborated for a number of years on this topic, and he read the entire manuscript of this book, providing many insightful comments. Steve Coté and Jim Grace also read parts of this book. Jim, in particular, provided some important counterpoint to my thoughts on latent variable models. Marco Festa-Bianchet provided the unpublished data that are reported in Chapter 5. I must also acknowledge my graduate students, Margaret McKenna, Driss Meziane, Jarceline Almeida-Cortez, Luc St-Pierre and Muhaymina Sari, as well as the many members of the SEMNET Internet discussion group.

Finally, I want to thank Judea Pearl for kindly responding to my many e-mails about d-separation and basis sets and to Clark Glymour, Richard Scheines and Peter Spirtes of Carnegie Mellon University for their generosity in extending an invitation to visit with them and for patiently answering my many questions about their discovery algorithms. Clark Glymour read and commented on some of the manuscript chapters.

I hope that you find this book to be useful, interesting and readable. I welcome your comments and feedback – especially if you don't agree with me.

Sherbrooke, 1999

Preface to the second edition

I had two motives, one positive and one more selfish, in writing the first edition of this book. The positive motive was to provide a detailed introduction of these methods to practising biologists, since they were largely unknown to students and researchers in this discipline. The more selfish motive was to provide a detailed *justification* of these methods to practising biologists. You see, I was frustrated. My research manuscripts that included these methods were being rejected by reviewers, who viewed the analyses as the statistical equivalents of conjurer's tricks. I concluded that a book-length explanation that was written specifically for biologists would provide such a justification. Now, writing fifteen years later, the situation is quite different. These methods have been increasingly adopted by biologists working in ecology, evolution, genetics and molecular biology. I hope that the first edition of this book, as well as Jim Grace's (2006) very fine book, have contributed to this change. Virtually every chapter has been updated in this second edition. These changes include, inter alia, new additions to the d-sep test, the inclusion of phylogenetic information and an expanded treatment of latent variables. The most extensive change is the detailed explanation for implementing these methods using the R programming language. The only computer programs for structural equation modelling that were available when I wrote the first edition were commercial ones. Since I didn't want to become a salesman for any particular commercial package, I didn't include the actual code and steps for carrying out the analyses. However, a 'user's guide' that omits such vital information is clearly lacking. Now that the freely available R program has become so ubiquitous for statistical analysis by biologists, and now that the methods in this book have been included in several R libraries, I have included detailed instructions in this second edition for carrying out the analyses.

Sherbrooke, 2014

1 Preliminaries

1.1 The shadow's cause

The *Wayang Kulit* is an ancient theatrical art, practised in Malaysia and throughout much of the Orient. The stories are often about battles between good and evil, as told in the great Hindu epics. What the audience actually see are not actors, nor even puppets, but, instead, the shadows of puppets projected onto a canvas screen. Behind the screen is a light. The puppet master creates the action by manipulating the puppets and props so that they will intercept the light and cast shadows. As these shadows dance across the screen the audience must deduce the story from these two-dimensional projections of the hidden three-dimensional objects. However, shadows can be ambiguous. In order to imply the three-dimensional action, the shadows must be detailed, with sharp contours, and they must be placed in context.

Biologists are unwitting participants in nature's shadow play. These shadows are cast when the causal processes in nature are intercepted by our measurements. Like the audience at the *Wayang Kulit*, the biologist cannot simply peek behind the screen and directly observe the actual causal processes. All that can be directly observed are the consequences of these processes in the form of complicated patterns of association and independence in the data. As with shadows, these correlational patterns are incomplete – and potentially ambiguous – projections of the original causal processes. As with shadows, we can infer much about the underlying causal processes if we can learn to study their details and sharpen their contours, especially if we can study them in context.

Unfortunately, unlike the puppet master in a *Wayang Kulit*, who takes care to cast informative shadows, nature is indifferent to the correlational shadows that it casts. This is the main reason why researchers go to such extraordinary lengths to randomise treatment allocations and to control variables. These methods, when they can be properly done, simplify the correlational shadows to manageable patterns that can be more easily mapped onto the underlying causal processes.

It is uncomfortably true, though rarely admitted in statistics texts, that many important areas of science are stubbornly impervious to experimental designs based on the randomisation of treatments to experimental units. Historically, the response to this embarrassing problem has been either to ignore it or to banish the very notion of causality from the language and to claim that the shadows dancing on the screen are all that exists. Ignoring a problem doesn't make it go away, and defining a problem out of existence doesn't make it so. We need to know what we can safely infer about causes from their observational shadows, what we can't infer and the degree of ambiguity that remains.

I wrote this book to introduce biologists to some very recent, and intellectually elegant, methods that help in the difficult task of inferring causes from observational data. Some of these methods, such as structural equation modelling (SEM), are well known to researchers in other fields, though largely unknown to biologists. Other methods, such as those based on causal graphs, are unknown to almost everyone but a small community of researchers. These methods help both to test pre-specified causal hypotheses and to help discover potentially useful hypotheses concerning causal structures.

This book has three objectives. First, it was written to convince biologists that inferring causes without randomised experiments is possible. If you are a typical reader then you are already more than a little sceptical. For this reason I devote the first two chapters to explaining why these methods are justified. The second objective is to produce a user's guide, devoid of as much jargon as possible, that explains how to use and interpret these methods. In the service of this second objective I will explain, when appropriate, how to do this using the open source statistical program R.[1] The third objective is to exemplify these methods using biological examples, taken mostly from my own research and from that of my students. Since I am an organismal biologist whose research deals primarily with plant physiological ecology, most of the examples will be from this area, but the extensions to other fields of biology should be obvious.

I came to these ideas unwillingly. In fact, I find myself in the embarrassing position of having claimed publicly that inferring causes without randomisation and experimental control is probably impossible and, if possible, is not to be recommended (Shipley and Peters 1990). I expressed such an opinion in the context of determining how the different traits of an organism interact as a causal system. I will return to this theme repeatedly in this book, because it is so basic to biology,[2] and yet it is completely unamenable to the one method that most modern biologists and statisticians would accept as providing convincing evidence of a causal relationship: the randomised experiment. However, even as I advanced the arguments in 1990, I was dissatisfied with the consequences that such arguments entailed. I was also uncomfortably aware of the logical weakness of such arguments; the fact that I did not know of any provably correct way of inferring causation without the randomised experiment did not mean that such a method cannot exist. In my defence, and beyond the folly of youth, I could point out that I was saying nothing original; such an opinion was (and still is) the position of most statisticians and biologists. This view is summed up in the mantra that is learned by almost every student who has ever taken an elementary course in statistics: *correlation does not imply causation.*

In fact, with few exceptions,[3] correlation does imply causation. If we observe a systematic relationship between two variables, and we have ruled out the likelihood that

---

[1] See www.r-project.org.

[2] This is also the problem that inspired Sewall Wright, one the most influential evolutionary biologists of the twentieth century, the inventor of path analysis and the intellectual grandparent of the methods described in this book. The history of path analysis is explored in more detail in Chapter 3.

[3] It could be argued that variables that covary only because they are time-ordered have no causal basis.

this is simply due to a random coincidence, then *something* must be causing this relationship. When the audience at a Malay shadow theatre see a solid round shadow on the screen they know that some three-dimensional object has cast it, though they may not know if the object is a ball or a rice bowl in profile. A more accurate sound bite for introductory statistics would be that a simple correlation implies an *unresolved* causal structure, since we cannot know which is the cause and which is the effect, or if both are common effects of other unmeasured variables.

Although correlation implies an unresolved causal structure the reverse is not true: causation implies a completely resolved correlational structure. By this I mean that, once a causal structure has been proposed, the complete pattern of correlation and partial correlation is unambiguously fixed. This point is developed more precisely in Chapter 2, but it is so central to this book that it deserves repetition: the causal relationships between objects or variables determine the correlational relationships between them. Just as the shape of an object fixes the shape of its shadow, the patterns of direct and indirect causation fix the correlational 'shadows' that we see in observational data. The causal processes generating our observed data impose constraints on the patterns of correlation that such data display. This is the central insight underlying the methods described in this book.

The term 'correlation' evokes the notion of a probabilistic association between random variables. One reason why statisticians rarely speak of causation, except to distance themselves from it, is that there did not exist, until very recently, any rigorous translation between the language of causality (however defined) and the language of probability distributions (Pearl 1988). It is therefore necessary to link causation to probability distributions in a very precise way. Such rigorous links are now being forged. It is now possible to give mathematical proofs that specify the correlational pattern that must exist given a causal structure. These proofs also allow us to specify the class of causal structures that must include the causal structure that generates a given correlational pattern. The methods described in this book are justified by these proofs. Since my objective is to describe these methods and show how they can help biologists in practical applications, I won't present these proofs but will direct the interested reader to the relevant primary literature as each proof is needed.

Another reason why some prefer to speak of associations rather than causes is perhaps that causation is seen as a metaphysical notion that is best left to philosophers. In fact, even philosophers of science cannot agree on what constitutes a 'cause'. I have no formal training in the philosophy of science and am neither able nor inclined to advance such a debate. This is not to say that philosophers of science have nothing useful to contribute. When it is directly relevant I will outline the development of philosophical investigations into the notion of 'causality' and place these ideas into the context of the methods that I will describe. However, I won't insist on any formal definition of 'cause', and will even admit that I have never seen anything in the life sciences that resembles the 'necessary and sufficient' conditions for causation that are so beloved of logicians.

You probably already have your own intuitive understanding of the term 'cause'. I won't take it away from you, though I hope it will be more refined after reading this book. When I first came across the idea that one can study causes without defining them,

I almost stopped reading the book (Spirtes, Glymour and Scheines 1993). I can advance three reasons why you should not follow through on this same impulse. First, and most important, the methods described here are not logically dependent on any particular definition of causality. The most basic assumption that these methods require is that causal relationships exist in relation to the phenomena that are studied by biologists.[4]

The second reason why you should continue reading even if you are sceptical is more practical and, admittedly, rhetorical: scientists commonly deal with notions whose meaning is somewhat ambiguous. Biologists are even more promiscuous than most with one notion that can still raise the blood pressure of philosophers and statisticians. This notion is 'probability', for which there are frequentist, objective Bayesian and subjective Bayesian definitions. In the 1920s von Mises is reported to have said: 'Today, probability theory is not a mathematical science' (Rao 1984). Mayo (1996) gives the following description of the present degree of consensus concerning the meaning of 'probability': 'Not only was there the controversy raging between the Bayesians and the error [i.e. frequentist] statisticians, but philosophers of statistics of all stripes were full of criticisms of Neyman–Pearson error [i.e. frequentist-based] statistics.' Needless to say, the fact that those best in a position to produce a definition of 'probability' cannot agree on one does not prevent biologists from effectively using probabilities, significance levels, confidence intervals and the other paraphernalia of modern statistics.[5] In fact, insisting on such an agreement would mean that modern statistics could not even have begun.

The third reason why you should continue reading, even if you are sceptical, is eminently practical. Although the randomised experiment is inferentially superior to the methods described in this book when randomisation can be properly applied, it cannot be properly applied to many (perhaps most) research questions asked by biologists. Unless you are willing simply to deny that causality is a meaningful concept then you will need some way of studying causal relationships when randomised experiments cannot be performed. Maintain your scepticism if you wish, but grant me the benefit of your doubt. A healthy scepticism while in a car dealership will keep you from buying a lemon. An unhealthy scepticism might prevent you from obtaining reliable transportation.

I said that the methods in this book are not logically dependent on any particular definition of causality. Rather than *defining* causality, the approach is to *axiomise* causality (Spirtes, Glymour and Scheines 1993). In other words, one begins by determining those attributes that scientists view as necessary for a relationship to be considered 'causal' and then develops a formal mathematical language that is based on such attributes. First, these relationships must be *transitive*: if A causes B and B causes C then it must also be true that A causes C. Second, such relationships must be 'local'; the technical term for this is that the relationships must obey the *Markov condition*, of which there are local and global versions. This is described in more detail in Chapter 2, but it can be intuitively understood to mean that events are caused only by their proximate causes. Thus,

---

[4] Perhaps quantum physics does not need such an assumption. I will leave this question to people better qualified than I. The world of biology does not operate at the quantum level.

[5] The perceptive reader will note that I have now compounded my problems. Not only do I propose to deal with one imperfectly defined notion – causality – but I will do it with reference to another imperfectly defined notion: a probability distribution.

if event A causes event C *only* through its effect on an intermediate event B (A→B→C) then the causal influence of A on C is blocked if event B is prevented from responding to A. Third, these relationships must be *irreflexive*: an event cannot cause itself. This is not to say that every event must be causally explained; to argue in this way would lead us directly into the paradox of infinite regress. Every causal explanation in science includes events that are accepted (measured, observed...) without being derived from previous events.[6] Finally, these relationships must be *asymmetric*: if A is a cause of B, B cannot simultaneously be a cause of A.[7] In my experience, scientists generally accept these four properties. In fact, so long as I avoid asking for definitions, I find that there is a large degree of agreement between scientists on whether any particular relationship should be considered causal or not. It might be of some comfort to empirically trained biologists that the methods described in this book are based on an almost empirical approach to causality. This is because deductive definitions of philosophers are replaced with attributes that working scientists have historically judged to be necessary for a relationship to be causal. However, this change of emphasis is, by itself, of little use.

Next, we require a new mathematical language that is able to express and manipulate these causal relationships. This mathematical language is that of directed graphs[8] (Pearl 1988; Spirtes, Glymour and Scheines 1993). Even this new mathematical language is not enough to be of practical use. Since, in the end, we wish to infer causal relationships from correlational data, we need a logically rigorous way of translating between the causal relationships encoded in directed graphs and the correlational relationships encoded in probability theory. Each of these requirements can now be fulfilled.

1.2   Fisher's genius and the randomised experiment

Since this book deals with causal inference from observational data, we should first look more closely at how biologists infer causes from experimental data. What is it about these experimental methods that allows scientists to speak comfortably about causes? What is it about inferring causality from non-experimental data that makes them squirm in their chairs? I will distinguish between two basic types of experiments: the controlled experiment and the randomised experiment. Although the controlled experiment takes

---

[6] The paradox of infinite regress is sometimes 'solved' by simply declaring a first cause: that which causes but which has no cause. This trick is hardly convincing, because, if we are allowed to invent such things by fiat, then we can declare them anywhere in the causal chain. The antiquity of this paradox can been seen in the first sentence of the first verse of Genesis: 'In the beginning God created the heavens and the earth.' According to the Confraternity Text of the Holy Bible, the Hebrew word that has been translated as 'created' was used only with reference to divine creation and meant 'to create out of nothing'.

[7] This does not exclude feedback loops so long as we understand these to be dynamic in nature: A causes B at time t, B causes A at time t+Δt, and so on. This is discussed more fully in Chapter 2.

[8] Biologists will find it ironic that this graphical language was actually proposed by Wright (1921), one of the most influential evolutionary biologists of the twentieth century, but his insight was largely ignored. This history is explored in Chapters 3 and 4.

historical precedence, the randomised experiment takes precedence in the strength of its causal inferences.

Fisher[9] described the principles of the randomised experiment in his classic *Design of Experiments* (Fisher 1926). Since he developed many of his statistical methods in the context of agronomy, let us consider a typical randomised experiment designed to determine if the addition of a nitrogen-based fertiliser can cause an increase in the seed yield of a particular variety of wheat. A field is divided into 30 plots of soil and the seed is sown. The treatment variable consists of the fertiliser, which is applied at either 0 or 20 kg/hectare. For each plot we place a small piece of paper in a hat. One-half of the pieces of paper have a '0' and the other half have a '20' written on them. After thoroughly mixing the pieces of paper, we randomly draw one for each plot to determine the treatment level that each plot is to receive. After applying the appropriate level of fertiliser independently to each plot, we make no further manipulations until harvest day, at which time we weigh the seed that is harvested from each plot.

The seed weight per plot is normally distributed within each treatment group. Those plots receiving no fertiliser produce 55 g of seed with a standard error of six. Those plots receiving 20 kg/hectare of fertiliser produce 80 g of seed with a standard error of six. Excluding the possibility that a very rare random event has occurred (with a probability of approximately 5×10^{-8}), we have very good evidence that there is a positive *association* between the addition of the fertiliser and the increased yield of the wheat. Here we see the first advantage of randomisation. By randomising the treatment allocation, we generate a sampling distribution that allows us to calculate the probability of observing a given result by chance if, in reality, there is no effect from the treatment. This helps us to distinguish between chance associations and systematic ones. Since one error that a researcher can make is to confuse a real difference with a difference due to sampling fluctuations, the sampling distribution allows us to calculate the probability of committing such an error.[10] However, Fisher, and many other statisticians (Kempthorpe 1979; Kendall and Stuart 1983),[11] go further by claiming that the process of randomisation allows us to differentiate between associations due to causal effects of the treatment and associations due to some variable that is a common cause both of the treatment and response variables. What allows us to move so confidently from this conclusion about an *association* (a 'co-relation') between fertiliser addition and increased seed yield to the claim that the added fertiliser actually *causes* the increased yield?

Given that two variables (X and Y) are associated, there can be only three elementary, but not mutually exclusive, causal explanations; either X causes Y, Y causes X or there are some other causes that are common to both X and Y. Here, I am making

---

[9] Sir Ronald A. Fisher (1890–1962) was chief statistician at the Rothamsted Agricultural Station. He was later Galton Professor at the University of London and Professor of Genetics at the University of Cambridge.

[10] It is for this reason that Mayo (1996) calls such frequency-based statistical tests 'error probes'.

[11] 'Only when the treatments in the experiment are applied by the experimenter using the full randomisation procedure is the chain of inductive inference sound; it is only under these circumstances that the experimenter can attribute whatever effect he observes to the treatment and to the treatment only' (Kempthorpe 1979).

no distinctions between 'direct' and 'indirect' causes; I argue in Chapter 2 that such terms have no meaning except relative to the other variables in the causal explanation. Remembering that transitivity is a property of causes, to say that X causes Y does not exclude the possibility that there are intervening variables $(X \rightarrow Z_1 \rightarrow Z_2 \rightarrow \ldots \rightarrow Y)$ in the causal chain between them. We can confidently exclude the possibility that the seed produced by the wheat caused the amount of fertiliser that was added. First, we already know the only cause of the amount of fertiliser that was added to any given plot: the number on the piece of paper that was drawn from the hat. Second, the fertiliser was added before the wheat plants began to produce seed.[12] What allows us to exclude the possibility that the observed association between fertiliser addition and seed yield is due to some unrecognised common cause of both? This was Fisher's genius; the treatments were randomly assigned to the experimental units (i.e. the plots with their associated wheat plants). By definition, such a random process ensures that the order in which the pieces of paper are chosen (and therefore the order in which the plots receive the treatment) is causally independent of any attributes of the plot, its soil or the plant at the moment of randomisation.

Let's retrace the logical steps. We began by asserting that, if there was a causal relationship between fertiliser addition and seed yield, there would also be a systematic relationship between these two variables in our data: *causation implies correlation*. When we observe a systematic relationship that cannot reasonably be attributed to sampling fluctuations, we conclude that there was some causal mechanism responsible for this association. Correlation does not necessarily imply a causal relationship from the fertiliser addition to the seed yield, but it does imply *some* causal relationship that is responsible for this association. There are only three such elementary causal relationships, and the process of randomisation has excluded two of them. We are left with the overwhelming likelihood that the fertiliser addition caused the increased seed yield. We cannot categorically exclude the two alternative causal explanations, since it is always possible that we were incredibly unlucky. Perhaps the random allocations resulted, by chance, in those plots that received the 20 kg/hectare of fertiliser having soil with a higher moisture-holding capacity or some other attribute that actually caused the increased seed yield? In any empirical investigation, experimental or observational, all we can do is to advance an argument that is beyond reasonable doubt, not a logical certainty.

The key role played by the process of randomisation seems to be what ensures, up to a probability that can be calculated from the sampling distribution produced by the randomisation, that no uncontrolled common cause of both the treatment and the response variables could produce a spurious association. Fisher said as much himself when he stated that randomisation 'relieves the experimenter from the anxiety of considering and estimating the magnitude of the innumerable causes by which his data may be

---

[12] Unless your meaning of 'cause' is very peculiar, you will not have objected to the notion that causal relationships cannot travel backwards in time. Despite some ambiguity in its formal definition, scientists would agree on a number of attributes associated with causal relationships. As with pornography, we have difficulty defining it but we all seem to know it when we see it.

disturbed'. Is this strictly true? Consider again the possibility that soil moisture content affects seed yield. By randomly assigning the fertiliser to plots, we ensure that, *on average*, the treatment and control plots have soil with the same moisture content, therefore removing any chance correlation between the treatment received by the plot and its soil moisture.[13] But the number of attributes of the experimental units (i.e. the plots with their attendant soil and plants) is limited only by our imagination. Let's say that there are 20 different attributes of the experimental units that could cause a difference in seed yield. What is the probability that at least one of these was sufficiently concentrated, by chance, in the treatment plots to produce a significant difference in seed yield even if the fertiliser had no causal effect? If this probability is not large enough for you then I can easily posit 50 or 100 different attributes that could cause a difference in seed yield. Since there are a large number of potential causes of seed yield, the likelihood that at least one of them was concentrated, by chance, in the treatment plots is not negligible even if we had used many more than the 30 plots.

Randomisation therefore serves two purposes in causal inference. First, it ensures that there is no causal effect coming from the experimental units to the treatment variable or from a common cause of both. Second, it helps to reduce the likelihood in the sample of a chance correlation between the treatment variable and some other cause of the treatment, but doesn't completely remove it. To cite Howson and Urbach (1989: 152, emphasis in original): 'Whatever the size of the sample, two treatment groups are *absolutely certain* to differ in some respect, indeed, in infinitely many respects, any of which might, unknown to us, be causally implicated in the trial outcome. So randomisation cannot possibly guarantee that the groups will be free from bias by unknown nuisance factors [i.e. variables correlated with the treatment]. And since one obviously doesn't know what those unknown factors are, one is in no position to calculate the probability of such a bias developing either.' This should not be interpreted as a severe weakness of the randomised experiment in any practical sense, but it does emphasise that even the randomised experiment does not provide any automatic assurance of causal inference, free of subjective assumptions.

Equally important is what is not required by the randomised experiment. The logic of experimentation up to Fisher's time was that of the controlled experiment, in which it was crucial that all other variables be experimentally fixed to constant values;[14] see, for example, Feiblman (1972: 149). Fisher (1970) explicitly rejects this as an inferior method, pointing out that it is logically impossible to know if 'all other variables' have been accounted for. This is not to say that Fisher does not advocate physically controlling for other causes in addition to randomisation. In fact, he explicitly recommends that

---

[13] More specifically, these two variables, being causally independent, are also probabilistically independent in the statistical population. This is not necessarily true in the sample due to sampling fluctuations.

[14] Clearly, this cannot be literally true. Consider a case in which the causal process is A→B→C, and we want to experimentally test whether A causes C. If we hold variable B constant then we would incorrectly surmise that A has no causal effect on C. It is crucial that common causes of A and C be held constant in order to exclude the possibility of a spurious relationship. It is also a good idea, though not crucial for the causal inference, that causes of C that are independent of A also be held constant, in order to reduce the residual variation of C.

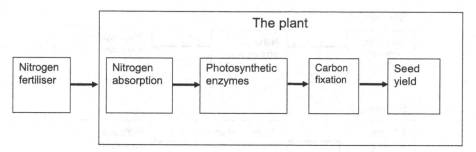

Figure 1.1 A hypothetical causal scenario that is not amenable to a randomised experiment

the researcher do this whenever possible. For instance, in discussing the comparison of plant yields of different varieties, he advises that they be planted in soil 'that appears to be uniform'. In the context of pot experiments he recommends that the soil be thoroughly mixed before putting it in the pots, that the watering be equalised, that the pots receive the same amount of light, and so on. The strength of the randomised experiment lies in the fact that we do not have to physically control – or even be aware of – other causally relevant variables in order to reduce (but not logically exclude) the possibility that the observed association is due to some unmeasured common cause in our sample.

Yet strength is not the same as omnipotence. Some readers will have noticed that the logic of the randomised experiment has, hidden within it, a weakness not yet discussed that severely restricts its usefulness to biologists; a weakness that is not removed even with an infinite sample size. In order to work, one must be able to randomly assign values of the hypothesised 'cause' to the experimental units independently of any attributes of these units. This assignment must be direct and not mediated by other attributes of the experimental units. However, a large proportion of biological studies involve relationships between different attributes of such experimental units.

In the experiment described above, the experimental units are the plots of ground with their wheat plants. The attributes of these units include those of the soil, the surrounding environment and the plants. Imagine that the researcher wants to test the following causal scenario: the added fertiliser increases the amount of nitrogen absorbed by the plant. This increases the amount of nitrogen-based photosynthetic enzymes in the leaves and therefore the net photosynthetic rate. The increased carbon fixation due to photosynthesis causes the increased seed yield (Figure 1.1).

The first part of this scenario is perfectly amenable to the randomised experiment since the nitrogen absorption is an attribute of the plant (the experimental unit) while the amount of fertiliser added is controlled completely by the researcher independently of any attribute of the plot or its wheat plants. The rest of the hypothesis is impervious to the randomised experiment. For instance, both the rate of nitrogen absorption and the concentration of photosynthetic enzymes are attributes of the plant (the experimental unit). It is impossible to randomly assign rates of nitrogen absorption to each plant independently of any of its other attributes, yet this is the crucial step in the randomised experiment that allows us to distinguish correlation from causation. It is true that the researcher can induce a *change* both in the rate of nitrogen absorption by the plant and

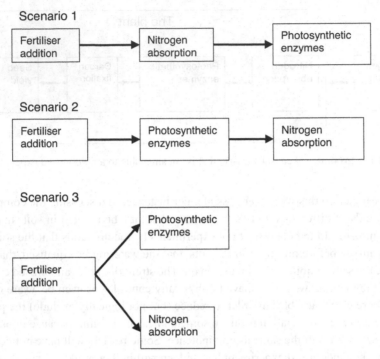

Figure 1.2 Three different causal scenarios that could generate an association between an increased nitrogen absorption and an increased enzyme concentration in the plant following the addition of fertiliser in a randomised experiment

in the concentration of photosynthetic enzymes in its leaves, but in each case these changes are due to the addition of the fertiliser. After observing an association between the increased nitrogen absorption and the increased enzyme concentration the randomisation of fertiliser addition does not exclude different causal scenarios, only some of which are shown in Figure 1.2.

When one is reading books about experimental design one's eyes often skim across the words 'experimental unit' without pausing to consider what these words mean. The experimental unit is the 'thing' to which the treatment levels are randomly assigned. The experimental unit is also an experimental *unit*. The causal relationships, if they exist, are between the external treatment variable and each of the attributes of the experimental unit that show a response. In biology, the experimental units (e.g. plants, leaves or cells) are integrated wholes whose parts cannot be disassembled without affecting the other parts. It is often not possible to randomly 'assign' values of one attribute of an experimental unit independently of the behaviour of its other attributes.[15] When such random assignments cannot be done, one cannot infer causality from a random experiment. A moment's reflection will show that this problem is very common in biology.

---

[15] This is not to say that it is always impossible. For instance, one can randomly add levels of insulin to the blood because the only cause of these changes (given proper controls) is the random numbers assigned to the animal. One cannot randomly add different numbers of functioning chloroplasts to a leaf.

Organismal, cell and molecular biology are rife with it. Physiology is hopelessly entangled. Evolution and ecology, dependent as they are on physiology and morphology, are often beyond its reach. If we accept that one cannot study causal relationships without the randomised experiment then a large proportion of biological research will have been gutted of any demonstrable causal content.

The usefulness of the randomised experiment is also severely reduced because of practical constraints. Remember that the inference is from the randomised treatment allocation to the experimental unit. The experimental unit must be the one that is relevant to the scientific hypothesis of interest. If the hypothesis refers to large-scale units (populations, ecosystems, landscapes) then the experimental unit must consist of such units. Someone wishing to know if increased carbon dioxide concentrations will change the community structure of forests will have to use entire forests as the experimental units. Such experiments are never done, and there is nothing in the inferential logic of randomised experiments that allows one to scale up from different (small-scale) experimental units. Even when proper randomised experiments can be done in principle, they cannot be done in practice due to financial or ethical constraints.

The biologist who wishes to study causal relationships using the randomised experiment is therefore severely limited in the questions that can be posed. The philosophically inclined scientist who insists that a positive response from a randomised experiment is an operational *definition* of a causal relationship would have to conclude that causality is irrelevant to much of science.

1.3   The controlled experiment

The currently prevalent notion that scientists cannot convincingly study causal relationships without the randomised experiment would have seemed incomprehensible to scientists before the twentieth century. Certainly, biologists *thought* that they were demonstrating causal relationships long before the invention of the randomised experiment. A wonderful example of this can be found in *An Introduction to the Study of Experimental Medicine* by the great nineteenth-century physiologist Claude Bernard (Bernard 1865).[16] I will cite a particularly interesting passage (Rapport and Wright 1963), and I ask that you pay special attention to the ways in which he tries to control variables. I will then develop the connection between the controlled experiment and the statistical methods described in this book.

In investigating how the blood, leaving the kidney, eliminated substances that I had injected, I chanced to observe that the blood in the renal vein was crimson, while the blood in the neighbouring veins was dark like ordinary venous blood. This unexpected peculiarity struck me, and I thus made observation of a fresh fact begotten by the experiment, but foreign to the experimental aim pursued at the moment. I therefore gave up my unverified original idea, and directed my attention to the singular colouring of the venous renal blood; and when I had noted it well and

---

[16] Rapport and Wright (1963) describe Claude Bernard (1813–1878) as an experimental genius and 'a master of the controlled experiment'.

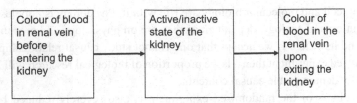

Figure 1.3 The hypothetical causal explanation invoked by Claude Bernard

assured myself that there was no source of error in my observation, I naturally asked myself what could be its cause. As I examined the urine flowing through the urethra and reflected about it, it occurred to me that the red colouring of the venous blood might well be connected with the secreting or active state of the kidney. On this hypothesis, if the renal secretion was stopped, the venous blood should become dark: that is what happened; when the renal secretion was re-established, the venous blood should become crimson again; this I also succeeded in verifying whenever I excited the secretion of urine. I thus secured experimental proof that there is a connection between the secretion of urine and the colouring of blood in the renal vein.

Our knowledge of human physiology has progressed far from the experiments of Bernard (physiologists might find it strange that he spoke of renal 'secretions'), yet his use of the controlled experiment would be immediately recognisable and accepted by modern physiologists. Fisher was correct in describing the controlled experiment as an inferior way of obtaining causal inferences, but the truth is that the randomised experiment is unsuited for much of biological research. The controlled experiment consists of proposing a hypothetical structure of cause–effect relationships, deducing what would happen if particular variables are controlled, or 'fixed' in a particular state, and then comparing the observed result with its predicted outcome. In the experiment described by Bernard, the hypothetical causal structure could be conceptualised as shown in Figure 1.3.

The key notion in Bernard's experiment was the realisation that, if his causal explanation was true, the type of *association* between the colour of the blood in the renal vein as it enters and leaves the kidney would change depending on the state of the hypothesised cause – i.e. whether the kidney was secreting or not. It is worth returning to his words: 'On this hypothesis, if the renal secretion was stopped, the venous blood should become dark: that is what happened; when the renal secretion was re-established, the venous blood should become crimson again; this I also succeeded in verifying whenever I excited the secretion of urine. I thus secured experimental proof that there is a connection between the secretion of urine and the colouring of blood in the renal vein.' Since he had explicitly stated earlier in the quote that he was enquiring into the 'cause' of the phenomenon, it is clear that he viewed the result of his experiments as establishing a *causal connection* between the secretion of urine and the colouring of blood in the renal vein.

Although the controlled experiment is an inferior method of making causal inferences relative to the randomised experiment, it is actually responsible for most of the causal knowledge that science has produced. The method involves two basic parts. First, one must propose an hypothesis stating how the measured variables are linked in the causal process. Second, one must deduce how the associations between the observations must change once particular combinations of variables are controlled so that they can no longer vary naturally – i.e. once particular combinations of variables are 'blocked'. The final step is to compare the patterns of association after such controls are established with the deductions. Historically, variables have been blocked by physically manipulating them. However (this is an important point that will be more fully developed and justified in Chapter 2), it is the control of variables, not how they are controlled, that is the crucial step. The weakness of the method, as Fisher pointed out, is that one can never be sure that all relevant variables have been identified and properly controlled. One can never be sure that, in manipulating one variable, one has not also changed some other, unknown variable. In any field of study, as Bernard documents in his book, the first causal hypotheses are generally wrong, and the process of testing, rejecting and revising them is what leads to progress in the field.

1.4   Physical controls and observational controls

It is the control of variables, not how they are controlled, that is the crucial step in the controlled experiment. What does it mean to 'control' a variable? Can such control be obtained in more than one way? In particular, can one control variables based on observational, rather than experimental, observations? The link between a physical control through an experimental manipulation and a statistical control through conditioning will be developed in the next chapter, but it is useful to provide an informal demonstration here using an example that should present no metaphysical problems to most biologists.

Body size in large mammals seems to be important in determining much of their ecology. In populations of bighorn sheep in the Rocky Mountains, it has been observed that the probability of survival of an individual through the winter is related to the size of the animal in the fall. However, this species has a strong sexual dimorphism, with males being up to 60 per cent larger than females. Perhaps the association between body size and survival is simply due to the fact that males have a better probability of survival than females, and this is unrelated to their body size? In observing these populations over many years, perhaps the observed association arises because those years showing better survival also have a larger proportion of males? Figure 1.4 shows these two alternative causal hypotheses. I have included boxes labelled 'Other causes' to emphasise that we are not assuming the chosen variables to be the only causes of body size or of survival.

Notice the similarity to Claude Bernard's question concerning the cause of blood colour in the renal vein. The difference between the two alternative causal explanations in Figure 1.4 is that the second assumes that the association between spring survival and autumn body size is due only to the sex ratio of the population. Thus, if the sex ratio could be held constant, the association would disappear. Since adult males and females of this species live in separate groups, it would be possible to physically separate them in their range and, in this way, physically control the sex ratio of the population. However, it is much easier to simply sort the data according to sex and then look for an

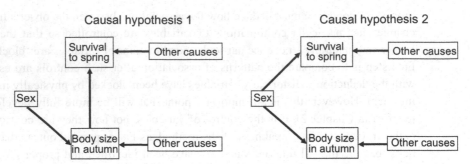

Figure 1.4 Two alternative causal explanations for the relationship between the sex and body size of bighorn sheep in the autumn and the probability of survival until the spring

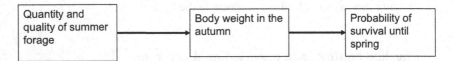

Figure 1.5 A hypothetical causal explanation for the relationship between the quality and quantity of summer forage, the body weight of bighorn sheep in the autumn and the probability of survival until the spring

association within each homogeneous group. The act of separating the data into two groups such that the variable in question – the sex ratio – is constant within each group represents a *statistical control*. We could imagine a situation in which we instruct one set of researchers to physically separate the original population into two groups based on sex, after which they test for the association within each of their experimental groups, and then ask them to combine the data together and give them to a second team of researchers. The second team would analyse the data using the statistical control. Both groups would come to identical conclusions.[17] In fact, using statistical controls might even be preferable in this situation. Simply observing the population over many years and then statistically controlling for the sex ratio on paper does not introduce any physical changes in the field population. It is likely that the act of physically separating the sexes in the field might introduce some unwanted, and potentially uncontrolled, change in the behavioural ecology of the animals during the rut that might bias the survival rates during the winter quite independently of body size.

Let's further extend this example to look at a case in which it is not as easy to separate the data into groups that are homogeneous with respect to the control variable. Perhaps the researchers have also noticed an association between the amount and quality of the rangeland vegetation during the early summer and the probability of survival during the next winter. They hypothesise that this pattern is caused by the animals being able to eat more during the summer, which increases their body size in the autumn, which in turn increases their chances of survival during the winter (Figure 1.5).

---

[17] It is not true that statistical and physical controls will always give the same conclusion. This is discussed in Chapter 2.

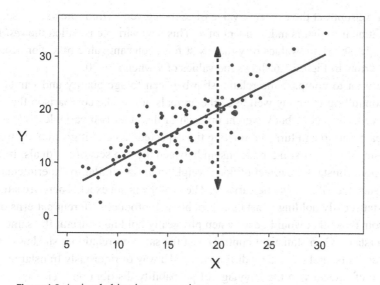

Figure 1.6 A simple bivariate regression
Notes: The solid line shows the expected value of y_i given the value of x_i ($E[y_i|x_i]$). The dotted line shows the possible values of y_i that are independent of x_i (the residuals).

The logic of the controlled experiment requires us to be able to compare the relationship between forage quality and winter survival after physically preventing body weight from changing, which we cannot do.[18] Since body weight is a continuous variable, we can't simply sort the data and then divide it into groups that are homogeneous for this variable. This is because each animal will have a different body weight. Nonetheless, there is a way of comparing the relationship between forage quality and winter survival while controlling the body weight of the animals during the comparison. This involves the concept of statistical conditioning, which will be more rigorously developed in Chapters 2 and 3. An intuitive understanding can be had with reference to a simple linear regression (Figure 1.6).

The formula for a linear regression is $y_i = \alpha + \beta x_i + N(0, \sigma)$. Here, the notation $N(0, \sigma)$ means 'a normally distributed random variable with a population mean of zero and a population standard deviation of σ'. As the formula makes clear, the observed value of y consists of two parts: one part that depends on x and one part that doesn't. If we let E(y|x) represent the expected value of y given x, we can write

$$E(y|x_i) = \alpha + \beta x_i$$
$$y_i = E(y|x_i) + N(0, \sigma)$$
$$(y_i - E(y|x_i)) = N(0, \sigma)$$

---

[18] It is actually possible, in principle if not in practice, to conduct a randomised experiment in this case, so long as we are interested only in knowing if summer forage quality causes a change in winter survival. This is because the hypothetical cause (vegetation quality and quantity) is not an attribute of the unit possessing the hypothetical effect (winter survival). Again, it is impossible to use a randomised experiment to determine if body size in the autumn is a cause of increased survival during the winter.

Thus, if we subtract the expected value of each y, given x, from the value itself, we get the variation in y that is independent of x. This new variable is called the *residual* of y given x. These are the values of y that exist for a constant value of x. For instance, the vertical arrow in Figure 1.6 shows the values of y when x = 20.

If we want to compare the relationship between forage quality and winter survival while controlling the body weight of the animals during the comparison then we have to remove the effect of body weight on each of the other two variables. We do this by taking each variable in turn, subtracting the expected value of it given body weight and then seeing if there is still a relationship between the two sets of residuals. In this way, we can hold constant the effect of body weight in a way similar to experimentally holding constant the effect of some variable. The analogy is not exact. There are situations in which statistically holding constant a variable will produce different patterns of association from those that would occur when physically holding constant the same variable. To understand when statistical controls cast the same correlational shadows as experimental controls, and when they differ, we need a way of rigorously translating from the language of causality to the language of probability distributions. This is the topic of the next chapter.

2 From cause to correlation and back

2.1 Translating from causal to statistical models

The official language of statistics is the probability calculus, based on the notion of a probability distribution. For instance, if you conduct an analysis of variance (ANOVA) then the key piece of information is the probability of observing a particular value of Fisher's F statistic in a random sample of data, given a particular hypothesis or model. To obtain this crucial piece of information, you (or your computer) must know the probability density function of the F statistic. Certain other (mathematical) languages are tolerated within statistics but, in the end, one must link one's ideas to a probability distribution in order to be understood. If we wish to study causal relationships using statistics, it is necessary that we translate, without error, from the language of causality to the only language that statistics can understand: probability theory.

Such a rigorous translation device did not exist until very recently (Pearl 1988). It is no wonder that statisticians have virtually banished the word 'cause' from statistics; such a word has no equivalent in their language.[1] Within the world of statistics the scientific notion of causality has, until recently, been a stranger in a strange land. Posing causal questions in the language of the probability calculus is like a unilingual Englishman asking for directions to the Louvre from a Frenchman who can't speak English. The Frenchman might understand that directions are being requested, and the Englishman might see fingers pointing in particular directions, but it is not at all certain that works of art will be found. Imperfect translations between the language of causality and the language of probability theory are equally disorienting.

Mistakes in translation come in all kinds. The most dangerous ones are the subtle errors in which a slight change in inflection or context of a word can change the meaning in disastrous ways. Because the French word *demande* both sounds like the English word 'demand' and has roughly the same meaning (it simply means 'to ask for', without any connotation of obligation), I have seen French-speaking people come up to a shop assistant and, while speaking English, 'demand service'. They think that they are politely asking for help, while the assistant thinks they are issuing an ultimatum. I once came close to being beaten by an enraged boyfriend simply because (I thought) I was complimenting his girlfriend on her long hair, which was drawn in a ponytail. The word

---

[1] Fisherian statistics does deal with causal hypotheses, but the causal inferences come from the experimental design, not from the mathematical details; see Chapter 1.

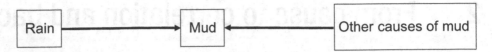

Figure 2.1 The causal relationships between rain, mud and other causes of mud

for 'tail' in French is *queue*, which takes a feminine gender. There is another word in colloquial French, *cul* (the 'l' is silent), that sounds almost the same. It takes a masculine gender, is pronounced only slightly differently and refers to a person's rear end; the correctly translated word rhymes with 'pass', but the reader will understand if I don't give the literal translation. So, while trying to make conversation with the boyfriend, I told him that his girlfriend had a nice *cul* instead of a nice *queue*. I immediately knew, from the look of rage on his face, that I had chosen the wrong word.

The same subtle mistakes of translation can occur when translating between the language of causality and the mathematical language of probability distributions. I began the first chapter by comparing causes and correlations to three-dimensional objects and their two-dimensional shadows. Clearly, there is a close relationship between the object and its shadow. Just as clearly, they are not the same thing. The goal of this chapter is to describe the relationship between variables involved in a causal process and the probability distribution of these variables that the causal process generates. Causal processes cast probability shadows, but 'causes' and 'probability distributions' are not the same thing either. It is important to understand exactly how the translation is made between causal processes and probability distributions in order to avoid the scientific equivalent of a punch in the nose from an enraged boyfriend.

I will make the distinction between a causal model, an observational model and a statistical model. Since every child knows that rain causes mud,[2] I will illustrate the difference between these three types of models with this analogy. The statement 'Rain causes mud' implies an asymmetric relationship: the presence of rain will create mud, but the presence of mud will not create rain. I will use the symbol '→' when I want to refer to such causal relationships. This leads naturally to the sort of 'box and arrow' diagrams with which most biologists are familiar (Figure 2.1).

To complete the description it is necessary to add the convention that, unless a causal relationship is explicitly included, it is understood not to exist. So, in Figure 2.1, the fact that there are no arrows between 'Rain' and 'Other causes of mud' means that there is no direct causal relationship between them; in fact, there is no causal relationship of any kind in this example, since the two are causally independent.

The observational model that is related to this causal model is the statement 'Having observed rain will give us information about what we will observe concerning mud'. Notice that this observational statement deals with information, not causes, and is not asymmetric. If we learn that it has rained then we will have added information concerning the presence of mud in our yard, but observing mud in our yard will also give us

---

[2] My children seem to have mastered this metaphysical concept well before age five. This is another example of how deeply ingrained the notion of causality is.

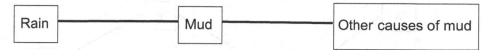

Figure 2.2 The observational relationships between rain, mud and other causes of mud

Mud (cm) = 0.1Rain (cm) + N(0,0.1)

Figure 2.3 A statistical model relating rain and mud

Rain (cm) = 10Mud (cm) + N(0,1)

Figure 2.4 Another statistical model relating rain and mud

information about whether or not it has rained. I will use the symbol '−' when I refer to such observational relationships. This leads to the model in Figure 2.2.

Notice that, although rain and other causes of mud are causally independent, they are not observationally independent given the state of mud; knowing that it has not rained but that there is mud in the front yard gives you information on the existence of other causes of mud.

The statistical model differs only in degree, not in kind, from the observational model. The statistical model (Figure 2.3) specifies the mathematical relationship between the variables as well as the probability distributions of the variables. Now we can use the equivalence operator of algebra ($=$), since we are stating a quantitative equivalence.

This mathematical statement says that the value obtained by measuring the depth of the mud, in centimetres, is the same as (is 'equivalent to') the value that is obtained by measuring the amount of rain that falls, in centimetres, multiplying this value by 0.1, and adding another value (in centimetres) obtained from a random value taken from a normal distribution whose population mean is zero and whose population standard deviation in 0.1.

What is the point of all this? According to Pearl (1997), a century of confusion between correlation and causation can be traced, in part, to a mistranslation of the word *cause*. When scientists and statisticians attempt to express notions of causality using mathematics they mistranslate 'cause', a word having connotations of asymmetry and all the other properties discussed in Chapter 1, as the algebraic notion "$=$" used in the language of probability theory. The symbols $\rightarrow$ and $=$ do not mean the same thing, because the algebraic concept of 'equivalence' and its symbol ($=$) do not have the properties of causes discussed in Chapter 1. It is perfectly correct to rearrange the equation in Figure 2.3 in order to imply that the amount of rain can be predicted from the amount of mud (Figure 2.4), even though any five-year-old child would recognise this as causally nonsensical.

This mistake is the scientific equivalent of telling a boyfriend that his girlfriend has *un beau cul* rather than *une belle queue*. The conceptual error occurs because we have replaced $\rightarrow$ with $=$. After translating from the language of causality to the language of observations, we have used the syntax of this observational language to produce a

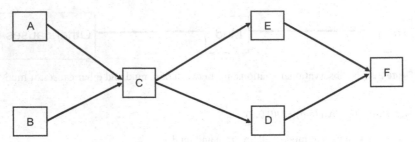

Figure 2.5 A directed graph describing the causal relationships between six variables or vertices (A to F)

perfectly reasonable statement for this observational language, but then we have performed a literal translation back into the language of causality without recognising the difference in syntax. There are computer programs that attempt to translate between human languages, and those that use literal word-by-word translations run into the same problems.[3] A newspaper headline such as 'Bill Gates worth $1,000,000,000', after being literally translated (word for word) into a different language and then retranslated back into English, might come up with a phrase such as 'payment request ["bill"] for doors in the fence ["gates"] costs $1,000,000,000'!

In the next few sections I develop a translation device to move between causal models and observational (statistical) models. To do this we require the necessary and sufficient conditions needed to specify a joint probability distribution that must exist given a causal process. Put another way, we require the necessary and sufficient conditions needed to specify the correlational shadow that will be cast by a causal process. This provides the key to translating between causal and statistical models. These sections require more effort to understand but in each case I will also provide a more intuitive description and some worked examples.

The strategy for translation from the physical world, in which the notion of causation is applicable, to the mathematical world of probability theory, in which the abstract notion of algebraic equivalence is applicable, involves two steps. First, since algebra cannot express the sort of relationships that we consider 'causal', we need a new mathematical language that can; this language is that of directed graphs. Second, we need a translation device that can unambiguously convert the statements expressed in such directed graphs into statements concerning the conditional independence of random variables obeying a particular probability distribution. This translation device is called 'd-separation' (short for '*directed* separation').

2.2   Directed graphs

It is now time to introduce some terminology concerning directed (sometimes called causal) graphs. These terms, although unfamiliar to most biologists, are quite easy to grasp and use. These terms will be defined using the causal graph shown in Figure 2.5.

---

[3] This was written in the first edition. Current translation programs are much better!

Here is a *partial* verbal (as opposed to mathematical) description of what Figure 2.5 means. Two of the six variables (A and B) are *causally independent*, meaning that changes in either will not affect the value of the other. Each of the four other variables (C, D, E and F) are *causally dependent* on A and B, either directly (C) or indirectly (D, E and F). By 'causally dependent' I mean that changes in either A or B will provoke changes in each of C, D, E and F, but changes in any of these will not provoke changes in either A or B. A and B are *direct* causes of C because changes in A or B will provoke changes in C irrespective of the behaviour of either D, E or F. A and B are *indirect causes* of D, E and F because changes in A or B will provoke changes in these variables only by causing changes in C; if C is prevented from changing then A and B will no longer cause changes in these three other variables. C is a direct *common* cause of D and E and an indirect cause of F through its effects on D and E. Finally, both D and E are direct causes of F, though they are not themselves causally independent.

It is clear that this directed graph is a very compact way of expressing even the previous incomplete verbal description of this causal system. This economy of description is a major reason why researchers in artificial intelligence adopted directed graphs as a way of economically programming causal knowledge (Pearl 1988).[4] In order to better use and interpret directed graphs, a few definitions are needed.

In graph theory, a directed graph is a set of vertices, represented by letters enclosed in boxes in Figure 2.5, and a set of edges, represented by lines; these lines can have either no, single or double arrowheads. The arrowheads denote the direction of the functional relationship between the vertices at either end of the line.[5] Since biologists use directed graphs to represent causal relationships between variables, you can replace the abstract term 'vertex' with the more familiar word 'variable' and the abstract term 'edge' with the more familiar word 'effect'. The symbols at the ends of the lines can be either an arrowhead or a 'missing' mark. Thus, the notation X→Y means 'X is a direct cause of Y'. The notation X←Y means 'Y is a direct cause of X'. Finally, the notation X ↔ Y means 'neither X nor Y is a cause of the other but both share common unknown causes represented by some unknown vertex not included in the causal graph'. This last notation is needed later when we use incomplete causal graphs with unspecified latent vertices.

A *direct cause* is a causal relationship between two vertices that exists independently of any other vertex in the causal explanation. This is denoted by an arrow (→) whose tail is at the cause and whose head is pointing to its direct effect. For instance, both A and B are direct causes of C in Figure 2.5. Furthermore, A and B are the *causal parents* of C, and C is their *causal child*. A cause is 'direct' only in relation to the other vertices in the causal explanation. This point is important, because a common error is to incorrectly equate a 'direct' cause relative to others in the causal graph with the more fundamental

---

[4] More accurately, directed graphs can economically store the conditional independence constraints implied by a causal system of an arbitrary joint probability distribution. This is explained in more detail below.

[5] In the jargon of graph theory, an undirected graph consists of a set of vertices {A,B,C, . . . } and a binary set denoting the presence or absence of edges (lines) between each pair of vertices. The graph becomes directed when we include a set of symbols for each edge showing direction. It is also possible to construct partially directed graphs. A graph is acyclic if there are no paths that lead a vertex back onto itself; otherwise it is cyclic. The causal graph in Figure 2.5 is therefore a directed acyclic graph (DAG).

claim that the cause is somehow 'direct' with respect to any other variable that might exist. Whenever you read the words 'direct cause' you should mentally add the words 'relative to the other variables that are explicitly invoked in the causal explanation'.

An *indirect cause* is a causal relationship between two vertices that is conditional on the behaviour of other vertices in the causal explanation. Again, a cause is 'indirect' only in relation to the other vertices in the causal explanation. For instance, in Figure 2.5 the vertex A is an indirect cause of vertex D (A→C→D) because its causal effect is conditional on the behaviour of vertex C. Furthermore, A and B are *causal ancestors* of D in Figure 2.5 and D is a *causal descendant* of both A and B.

Perhaps an example would help at this point. If we wish to give a causal description of the murder of a victim by a gunman and this explanation involves only these two 'variables' then we would say that the gunman's actions were the direct cause of the victim's death, and write 'Gunman's actions→Murder of victim'. On the other hand, if we also include the presence of the bullet penetrating the victim's heart in our causal explanation then we would say that the bullet was the direct cause of death and the gunman was an indirect cause, and write 'Gunman's actions→Bullet→Murder of victim'. If we wish to go into more gruesome physiological detail then we would describe how the bullet interrupts the functioning of the heart and the bullet would no longer be a direct cause of the victim's death. Virtually any causal mechanism can be further decomposed into a more detailed causal mechanism, and so describing a cause as 'direct' or 'indirect' can be meaningful only in relative terms in the context of the other variables that make up the causal explanation. This is simply the reductionist method common in science, and the trick is always to choose a level of causal complexity that is sufficiently detailed that it meets the goals of the study while remaining applicable in practice.

A *directed path* between two vertices in a causal graph exists if it is possible to trace an ordered sequence of vertices that must be traversed, when following the direction of the edges (head to tail), in order to travel between the first and the second. If no such directed path exists then the two vertices *are causally independent*; causal conditional independence is defined below. It is possible for there to be more than one directed path linking two vertices. In Figure 2.5 there are two different directed paths between A and F: A→C→D→F and A→C→E→F.

An *undirected path* between two vertices in a causal graph exists if it is possible to trace an ordered sequence of vertices than must be traversed, *ignoring* the direction of the edges (head to tail), in order to travel from the first to the second. Be careful! An undirected path is *not* one having no arrowheads; rather, it is simply one resulting from ignoring them. In other words, an undirected path can also be a directed path, but this is not necessarily the case. For instance, there is an undirected path between A and B in Figure 2.5 (A→C←B) that is not also a directed path.

A *collider* vertex on an undirected path is a vertex with arrows pointing into it from both directions. The vertex F in the undirected path D→F←E in Figure 2.5 is a collider along this undirected path. It is possible for the same vertex to be a collider along one path and a non-collider along another path. A vertex that is a collider along an undirected path is *inactive* in its normal (unconditioned) state. This means that, in its normal (unconditioned) state, a collider blocks (prevents) the transmission of causal effects along such a path. The contrary of a collider is a *non-collider*. The vertex C in the path A→C→D in Figure 2.5 is a non-collider. A vertex that is a non-collider along a path is said to be active in its normal (unconditioned) state. This means that, in its normal (unconditioned) state, a non-collider permits the transmission of causal effects along such a path. It is sometimes easier to imagine a path as an electrical circuit and the variables (vertices) along the path as switches. A variable along a path that is a collider is like a switch that is normally OFF and a variable along a path that is a non-collider is like a switch that is normally ON.

An *unshielded collider* vertex is a set of three vertices A→B←C along a path such that B is a collider and, additionally, there is no edge between A and C. In Figure 2.5 the vertex F in the undirected path D→F←E is not only a collider but also an unshielded collider, since there is no edge between D and E. The contrary of an unshielded collider is a *shielded collider*.

2.3   Causal conditioning

I have been referring to the letters in the causal graph as 'vertices'. Once we include the notion of a probability distribution that is generated by the causal graph, these vertices will also represent random variables. These vertices can be conceived to exist in one of two binary states along a given path: active or inactive. As stated above, the natural state of a non-collider is the active (ON) state and the natural state of a collider is the inactive (OFF) state. Again, it is possible for a vertex to be active along one path and inactive along another. Intuitively, one can think of the arrows as pointing out the direction of causal influence. Thus, a vertex that is both an effect and a cause (a type of non-collider), such as vertex C along the path A→C→D in Figure 2.5, is active, because it allows the causal influence of A to be transmitted to D. In the same way, a vertex that is an effect of two vertices and therefore a cause to neither (a collider) is inactive, because it blocks the causal influence from being transmitted along the path. An example is the vertex F along the path D→F←E in Figure 2.5. *Conditioning* on a vertex in a causal graph means to change its state; if it was active then conditioning inactivates it, but if it was inactive then conditioning activates it. So, since vertex C along the path A→C→D is naturally active (ON), conditioning on it changes its state to inactive (OFF), thus blocking any indirect causal influence of A on D.

2.4   D-separation

Remembering that we are still not discussing probability distributions or statistical models, and are still concerned only with the properties of directed acyclic graphs, we can now define what is meant by the 'independence' of vertices, or groups of vertices, in a causal graph upon conditioning on some other set of vertices. This property is called *d-separation* ('directed separation': Verma and Pearl 1988; Pearl 1988; Geiger, Verma and Pearl 1990). The definition of d-separation uses the definitions above and, although

> **Box 2.1** Formal definition of *d-separation*[6]
>
> Given a causal graph G, if X and Y are two different vertices in G, and **W** is a set of vertices in G that does not contain X or Y, then X and Y are d-separated given **W** in G if and only if there exists no undirected path U between X and Y such that (a) every collider on U is either in **W** or else has a descendant in **W** and (b) no other vertex on U is in **W**.

it is awkward to define in words, it is very easy to understand when looking at a causal graph. The formal definition is given in Box 2.1. I then give a more informal definition, and finally I illustrate it using figures.

Informally, d-separation gives the necessary and sufficient conditions for two vertices in a directed acyclic (causal) graph to be observationally (probabilistically) independent upon conditioning on some other set of vertices. Stated a different way, d-separation specifies how causal information about the effects of statistical conditioning on a causal system flows between vertices in the causal graph. Notice that I say causal *information* (what we can know about the result of statistical conditioning in a causal system) and not causal *effects*. D-separation is the translation device between the language of causality and the language of probability distributions that we have been searching for. In a few pages you will understand how this translation device works. Just be patient for a few more pages while I explain how to obtain d-separation claims from a causal graph.

To know if two vertices (X, Y) are d-separated given some set of other vertices in the causal graph, which we will call **W**, do the following.

(1) List every undirected path between X and Y. In other words, find every unique way that you can get to both X and Y in the causal graph if you ignore the direction of the arrowheads and then write down these paths.

(2) For every such undirected path between X and Y (which is an ordered sequence of vertices that must be traversed, ignoring the directions of the arrows), see if *any* non-colliding vertices in this path are in the conditioning set **W**. If so, then the path is blocked and there is no causal influence between X and Y along this path. Remembering that conditioning on a non-collider changes its state to inactive, at least one of the vertices in **W** blocks any causal influence between X and Y along this undirected path.

(3) For every such undirected path between X and Y, see if *every* collider vertex along this path is either a member of the conditioning set **W** or else has a causal descendant that is a member of the conditioning set **W**. If not, then the path is blocked and there is no causal influence between X and Y along this path. Remembering that conditioning on a collider changes its state from inactive to active, there is at least

---

[6] D-separation can also be extended to determining the causal independence of two sets of vertices **A** and **B**, upon conditioning on a third set **W**.

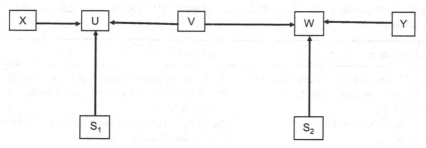

Figure 2.6 A directed graph used to illustrate the notion of d-separation

one collider along this undirected path that remains inactive, and so this path cannot transmit causal influence between X and Y.

The use of d-separation to deduce probabilistic independence upon conditioning from a causal system is best understood using a diagram (Figure 2.6) given by Spirtes, Glymour and Scheines (1993). First, we need some notation to speed things up. Whenever you see a d-separation (or 'd-sep') claim using the notation $X \perp\!\!\!\perp Y | \{\mathbf{W}\}$ you should read this as 'vertices X and Y are d-separated ("$\perp\!\!\!\perp$") given ("|") the set $\mathbf{W} = \{v_1, v_2, \dots\}$ of conditioning vertices in the causal graph'. So, are vertices X and V d-separated in the causal graph in Figure 2.6 if we don't condition on any other vertices? Using our notation, and letting $\{\phi\}$ mean an empty set, is it true that $X \perp\!\!\!\perp V | \{\phi\}$? To get an answer, we first write out all undirected paths linking vertices X and V. There is only one, $X \rightarrow U \leftarrow V$. Since the causal information collides at U along this path, its natural (unconditioned) state is OFF and causal information cannot pass between X and V along this path. Next, we look at each non-collider vertex along each undirected path (there are none in our single path) and then check to see if any of these non-colliders are in our conditioning set (which is empty in this case). If we had found any non-colliders then our undirected path would be blocked, but this is not the case. Next, we look at each collider vertex along each undirected path (there is one: vertex U) along our single undirected path and check to see if all these colliders are in our conditioning set or else are ancestors of all the vertices in the conditioning set. In our case, our conditioning set is empty, and so our single undirected path ($X \rightarrow U \leftarrow V$) remained blocked. Therefore, X is d-separated from V if we don't condition on any variables. How about $X \perp\!\!\!\perp V | \{U\}$? We have the same list of undirected paths as previously but now we condition on vertex U. Since U is a collider vertex along this undirected path, and U is also in the conditioning set, we change the status of this collider vertex from blocked (its natural state) to open. Therefore, information about the state of vertices X and V can flow between them when conditioning on vertex U. If this seems counter-intuitive to you then wait a few pages for the explanation.

It is important to be able to use this d-sep operation when using causal graphs. Table 2.1 lists some of the d-separation statements that can be obtained from Figure 2.6; the negation $\sim X \perp\!\!\!\perp Y | \{\mathbf{W}\}$ means 'vertices X and Y are *not* d-separated given the conditioning set $\mathbf{W}$'.

Table 2.1 Various probabilistic independence relationships of the directed graph in Figure 2.6 that can be deduced using d-separation

Independence relation	Explanation
$X \perp\!\!\!\perp V \vert^\phi$. X and V unconditionally independent.	There are no directed paths between X and V.
$\sim X \perp\!\!\!\perp V \vert U$. X and V not independent, conditioned on U.	Since $X \rightarrow U \leftarrow V$ collides at U, conditioning on U activates this path.
$\sim X \perp\!\!\!\perp V \vert S_1$. X and V not independent, conditioned on S_1.	Since S_1 is a causal ancestor of U, conditioning on S_1 activates U along path $X \rightarrow U \leftarrow V$.
$\sim U \perp\!\!\!\perp W \vert^\phi$. U and W are not unconditionally independent.	The path $U \leftarrow V \rightarrow W$ is naturally active. U and W share a common cause: (V) and V is not in the conditioning set $\{\phi\}$.
$U \perp\!\!\!\perp W \vert V$. U and W are independent, conditioned on V.	There is only one naturally active path between U and W: $U \leftarrow V \rightarrow W$. Conditioning on V inactivates V, blocking this path.
$X \perp\!\!\!\perp Y \vert^\phi$. X and Y are unconditionally independent.	The only undirected path between X and Y is naturally blocked by both U and W.
$\sim X \perp\!\!\!\perp Y \vert \{U,W\}$. X is not independent of Y, conditioned simultaneously on U and W.	The only undirected path between X and Y has two colliders, and both are in the conditioning set. This activates the undirected path.
$\sim X \perp\!\!\!\perp Y \vert \{S_1,S_2\}$. X is not independent of Y, conditioned simultaneously on S_1 and S_2.	The only undirected path between X and Y has two colliders, and the causal ancestors of both are in the conditioning set. This activates the undirected path.
$X \perp\!\!\!\perp Y \vert \{U,W,V\}$. X is independent of Y, conditioned simultaneously on U, W and V.	Although conditioning on both U and W activates these two colliders, conditioning on V disactivates this non-collider.

The causal inferences about this graph that are listed in Table 2.1 are not exhaustive. After a few minutes of practice it is easy to simply read off the conditional independence relations from such a causal graph. However, there also exists a function (dSep) in the ggm (graphical Gaussian model) library of R that can give the answers to d-sep claims.

The first step (after loading the ggm library) is to input your DAG. This is done using the DAG function and the ~ operator of R that is commonly used for specifying model formulae. The syntax of the DAG function is DAG(..., order = FALSE). The first argument (...) is the sequence of model formulae that specifies the DAG. The second argument (order) specifies if you want to keep the model formulae in the order in which you entered them (order = FALSE), which is the default, or if you want the order rearranged to respect the topological of cause–effect ordering. Using Figure 2.6 as an example, here is how you would use the DAG function to specify this causal graph:

```
Figure. 2.6←DAG(
U~X+V,
S1~U,
W~V+Y,
S2~W,order = FALSE)
```

There is some flexibility in the specification of the DAG. You can enter each parent–child node separately. For instance, rather than entering U~X+Y you could enter U~X,

U~Y. If you have a vertex (say 'v') that is isolated (i.e. one that neither causes nor is caused by another variable) then you would specify this as v~v. The output of the DAG function is a square Boolean matrix in which the rows represent the causal parents, the columns represent the causal children and a '1' in cell (i,j) means that there is an arrow going from row i to column j. The row names attribute of this matrix are the names that you input for your vertices.

After you have input your causal graph using the DAG function and saved it, then you can input d-sep queries using the dSep function. The syntax of the dSep function is dSep(amat,first,second,cond). The first argument, 'amat', is the Boolean matrix that you would normally create using the DAG function. The next three arguments are the names of the two vertices whose d-separation status you wish to know ('first' and 'second') and the names of the conditioning vertices ('cond'). Each of these last three arguments can be specified as single names or as a vector of names. The name of an empty set is 'NULL'.

The first d-sep claim in Table 2.1 is $X \perp\!\!\!\perp V | \{\emptyset\}$. Here is how to use the dSep function to ask if X is d-separated from V in Figure 2.6 if no other vertices are fixed:

```
dSep(Figure. 2.6,first = "X",second = "V",cond = NULL)
[1] TRUE
How about X⊥⊥Y|{U,W}?
dSep(Figure. 2.6,first = "X",second = "Y",cond = c("U","W"))
[1] FALSE
```

D-separation leads to a wealth of very useful results involving causal inference, many of which will be described in later chapters. However, until d-separation is related to probability distributions it provides no way of inferring causal relationships from observational data. Before making this link explicit, we first need some notions from probability theory.

2.5   Probability distributions

The vertices of a causal graph represent attributes in a causal system (e.g. the nitrogen concentration in a leaf or the body mass of a sheep). When we randomly sample observational units (leaves, sheep) possessing these attributes (nitrogen concentration, body mass) from some statistical population that is governed by this causal system, the vertices of the causal graph are also random variables that obey a probability distribution. Since causal relationships involve at least two such random variables, we must deal with joint probability distributions.

As I have already briefly mentioned, the notion of 'probability' differs depending on whether one subscribes to a frequentist, objective Bayesian or subjective Bayesian school of statistics. Since most statistical methods familiar to biologists derive from a frequentist perspective, I will use this definition. One begins with a hypothetical statistical population (say, all wheat plants grown in Europe) that contains all the observational

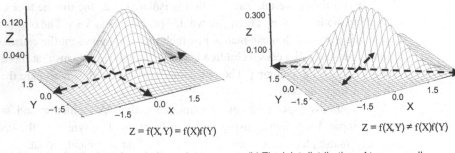

$$Z = f(X, Y) = f(X)f(Y)$$ $$\qquad\qquad Z = f(X, Y) \neq f(X)f(Y)$$

(a) The joint distribution of two independent, normally distributed random variables

(b) The joint distribution of two normally distributed random variables that are not independent

Figure 2.7 Two different versions of a bivariate normal probability distribution

units (individual plants) of interest. Each observational unit has a variable (say, the protein content of a seed) that can take different values (1.2 mg, 3.1 mg, etc.). The proportion of observational units (individual plants) in the statistical population (wheat grown in Europe) taking different values of the variable of interest (seed protein content) is the probability of this variable in this statistical population. Another way of saying this is that the probability of a random variable (X) taking a value $X = x_i$ (or having a value within an infinitesimal interval around x_i) in a statistical population of size N is the limiting frequency of $X = x_i$ in a random sample of size n as n approaches N.

A probability distribution is the distribution of the limiting (relative) frequencies of $X = x_1, x_2, \ldots$ in such a statistical population. Happily, it is an empirical fact that the distribution of many variables, when randomly sampled, can be closely approximated by various mathematical functions. Many of these functions are well known to biologists (normal distribution, Poisson distribution, binomial distribution, Fisher's F-distribution, chi-square distribution), and there are many less well-known functions that can be used as well. It is always an empirical question whether or not one of these mathematical distributions is a sufficiently close approximation of one's data to be acceptable. For instance, the relative frequency of the seed protein content per plant is likely to follow a normal distribution. The formula for the normal distribution is

$$f(x; \mu, \sigma) = \frac{1}{\sqrt{2\pi\sigma^2}} e^{\frac{-(x-\mu)^2}{2\sigma^2}}$$

When only one variable is measured on each observational unit then one obtains a *univariate* distribution. When one measures more than one variable on each observational unit (say, both the protein content and the average seed weight per plant) then one obtains a *multivariate* distribution.[7] If one obtains the relative frequencies of values of each unique set of multivariate observations then one has a multivariate probability distribution. Again, there are many multivariate mathematical functions that approximate such multivariate probability distributions. Figure 2.7 shows two versions of a bivariate normal distribution.

---

[7] In this case, a bivariate normal distribution.

2.6   Probabilistic (conditional) independence

By definition, two random variables (X, Y) are (unconditionally) independent if the joint probability density of X and Y is the product of the probability density of X and the probability density of Y. Let's use the notation $I(X,\Phi,Y)$ to mean that random variables X and Y are independent, conditional on no other variables (i.e. our empty set Φ). Thus:

$$\text{if } I(X, \phi, Y) \text{ then } P(X, Y) = P(X) \cdot P(Y)$$

For instance, if X and Y are each distributed as a standard normal distribution and they are also independent (Figure 2.7(a)) then the joint probability distribution can be obtained as follows:

$$f(X; 0, 1) = \frac{1}{\sqrt{2\pi}} e^{\frac{-(X)^2}{2}}$$

$$f(Y; 0, 1) = \frac{1}{\sqrt{2\pi}} e^{\frac{-(Y)^2}{2}}$$

$$f(X, Y) = f(X; 0, 1) \cdot f(Y; 0, 1) = \frac{1}{2\pi} e^{\frac{-(X^2+Y^2)}{2}}$$

If two random variables (X, Y) are not (unconditionally) independent then the joint probability density of X and Y is not the product of the two univariate probability densities. If the variables are dependent then one cannot simply multiply one univariate probability density to the other, because we have to take into consideration the interaction of the two (Figure 2.7(a)).

Figure 2.7(a) shows the bivariate normal density function of two independent variables. Note that the mean value of Y is the same (zero) no matter what the value of X, and vice versa; the value of one variable doesn't change with changes in the average, or *expected*, value of the other variable. Knowing that the value of X is 1.5 rather than −1.5 in Figure 2.7(a) doesn't give us any additional information about the values of Y that we will encounter. Figure 2.7(b) shows the bivariate normal density function of two dependent variables. Here, the mean value of Y is not independent of the value of X because, as the value of X increases, the mean value of Y decreases. Now, knowing that the value of X is 1.5 rather than −1.5 tells us that the value of Y is more likely to be closer to −1.5 than to 1.5.

Similarly, X and Y are independent, conditional on ('given') a set of other variables **Z**, if the joint probability density of X and Y given **Z** equals the product of the probability density of X given **Z** and the probability density of Y given **Z** for all values of X, Y and **Z** for which the probability density of **Z** is not equal to zero.[8] The notion of conditional independence will be explained in more detail in Chapter 3. Thus:

$$\text{if } I(X,\mathbf{Z}, Y) \text{ then } P(X, Y|Z) = P(X|Z) \cdot P(Y|Z)$$

Because the notion of conditional independence is sometimes difficult to grasp, here is a verbal description. To say that a variable X is a random variable obeying a particular

---

[8] This can be generalised to joint distributions of sets of variables **X** and **Y** conditional on another set **Z**.

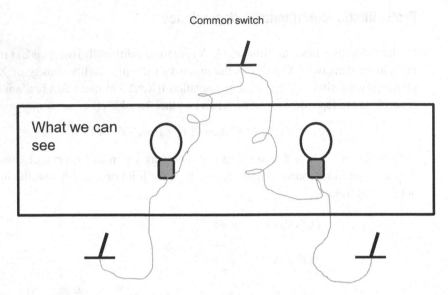

Figure 2.8 An electrical circuit in which two light bulbs are each controlled by the same ON/OFF switch; each bulb can also be turned ON independently with its own unique switch

probability distribution is to say that we do not know for certain what the value of this variable will be when we next measure it on an observational unit but that we do have information about how likely it is to take on different values. What does it mean to say that our random variable X is dependent on some other random variable Y, also obeying some particular probability distribution? This means that we will have even more information about the likely values of X if we are given the value of Y ('conditional on Y') than if we don't know the value of Y. What if our random variable X is dependent on random variable Y, but conditionally independent given the value of some third random variable, Z? This means that whatever information we have about X that is provided by Y (and vice versa) is also provided by Z, and so, once we know Z, Y provides no *additional* information.

Let's make this a bit less abstract. Imagine an electrical circuit in which two light bulbs are each controlled by the same ON/OFF switch (Figure 2.8). Each bulb can also be turned on or off by its own separate switch. We can see the light bulbs but can't see the switches. For each bulb, if either switch (its own unique one or the common one) is on, the bulb lights up. These switches turn ON and OFF at random intervals and are not connected together. Is the light bulb on the right lighted or dark? If I tell you that the light bulb on the left is alight, does this information make it more or less likely that the light bulb on the right is also alight? Since one cause of the left bulb being turned on is that the common switch is ON, knowing that the left bulb is alight increases the *chances* that the bulb on the right is also ON. This means that the two random variables describing the states of the two light bulbs are dependent random variables. Similarly, if I tell you only that the common switch is on, you also know something extra about the state of the bulb on the right: it is certainly alight. This means that these two random variables (the states of the common switch and the second light bulb) are also dependent random variables. Now, if I tell you not only that the common switch is on (information about this random variable) but also that the light bulb on the left is on (information about this second random variable), this second bit of information about the state of the left light bulb gives you no extra information about the state of the right light bulb (the third random variable) that was not already provided by the first bit of information (about the common switch). In other words, the two random variables referring to the two light bulbs are independent (provide no mutual information) given ('conditional on') the information contained in the third random variable.

2.7   The Markov condition

Many ecologists, especially those who study vegetation dynamics, are familiar with Markov chain models (van Hulst 1979). These models predict vegetation dynamics based on a 'transition matrix'. The transition matrix gives the probability that a location that is occupied by a species s_i at time t will be replaced by species s_j at time t+1. The model is 'Markovian' because of the assumption that changes in the vegetation at time t+1 depend at most on the state of the vegetation at time t, but not on states of the vegetation at earlier times. Stated another way, these models are Markovian because they assume that the more distant past (t–1) affects the immediate future (t+1) only indirectly through the present (t); thus, (t–1)→(t)→(t+1). Stated a third way, these models are Markovian because they assume that the state of the random variable representing the immediate future is independent of the more distant past, conditional on the present state.

In the context of causal models, the Markov condition is a property both of a directed acyclic (causal) graph and the joint probability distribution that is generated by the graph. The condition is satisfied if, given a vertex v_i in the graph, or a random variable v_i in the probability distribution, v_i is independent of all ancestral causes given its causal parents.[9] In the context of a causal model, this assumption is simply the claim that, once we know the direct causes of an event, knowledge of more distant (indirect) causes provides no new information. Notice that this is one of the properties stipulated in Chapter 1 for our mathematical language of causality. To use a previous example,[10] assume that the only cause of an increased concentration of photosynthetic enzymes in a leaf is the added fertiliser that was put on the ground, and that the only cause of an increased photosynthetic rate is the increased concentration of photosynthetic enzymes. Then, knowing how much fertiliser was added gives us no new information about the photosynthetic rate once we already know the concentration of photosynthetic enzymes in the leaf.

An important property of probability distributions that obey the Markov condition is that they can be decomposed into conditional probabilities involving only variables and their causal parents. For example, Figure 2.9 shows a causal graph and the joint

[9] $P(v_i)=\Pi P(v_i|parents(v_i))$.   [10] Fertiliser→photosynthetic enzymes→photosynthetic rate.

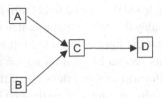

$$P(A,B,C,D) = P(A) \bullet P(B) \bullet P(C \mid \{A,B\}) \bullet P(D \mid C)$$

Figure 2.9 A causal graph involving four variables and the joint probability distribution that is generated by it

probability distribution that is generated by it. This decomposition states that, to know the probability distribution of D, we need only to know the value of C – i.e. P(D|C). To know the probability distribution of C we need only to know the values of A and B – i.e. P(C|{A,B}). A and B are independent, and so to know the joint probability distribution of A and B we need only to know the marginal distributions of A and B – i.e. P(A)P(B).

2.8   The translation from causal models to observational models

Although causal models and observational models are not the same thing, there is a remarkable relationship between the two. Consider the case of causal graphs that do not have feedback relationships – that is, directed paths from some vertex that do not lead back to the same vertex. Theorem 10 by Pearl (1988) states that, for any causal graph without feedback loops (a directed acyclic graph), every d-separation statement obtained from the graph implies an independence relation in the joint probability distribution of the random variables represented by its vertices.

This central insight has been a long time in coming, and I imagine that many readers will wonder whether the effort was worth the return, so let me rephrase it:

Once we have specified the acyclic causal graph, then every d-separation relation that exists in our causal graph must be mirrored in an equivalent statistical independency in the observational data if the causal model is correct.

The above statement is incredibly general. It does not depend on any distributional assumptions of the random variables or on the functional form of the causal relationships. In the same way, if even one statistical independency in the data disagrees with what d-separations of the causal graph predict, the causal model must be wrong. This is the translation device that we needed in order to properly translate the causal claims represented in the directed graph into the 'official' language of probability theory used by statisticians to express observational models. After wading through the jargon developed above, I hope that you will recognise the elegant simplicity of this strategy (Figure 2.10). First, express one's causal hypothesis in a mathematical language (directed graphs) that can properly express the asymmetric types of relationships that scientists imply when they use the language of causality. Second, use the translation device (d-separation) to translate from this directed graph into the well-known

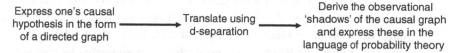

Figure 2.10 The strategy used to translate from a causal model to an observational model

mathematical language (probability theory) that is used in statistics to express notions of association. Finally, determine the types of (conditional) independence relationships that must occur in the resulting joint probability distribution. The hidden causal processes cast observational shadows in nature's shadow play. D-separation is the translation device by which one can predict the shape of these shadows given an hypothesised causal process. The shadows are in the form of conditional independence relationships that the joint probability distribution (and therefore the observational model) must possess if the data are really generated by the hypothesised directed graph.

2.9   Counter-intuitive consequences and limitations of d-separation: conditioning on a causal child

D-separation does not tell us how a causal system will respond following an external manipulation.[11] Rather, d-separation is a mathematical operation that gives the correlational *consequences* of conditioning on a variable in a causal system. One non-intuitive consequence is that two causally independent variables will be correlated if one conditions on any of their common descendants. This is because conditioning on a collider vertex along a path between vertices X and Y means that X and Y are not d-separated. This has important consequences for applied regression analysis and shows how such a method can give very misleading results if these are interpreted as giving information about causal relationships.

Consider a causal system in which two causally independent variables (X and Y) jointly cause variable Z: X→Z←Y. To be more specific, let's assume that the nitrogen content (X) and the stomatal density (Y) of the leaves of individuals of a particular species jointly cause the observed net photosynthetic rate (Z). Further, assume that leaf nitrogen content and stomatal density are causally independent. The causal graph is therefore leaf nitrogen→net photosynthetic rate←stomatal density. Let the functional relationships between these variables be as follows:

$$leaf\ nitrogen = N(0, 1)$$
$$stomatal\ density = N(0, 1)$$
$$net\ photosynthesis = 0.5\ leaf\ nitrogen + 0.5\ stomatal\ density + N(0, 0.707)$$

These three equations can be used to conduct numerical simulations[12] that can demonstrate the consequences of conditioning on a common causal child (the net

---

[11] This is explained later in this chapter.

[12] Such simulations are often called Monte Carlo simulations, after the famous gambling city, because they make use of random number generators to simulate a random process.

photosynthetic rate). Since I will use this method repeatedly in this book, I will explain how it is done in some detail. The first equation states that the leaf nitrogen concentration of a particular plant has causes not included in the model. Since the plant is chosen at random the leaf nitrogen concentration is simulated by choosing at random from a normal distribution whose population mean is zero and whose population standard deviation is 1. The second equation states that the stomatal density of the same leaf of this individual also has causes not included in the model (not the same unknown causes, since otherwise it would not be causally independent) and its value is simulated by choosing another (independent) number from the same probability distribution. The third equation states that the net photosynthetic rate of this same leaf is jointly caused by the two previous variables. The quantitative effect of these two causes on the net photosynthetic rate is obtained by adding 0.5 times the leaf nitrogen concentration plus 0.5 times the stomatal density plus a new (independent) random number taken from a normal distribution whose population mean is zero, whose population variance is 1–$2(0.5^2)$ and whose population standard deviation is therefore the square root of this value; this third random variable represents all those other causes of net photosynthetic rate other than leaf nitrogen and stomatal density, and these other unspecified causes are not causally connected to either of the specified causes.

By repeating this process a large number of times, one obtains a random 'sample' of 'observations' that agree with the generating process specified by the equations.[13] As will described in Chapter 3, this model is actually a very simple path model. Here is some simple R code to do this simulation:

```
leaf.nitrogen←rnorm(1000,0,1)
stomatal.density←rnorm(1000,0,1)
net.photosynthesis←0.5*leaf.nitrogen+0.5*stomatal.density+
rnorm(1000,0,0.707)
```

After generating 1,000 independent 'observations' that agree with these equations, and respecting the causal relationships specified by our causal system, these are the regression equations that are obtained:

leaf nitrogen $= N(0.002, 1.001)$

stomatal density $= N(-0.044, 1.000)$

net photosynthesis $= 0.001 + 0.470$ *leaf nitrogen* $+ 0.514$ *stomatal density* $+ N(0, 0.707)$

Happily, the partial regression coefficients as well as the means and standard deviations of the random variables are what we should find, given sampling variation with a sample size of 1,000.

What happens if we give these data to a friend who mistakenly thinks that leaf nitrogen concentration is actually caused by net photosynthetic rate and stomatal density? In other words, she mistakenly thinks that the causal graph is net photosynthetic rate→leaf

---

[13] Many commercial statistical packages can generate random numbers from specified probability distributions. A good reference, along with FORTRAN subroutines, is the book by Press et al. (1986). The R language incorporates most of these, of which the 'rnorm' function is one.

nitrogen←stomatal density. We know, because we generated the numbers, that leaf nitrogen and stomatal density are actually independent (the Pearson correlation coefficient between them is −0.037, which is not statistically significant at the traditional 0.05 confidence level), but this is the set of regression equations that results from this incorrect causal hypothesis:

net photosynthesis $= N(-0.002, 0.987)$

stomatal density $= N(-0.044, 1.000)$

leaf nitrogen $= 0.000 + 0.654$ *net photosynthesis* $- 0.350$ *stomatal density* $+ N(0, 0.834)$

Tests of significance for the two partial regression coefficients show that each is significantly different from zero at a probability of less than 1×10^{-6}. Why would the multiple regression mistakenly report a highly significant 'effect' of stomatal density on leaf nitrogen when we know that they are both statistically (remember that the correlation between the two was only −0.037) and causally independent (because we made them that way in the simulation)? There is no 'mistake' in the statistics; rather, it is due to our mistranslation between the language of probability and the language of causality. The regression equation is an observational model. It is simply telling us that knowing something about the stomatal density gives us extra information about (or helps to predict) the amount of nitrogen in the leaf, *when we compare leaves with the same net photosynthetic rate*.[14] This is exactly what d-separation, applied to the correct causal graph, tells us will happen: leaf nitrogen and stomatal density, while unconditionally d-separated, are not d-separated (therefore observationally associated) upon conditioning on their causal child (the net photosynthetic rate).

This counter-intuitive claim is easier to understand with an everyday example. Consider again the simple causal world consisting only of rain, watering cans and mud, related as rain→mud←watering cans. Now, in this world there are no causal links between watering cans and rain. Knowing that no one has dumped water from the watering can tells us nothing about whether or not it is raining; we can predict nothing about the occurrence of rain by knowing something about the watering can. On the other hand, if we see that there is mud (the causal child of the two independent causes) *and* we know that no one has dumped water from the watering can (i.e. conditional on this variable) then we can predict that it has rained. Conditioning on a common child of the two causally independent variables (rain and watering cans) renders them observationally dependent. This is because information, unlike causality, is symmetric.

Many researchers believe that the more variables that can be statistically controlled in a multiple regression, the less biased and the more reliable the resulting model. The above example shows this to be wrong and warns against such methods as stepwise multiple regression if the resulting model is to be interpreted as something more than simply a prediction device.[15] This point is rarely mentioned in most statistics texts.

---

[14] Remember that a partial regression coefficient is a function of the partial correlation coefficient. The partial correlation coefficient measures the degree of linear association between two variables upon conditioning on some other set of variables; see Chapter 3.

[15] Even as a prediction device, such models are valid only if no manipulations are done to the population.

2.10   Counter-intuitive consequences and limitations of d-separation: conditioning due to selection bias

There is also an interesting consequence of d-separation that might occur in selection experiments. 'Body condition' is a somewhat vague concept that is sometimes used to refer to the general health and vigour of an animal. It is occasionally operationalised as an index based on a weighting of such things as the amount of subcutaneous fat, the parasite load or other variables judged relevant to the health of the species. Imagine a wildlife manager who wants to select for an improved body condition of bighorn sheep. His measure of body condition is obtained by adding together the thickness of subcutaneous fat in the autumn (cm) and a score for parasite load (0 = none, 1 = average load, 2 = above-average load) as follows: body condition = 0.5fat + parasite load. These two components of body condition are causally unrelated. He decides to protect all individuals whose body condition is greater than 3 and removes all others from the population by allowing hunters to kill them. The causal graph of this process is fat thickness→body condition←parasite load. If someone else were to then measure the fat thickness and parasite load in the remaining population after the selective hunt, she would find that these two variables were correlated, even though there is, in reality, no causal link between the two.[16] This occurs because the selective hunt has removed all those individuals not meeting the selection criterion, and this effectively results in conditioning on body condition.

2.11   Counter-intuitive consequences and limitations of d-separation: feedback loops and cyclic causal graphs

The relationship between d-separation in an acyclic causal model (a DAG) and independencies in a probability distribution is therefore very general. What happens if there are feedback loops in the causal model? We don't know for sure, although this is an area of active research (Richardson 1996b). Spirtes (1995) has shown that d-separation in a cyclic causal model still implies independence in the joint probability distribution that it generates, but only if the relationships are linear. Pearl and Dechter (1996) have also shown that the relationship between d-separation and probabilistic independence also holds if all variables are discrete without any restriction on the functional form of the relationships. Unfortunately, Spirtes (1995) has also shown, by a counter-example, that d-separation does not always imply probabilistic independence when the functional relationships are non-linear and the variables are continuous. There are some grammatical constructs in the language of causality for which no one has yet found a good translation.

---

[16] On the other hand, if this process were to be repeated for a number of generations and the two attributes were heritable, there would develop a causal link, since the average values of the attributes in the next generation would depend on who survives, and this is a consequence of these same attributes in the previous generation.

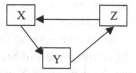

Figure 2.11 A cyclic causal graph that seemingly violates many of the properties of 'causal' relationships

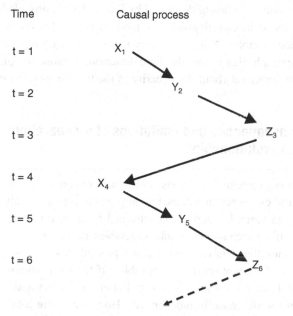

Figure 2.12 The causal relationships between X, Y and Z from Figure 2.11 when the time dimension is included in the causal graph

There are other curious properties of causal models with feedback loops. Consider Figure 2.11: such a causal model seems to violate many properties of causes. The relationship is no longer asymmetric, since X causes Z (indirectly through Y) and Z also causes X. The relationship is no longer irreflexive, since X seems to cause itself through its effects on Y and Z.

These counter-intuitive aspects of feedback loops can be resolved if we remember that causality is a process that must follow time's arrow but causal graphs do not explicitly include this time dimension. Causal graphs with feedback loops represent either a 'time slice' of an ongoing dynamic process or a description of this dynamic process at equilibrium, an interpretation that appears to have been first proposed by F. M. Fisher (1970). Thomas Richardson's very interesting PhD thesis (Richardson 1996b) provides a history of the use and interpretation of such cyclic, or 'feedback', models[17] in economics. A more complete causal description of the process shown in Figure 2.11 is given in Figure 2.12; the subscripts on the vertices index the state of that vertex at a

---

[17] In the literature of structural equation modelling, cyclic or feedback models are called 'non-recursive'. This whole subject area is replete with confusing and intimidating jargon.

given time. From Figure 2.12 we see that, once the explicit time dimension is included in the directed graph, the apparent paradoxes disappear. Rather than circles, when we ignore the time dimension (as in Figure 2.11), we have spirals that never close on themselves when the time dimension is included. Just as the one-year-old Bill Shipley is not the same individual as I am as I write these words, the 'X' that causes Y at time t = 1 will not be that same 'X' that is caused by Z at time t = 4 in Figure 2.12.

Conceived in this way, both acyclic and cyclic causal models represent 'time slices' of some causal process. Samuel Mason, described by Heise (1975), provided a general treatment of feedback loops in causal graphs over 60 years ago for the case of linear relationships between variables. Nonetheless, trying to model causal processes with feedback using directed graphs that ignore this time dimension is more complicated and requires one to make assumptions about the linearity of the functional relationships.

2.12   Counter-intuitive consequences and limitations of d-separation: imposed conservation relationships

Relationships derived from imposed (as opposed to dynamic) conservation constraints are superficially similar to cyclic relationships, but they are conceptually quite different. By 'conservation' I mean variables that are constrained to maintain some conserved property. For instance, if I purchase fruits and vegetables in a store and then count the total amount of money that I have spent, I can represent this as money spent on fruits→total money spend←money spent on vegetables. If the total amount of money that I can spend is not fixed then the amount that I spend on fruits and the amount that I spend on vegetables are causally independent. However, if the total amount of money is fixed, or *conserved*, due to some influence outside the causal system then every dollar that I spend on fruit causes a decrease in the amount of money that I spend on vegetables. There is now a causal link between the amount of money spent on fruits and on vegetables due only to the requirement that the total amount of money be conserved.

There is no obvious way to express such relationships in a causal graph. One might be tempted to modify our original acyclic graph by adding a cyclic path between 'Fruits' and 'Vegetables' but, if we do this, one cannot interpret such a cyclic graph as a static graph of a dynamic process; the conservation constraint is imposed from outside and is not due to a dynamic equilibrium that results from the prior interaction of 'Money spent on fruits' and 'Money spent on vegetables'. In other words, it is not as if I spend one dollar more on fruits at time t = 1, which causes me to spend one dollar less on vegetables at time t = 2, which then causes me to spend one dollar less on fruits at time t = 3, and so on until some dynamic equilibrium is attained. The conservation of the total amount of money spent is imposed from outside the causal system.

One might also be tempted to interpret the conservation requirement as equivalent to physically fixing the total amount of money at a constant value. If this were true then one could maintain the causal graph 'Money spent on fruits→Total money spent←Money spent on vegetables' but with the variable 'Total money spent' being fixed due to the imposed conservation requirement. Because 'Total money spent' is now viewed as being fixed rather than being allowed to randomly vary, 'Money spent on fruits' would not be d-separated from 'Money spent on vegetables' (remember d-separation); this is because 'Total money spent' is the causal child of each of 'Money spent on fruits' and 'Money spent on vegetables'. This would indeed imply a correlation between 'Fruits' and 'Vegetables'. Unfortunately, our causal system does not simply imply that the money spent on fruits is *correlated* with the money spent on vegetables, but that there is actually a causal connection between them that exists only when the conservation requirement is in place. D-separation upon conditioning on a common causal child does not imply that any new causal connections form between the causal parents. Perhaps the best causal representation is to consider that the causal graph 'Money spent on fruits→Total money spent←Money spent on vegetables' is actually replaced by the causal graph 'Money spent on fruits←Total money spent→Money spent on vegetables' upon conditioning on 'Total money spent'.

Systems that contain imposed conservation laws (conservation of energy, mass, volume, number, etc.) cannot yet be properly expressed using directed graphs and d-separation. In fact, such 'causal' relationships resemble Plato's notion of 'formal causes' rather that the 'efficient causes' with which scientists are used to working. However, it is important to keep in mind that this does not apply to conservation relationships that are due to a dynamic equilibrium, for which cyclic graphs can be used, but, rather, to conservation relationships that are imposed independently of the causal parents of the conserved variable.

2.13   Counter-intuitive consequences and limitations of d-separation: unfaithfulness

Let's go back to the relationship between d-separation and probabilistic independence. We now know that, once we have specified the acyclic causal model, every d-separation relation that exists in our causal model must be mirrored in an equivalent statistical independency in the observational data if the causal model is correct. This does not depend on any distributional assumptions of the random variables or on the functional form of the causal relationships. Is the contrary also true? If we find a (conditional) independency in our data then does this mean that there must always be a d-separation relationship in the causal process generating the data? The answer is: almost, but not, always. It is possible, as a limiting case, for there to be independencies in the data that are not predicted by the d-separation criterion. For instance, this can occur if the quantitative causal effect of two variables along different directed paths exactly cancel each other out. Two examples are shown in Figure 2.13. In these causal models we see that no vertex is unconditionally d-separated from any other vertex. Assume that the joint probability distribution over the three vertices is multivariate normal and that the functional relationships between the variables are linear. Under these conditions, we can

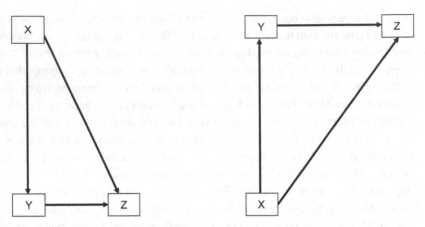

Figure 2.13 Two causal graphs for which special combinations of causal strengths can result in unfaithful probability distributions

use Pearson's partial correlation to measure probabilistic independence.[18] By definition, the partial correlation between X and Z, conditioned on Y, is given by:

$$\rho_{XY \cdot Z} = \frac{\rho_{XY} - \rho_{XY}\rho_{ZY}}{\sqrt{(1 - \rho_{XY}^2)(1 - \rho_{ZY}^2)}}$$

It can happen that $\rho_{XZ \cdot Y} = 0$ (i.e. $\rho_{XZ} = \rho_{XY}\rho_{ZY}$) even though X and Z are not d-separated given Y, if the correlations between each pair of variables exactly cancel each other. Using the rules of path analysis (Chapter 4), this will happen only if Y is perfectly correlated with X in the first model in Figure 2.13, or if the indirect effect of X on Z is exactly equal in strength but opposite in sign to the direct effect of X on Z.

When this occurs we say that the probability distribution is *unfaithful* to the causal graph (Pearl 1988; Spirtes, Glymour and Scheines 1993). I will call such probabilistic independencies that are not predicted by d-separation, and that depend on a particular combination of quantitative effects, *balancing independencies*, to emphasise that such independencies require a very peculiar balancing of the positive and negative effects between the variables along different paths. Clearly, this can occur only under very special conditions, and anyone who wanted to link a causal model with such an unfaithful probability distribution would require strong external evidence to support such a delicate balance of causal effects. This is not to say that such things are impossible. It sometimes happens that an organism attempts to maintain some constant set-point value by balancing different causal effects; an example is the control of the internal CO_2 concentration of a leaf, as described in Chapter 3. Essentially, in proposing such a claim we are saying that nature is conspiring to give the impression of independence by exactly balancing the positive and negative effects.

---

[18] Pearson partial correlations are explained more fully in Chapter 3.

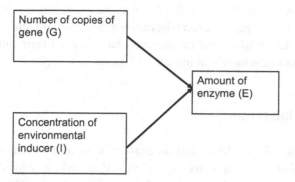

Figure 2.14 A biological example of a causal process that can potentially result in context-sensitive independence

2.14   Counter-intuitive consequences and limitations of d-separation: context-sensitive independence

Another way in which independencies can occur in the joint probability distribution without being mirrored in the d-separation criterion is as a result of *context-sensitive* independence. An example of this in biology is enzyme induction.[19] Imagine a case in which the number of functional copies of a gene (G) determines the rate at which some enzyme is produced (E). If there are no functional copies of the gene then the enzyme is never produced. However, the rate at which these genes are transcribed is determined by the amount of some environmental inducer (I). If the environment completely lacks the inducer then no genes are transcribed and the enzyme is still never produced. It is possible to arrange an experimental set-up in which the number of functional genes is causally independent of the concentration of the inducer in the environment.[20] The number of functional genes and the concentration of the inducer are both causes of enzyme production. We can construct a causal graph of this process (Figure 2.14).

Now, applying d-separation to the causal graph in Figure 2.14 predicts that G is independent of I but that E is dependent on both G and I. However, if there are no copies of G (i.e. G = 0) then the concentration of the inducer will be independent of the amount of enzyme that is produced (which will be zero). Similarly, if there is no inducer (i.e. I = 0) then the number of copies of the gene will be independent of the amount of enzyme that is produced (which will be zero). In other words, for the special cases of G = 0 and/or I = 0, d-separation predicts a dependence when, in fact, there is independence. Note that the d-separation theorem still holds; d-separation does not predict any *independence* relations that do not exist. So long as the experiment involves experimental

---

[19] A classic example is the *lac* operon of *E. coli*, whose transcription in the presence of lactose induces the production of β-galactosidase, lac permease and transacetylase, thus converting lactose into galactose and glucose (De Robertis and De Robertis 1980).

[20] Whether this would be true in the biological population is an empirical question. Perhaps the presence of a functional gene was selected on the basis of the presence of the inducer. In this case, the inducer would be a cause of the presence (and perhaps the number of copies) of the gene.

units, at least some of which include $G \neq 0$ and $I \neq 0$, the d-separation criterion still predicts both probabilistic independence and dependence. Similarly, if both G and I are true random variables (i.e. in which the experimenter has not fixed their values) then any reasonably large random sample will include such cases.

2.15   The logic of causal inference

Now that we have our translation device and are aware of some of the counter-intuitive results and limitations that can occur with d-separation, we have to be able to infer causal consequences from observational data by using this translation device. The details of how to carry out such inferences will occupy the rest of this book. However, before we look at the statistical details we must first consider the logic of causal and statistical inferences.

Since we are talking about the logic of inferences from empirical experience, it is useful to briefly look at what philosophers of science have had to say about valid inference. Logical positivism, itself rooted in the British empiricism of the nineteenth century that so influenced people such as Karl Pearson,[21] was dominant in the twentieth century up to the middle of the 1960s. This philosophical school was based on the verifiability theory of meaning; to be meaningful, a statement had to be of a kind that could be shown to be either true or false. For logical positivism, there were only two kinds of meaningful statements. The first kind was composed of *analytical* statements (tautologies, mathematical or logical statements) whose truth could be determined by deducing them from axioms or definitions. The second kind was composed of *empirical* statements that were either self-evident observations ('The water is 23°C') or could be logically deduced from combinations of basic observations whose truth was self-evident.[22] Thus, logical positivists emphasised the hypothetico-deductive method: a hypothesis was formulated to explain some phenomenon by showing that it followed deductively from the hypothesis. The scientist attempted to validate the hypothesis by deducing logical consequences of the hypothesis that were not involved in its formulation and testing these against additional observations. A simplified version of the argument goes like this:

- if my hypothesis is true then consequence C must also be true;
- consequence C is true;
- therefore, my hypothesis is true.

Readers will immediately recognise that such an argument commits the logical fallacy of affirming the consequent. It is possible for the consequence to be true even though the hypothesis that deduced it is false, since there can always be other reasons for the truth of C.

---

[21] This is explored in more detail in Chapter 3.

[22] That even such simple observational or experiential statements cannot be considered objectively self-evident was shown at the beginning of the twentieth century by Duhem (1914).

Popper (1980) has pointed out that, although we cannot use such an argument to verify hypotheses, we can use it to reject them without committing any logical fallacy:

- if my hypothesis is true then consequence C must also be true;
- consequence C is false;
- therefore, my hypothesis is false.

Practising scientists would quickly recognise that this argument, though logically acceptable, has important shortcomings when applied to empirical studies. It was recognised as long ago as the turn of the twentieth century (Duhem 1914) that no hypothesis is tested in isolation. Every time that we draw a conclusion from some empirical observation we rely on a whole set of auxiliary hypotheses (A_1, A_2, etc.) as well. Some of these have been repeatedly tested so many times and in so many situations that we scarcely doubt their truth. Other auxiliary assumptions may be less well established. These auxiliary assumptions will typically include ones concerning the experimental or observational background, the statistical properties of the data, and so on. Did the experimental control really prevent the variable from changing? Were the data really normally distributed, as the statistical test assumes? Such auxiliary assumptions are legion in every empirical study, including the randomised experiment, the controlled experiment or the methods described in this book involving statistical controls. A large part of every empirical investigation involves checking, as best as one can, such auxiliary assumptions so that, once the result is obtained, blame or praise can be directed at the main hypothesis rather than at the auxiliary assumptions.

Popper's process of inference might therefore be simplistically paraphrased[23] as:

- if auxiliary hypotheses A_1, A_2, ... A_n are true, and
- if my hypothesis is true then consequence C must be true;
- consequence C is false;
- therefore, my hypothesis is false.

Unfortunately, to argue in such a manner is also logically fallacious. Consequence C might be false not because the hypothesis is false but, rather, because one or more of the auxiliary hypotheses are false. The empirical researcher is now back where he or she started: there is no way of determining either the truth or falsity of his or her hypothesis in any absolute sense from logical deduction. This conclusion applies just as well to the randomised experiment, the controlled experiment or the methods described in this book. However, most biologists would recognise the falsifiability criterion as important to science, and would probably modify the simplistic paraphrase of Popper's inference by attempting to judge which – the auxiliary hypotheses and background conditions or the hypothesis under scrutiny – is on firmer empirical ground. If the auxiliary assumptions seem more likely to be true than the hypothesis under scrutiny, yet if the data do not accord with the predicted consequences then the hypothesis would be tentatively rejected. If there are no reasoned arguments to suggest that the auxiliary assumptions

---

[23] *Simplistic* because it is wrong. Popper did not make such a claim.

are false, and the data also accord with the predictions of the hypothesis under scrutiny, then the hypothesis would be tentatively accepted.

Pollack (1986) calls such reasoning *defeasible* reasoning.[24] Revealingly, practising scientists have explicitly described their inferences in such terms for a long time. At the turn of the twentieth century Thomas Huxley likened the decision to accept or reject a scientific hypothesis to a criminal trial in a court of law (reproduced by Rapport and Wright 1963) in which guilt must be demonstrated beyond reasonable doubt.

Let's apply this reasoning to the examples in Chapter 1 involving the randomised and the controlled experiments. Later, I will apply the same reasoning to the methods involving statistical control.

Here is the logic of causal inference with respect to the randomised experiment to test the hypothesis that fertiliser addition increases seed yield.

- If the randomisation procedure was properly done so that the alternate causal explanations were excluded;
- if the experimental treatment was properly applied;
- if the observational data do not violate the assumptions of the statistical test;
- if the observed degree of association was not due to sampling fluctuations;
- then by the causal hypothesis the amount of seed produced will be associated with the presence of the fertiliser.
- There is/is not an association between the two variables.
- Therefore, the fertiliser addition might have caused/did not cause the increased seed yield.

This list of auxiliary assumptions is only partial. In particular, we still have to make the basic assumption linking causality to observational associations, as described in Chapter 1. At this stage we must either reject one of the auxiliary assumptions or tentatively accept the conclusion concerning the causal hypothesis. If the probability associated with the test for the association is sufficiently large, traditionally above 0.05,[25] then we are willing to reject one of the auxiliary assumptions (the observed measure of

---

[24] *Defeasible* because it can be *defeated* with subsequent evidence.

[25] See Cowles and Davis (1982b) for a history of the 5 per cent significance level. The first edition of Fisher's classic book states (Fisher 1925: 47): 'It is convenient to take this point as a limit in judging whether a deviation is to be considered significant or not. Deviations exceeding twice the standard deviation are thus formally regarded as significant.' The words 'convenient' and 'formal' emphasise the somewhat arbitrary nature of this value. In fact, this level can be traced back even further to the use of three times the probable error (about two-thirds of a standard deviation). Strictly speaking, twice the standard deviation of a normal distribution gives a probability level of 0.0456; perhaps Fisher simply rounded this up to 0.05 for his tables. Pearson and Kendall (1970) record Pearson's reasons at the turn of the century: $p = 0.5586$, 'thus we may consider the fit remarkably good'; $p = 0.28$, 'fairly represented'; $p = 0.10$, 'not very improbable'; $p = 0.01$, 'this very improbable result'. Note that some doubt began at 0.1 and Pearson was quite convinced at $p = 0.01$. The midpoint between 0.1 and 0.01 is 0.05. Cowles and Davis (1982a) conducted a small psychological experiment by fooling students into believing that they were participating in a real betting game (with money) that was, in reality, fixed. The object was to see how unlikely a result people would accept before they began to doubt the fairness of the game. They found that, 'on average, people do have doubts about the operation of chance when the odds reach about 9 to 1 [i.e. 0.09], and are pretty well convinced when the odds are 99 to 1 [i.e. ~0.0101]...If these data are accepted, the 5% level would appear to have the appealing merit of having some grounding in common sense.'

association was not due to sampling fluctuations) rather than accept the causal hypothesis. Thus, we reject our causal hypothesis. This rejection must remain tentative. This is because another of the auxiliary assumptions (not listed above) is that the sample size is large enough to permit the statistical test to differentiate between sampling fluctuations and systematic differences. However, note that it is not enough to propose any old reason to reject one of the auxiliary assumptions; we must propose a reason that has empirical support. We must produce *reasonable* doubt; in the context of the assumption concerning sampling fluctuations, scientists generally require a probability above 0.05.

Here it is useful to quote from the first edition of Fisher's (1925: 504) influential *Statistical Methods for Research Workers*: 'Personally, the writer prefers to set a low standard of significance at the 5 per cent point, and ignore entirely all results which fail to reach this level. A scientific fact should be regarded as experimentally established only if a properly designed experiment rarely fails to give this level of significance.' It is clear that Fisher was demanding reasonable doubt concerning the null hypothesis, since he asked only that a result 'rarely fail' to reject it. What if the probability of the statistical test was sufficiently small, such as 0.01, that we do not have reasonable grounds to reject our auxiliary assumption concerning sampling fluctuations? What if we do not have reasonable grounds to reject the other auxiliary assumptions? What if the sampling variation was small compared to a reasonable effect size? Then we must tentatively accept the causal hypothesis.

Again, this acceptance must remain tentative, since new empirical data might provide such reasonable doubt. Is there any automatic way of measuring the relative support for or against each of the auxiliary assumptions and of the principal causal hypothesis? No. Although the support (in terms of objective probabilities) for some assumptions can be obtained – for instance, those concerning the normality or linearity of the data – there are many other assumptions that deal with experimental procedure or a lack of confounding variables for which no objective probability can be calculated. This is one reason why so many contemporary philosophers of science prefer Bayesian methods to frequency-based interpretations of probabilistic inference (see, for example, Howson and Urbach 1989). Such Bayesian methods suffer from their own set of conceptual problems (Mayo 1996). In the end, even the randomised experiment requires subjective decisions on the part of the researcher. This is why the independent replication of experiments in different locations, using slightly different environmental or experimental conditions and therefore having different sets of auxiliary assumptions, is so important. As the causal hypothesis continues to be accepted in these new experiments, it becomes less and less reasonable to suppose that incorrect auxiliary assumptions are conspiring to give the illusion of a correct causal hypothesis.

Here is the logic of our inferences with respect to the controlled experiment to test the hypothesis that renal activity causes the change in the colour of the venous renal blood, described in Chapter 1.

- If the activity of the kidney was effectively controlled;
- if the colour of the blood was accurately determined;

- if the experimental manipulation did not change some other uncontrolled attribute besides kidney function that is a common cause of the colour of blood in the renal vein before entering, and after leaving, the kidney;
- if there was not some unknown (and therefore uncontrolled) common cause of the colour of blood in the renal vein before entering, and after leaving, the kidney;
- if a rare random event did not occur;
- then, by the causal hypothesis, blood will change colour only when the kidney is active.
- The blood did change colour in relation to kidney activity.
- Therefore, kidney activity does cause the change in the colour of blood leaving the renal vein.

Again, this list of auxiliary assumptions is only partial. Again, one must either produce reasonable evidence that one or more of the auxiliary assumptions is false or tentatively accept the hypothesis. In particular, more of these auxiliary assumptions concern properties of the experiment or of the experimental units for which we cannot calculate any objective probability concerning their veracity. This was one of the primary reasons why Fisher rejected the controlled experiment as inferior. In the controlled experiment these auxiliary assumptions are more substantial, but it is still not enough to raise any old doubt; there must be some empirical evidence to support the decision to reject one of these assumptions. Since we want the data to cast doubt or praise on the principal causal hypothesis and not on the auxiliary assumptions, we will ask only for evidence that casts reasonable doubt. It is not enough to reject the causal hypothesis simply because 'experimental manipulation *might have* changed some other uncontrolled attribute besides kidney function that is a common cause of the colour of blood in the renal vein before entering, and after leaving, the kidney'. We must advance *some* evidence to support the idea that such an uncontrolled factor in fact exists. For instance, a critic might reasonably point out that some other attribute is also known to be correlated with blood colour and that the experimental manipulation was known to have changed this attribute. Although such evidence would certainly not be sufficient to demonstrate that this other attribute definitely was the cause, it might be enough to cast doubt on the veracity of the principal hypothesis.

This is the same criterion that we used beforehand to choose a significance level in our statistical test. Rejecting a statistical hypothesis because the probability associated with it was, say, 0.5 would not be reasonable. Certainly, this gives some doubt about the truth of the hypothesis, but our doubt is not sufficiently strong that we would have a clear preference for the contrary hypothesis. It is the same defeasible argument that might be raised in a murder trial. If the prosecution has demonstrated that the accused had a strong motive, if it produced a number of reliable eyewitnesses and if it produced physical evidence implicating the accused then it would not be enough for the defence to simply claim that 'maybe someone else did it'. However, if the defence could produce some contrary empirical evidence implicating someone else then reasonable doubt would be cast on the prosecution's argument. In fact, I think that the analogy between testing a scientific hypothesis and testing the innocence of the accused in a criminal trial can be stretched even further. There is no objective definition of reasonable doubt in a criminal trial; what is reasonable is decided by the jury in the context of legal precedence. In the same way, there is no objective definition of reasonable doubt in a scientific claim. In the first instance, reasonable doubt is decided by the peer reviewers of the scientific article, and, ultimately, reasonable doubt is decided by the entire scientific community. One should not conclude from this that such decisions are purely subjective acts and that scientific claims are therefore simply relativistic stories whose truth is decided by fiat by a power elite. Judgements concerning reasonable doubt and statistical significance are constrained in that they must deliver predictive agreement with the natural world in the long run.

Now let's look at the process of inference with respect to causal graphs.

- If the data were generated according to the causal model;
- if the causal process generating the data does not include non-linear feedback relationships;
- if the statistical test used to test the independence relationships is appropriate for the data;
- if a rare sampling fluctuation did not occur;
- then each d-separation statement will be mirrored by a probabilistic independence in the data.
- At least one predicted probabilistic independence did not exist.
- Therefore, the causal model is wrong.

By now, you should have recognised the similarity of these inferences. We can prove by logical deduction that d-separation implies probabilistic independence in such directed acyclic graphs. We can prove that, barring the case of non-linear feedback with non-normal data (an auxiliary assumption), every d-separation statement obtained from any DAG must be mirrored by a probabilistic independence in any data that were generated according to the causal process that was coded by this DAG. We can prove that, barring a non-faithful probability distribution (another auxiliary assumption, but one that is relevant only if the causal hypothesis is accepted, not if it is rejected), there can be no independence relation in the data that is not mirrored by d-separation. So, if we have used a statistical test that is appropriate for our data and have obtained a probability that is sufficiently low to reasonably exclude a rare sampling event, we must tentatively reject our causal model. As in the case of the controlled experiment, if we are led to tentatively accept our causal model then this will require that we can't reasonably propose an alternative causal explanation that also fits our data as well. As always, it is not sufficient to simply claim that '*maybe* there is such an alternative causal explanation'. One must be able to propose an alternative causal explanation that has at least enough empirical support to cast reasonable doubt on the proposed explanation.

2.16   Statistical control is not always the same as physical control

We have now seen how to translate from a causal hypothesis into a statistical hypothesis. First, transcribe the causal hypothesis into a causal graph showing how each variable is causally linked to other variables in the form of direct and indirect effects. Second, use

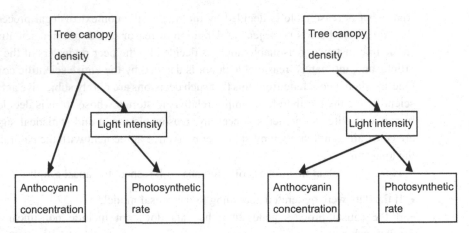

Figure 2.15 Two different causal scenarios linking the same four variables

the d-separation criterion to predict what types of probabilistic independence relationships must exist when we observe a random sample of units that obey such a causal process. In Chapter 1 I alluded to the fact that the key to a controlled experiment is *control* over variables, not how the control is produced. It is time to look at this more carefully. The relationship between control through external (experimental) manipulation and probability distributions is given by the manipulation theorem (Spirtes, Glymour and Scheines 1993). Let me introduce another definition.

Back-door path: given two variables, X and Y, and a variable F that is a causal ancestor of both X and Y, a back-door path goes from F to each of X and Y. Thus X← ← ←F→ → →Y.

Whenever someone directly physically controls some set of variables through experimental manipulation, he or she is changing the causal process that is generating the data. Whenever someone physically fixes some variable at a given level the variable stops being random[26] and is then under the complete control of the experimenter. In other words, whatever causes might have determined the 'random' values of the variable before the manipulation have been removed by the manipulation. The only direct cause of the controlled variable after the manipulation has been performed is the will of the experimenter.

Imagine that someone has randomly sampled herbaceous plants growing in the understorey of an open stand of trees. The measured variables are the light intensities experienced by the herbaceous plants, their photosynthetic rates and the concentration of anthocyanins (red-coloured pigments) in their leaves. Each of these three variables is random since they are outside the control of the researcher. One cause of variation in light intensity at ground level is the presence of trees. The researcher proposes two alternative causal explanations for the data (Figure 2.15).

---

[26] The notion of 'randomness' is another example of a concept that is regularly invoked in science even though it is extraordinarily difficult to define.

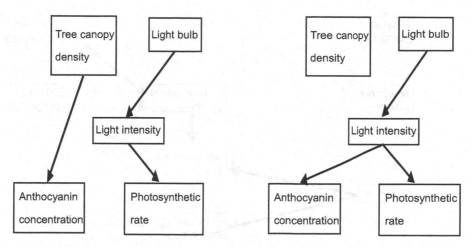

Figure 2.16 Experimental manipulation of the causal systems that are shown in Figure 2.15

To test between these two alternative explanations, the researcher experimentally manipulates light intensity by installing a neutral shade cloth between the trees and the herbs, and then adds an artificial source of lighting. Remembering that this is a controlled experiment, the researcher would want to take precautions to ensure that other environmental variables (temperature, humidity, etc.) are not changed by this manipulation. The manipulation theorem, in graphical terms,[27] states that the probability distribution of this new causal system can be described by taking the original (unmanipulated) causal graphs, removing any arrows leading into the manipulated variable (light intensity) and adding a new variable representing the new causes of the manipulated variable (Figure 2.16).

D-separation will predict the pattern of probabilistic independencies in this new causal system. Notice that anthocyanin concentration is d-separated from the photosynthetic rate according to the first hypothesis in both the manipulated system (Figure 2.16), when light intensity is experimentally fixed, and the unmanipulated system (Figure 2.15), when light intensity is statistically fixed by conditioning. The same d-connection relationships between anthocyanin concentration and the photosynthetic rate hold in the second scenario whether based on physically or statistically controlling light intensity. In other words, statistical and experimental controls are alternative ways of doing the same thing: predicting how the associations between variables will change once other sets of variables are 'held constant'. This does not mean that the two types of control always predict the same types of observational independencies in our data; remember the example of d-separation upon conditioning on a causal child, described previously. Once we have a way of measuring how closely the predictions agree with the observations, we have a way of testing, and potentially falsifying, causal hypotheses even in cases in which we cannot physically control the variables of interest.

---

[27] The manipulation theorem also predicts how the joint probability distribution in the new manipulated causal system differs, if at all, from the original distribution before the manipulation.

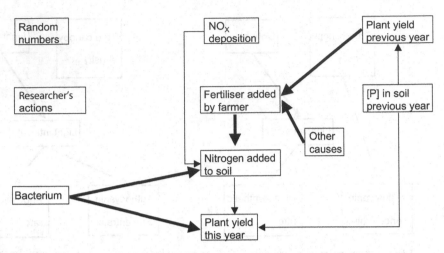

Figure 2.17 A hypothetical causal system before experimental manipulation

With these notions we can now go back and look again at the randomised experiment in Chapter 1. Let's consider an example involving an agricultural researcher who is interested in determining if, and how, the addition of a nitrate fertiliser can increase plant yield. To be more specific, imagine that the plant is alfalfa, which contains a bacterium in its roots that is capable of directly fixing atmospheric nitrogen (N_2). The researcher meets a farmer, who tells him that adding such a nitrate fertiliser in the past had increased the yield of alfalfa. After further questioning, the researcher learns that the farmer had tended to add more fertiliser to those parts of the field that, in previous years, had produced the lowest yields. The researcher knows that other things can also affect the amount of fertiliser that a farmer will add to different parts of a field. For instance, parts of the field that cause the farmer to slow down the speed of his tractor will therefore tend to receive more fertiliser, and so on.[28] Imagine that, unknown to the researcher, the actual causal processes are as shown in Figure 2.17. There are only three sources of nitrogen: the nitrate that is added to the soil by the fertiliser, by NO_X deposition and from N_2 fixation by the bacterium. The amount of fertiliser added by the farmer in different parts of the field is determined by the yield of plants the previous year as well as the contours of the field. In reality, all the sources of nitrogen and the soil phosphate level are causes of yield.

Before experimenting with this system, the researcher has previous causal knowledge of only part of it, shown by the thicker arrows in Figure 2.17. He knows that the bacterium will increase the alfalfa yield. He knows that the bacterium will increase the nitrate concentration in the soil. He knows that the yield of alfalfa in previous years has affected the amount of nitrate fertiliser that the farmer had added, and he knows that the amount of added nitrate fertiliser is *associated* with increased yields. What he doesn't know is whether the nitrogen added to the soil is the cause of the subsequent plant yield.

---

[28] This scenario will work only if the governor is not functioning!

Since the experiment has not yet begun, the 'Random numbers' in Figure 2.17 do not affect any actions by the researcher, and he has no causal effect on any variable in the system. The 'Random numbers' and the 'Researcher's actions' are therefore causally independent of each other and of every other variable in the system.

Based only on the *partial* knowledge shown by the thick arrows, can the researcher use d-separation and statistical control to confidently infer that the added nitrate fertiliser causes an increase in plant yield? No. He knows that the yields of previous years were a cause of the farmer's fertiliser addition and not vice versa; therefore, he knows that he can block any possible back-door path between the amount of fertiliser added and the plant yield that passes through the variable 'Plant yield previous year'. Unfortunately, he also knows that this was not the *only* possible cause of the amount of fertiliser added by the farmer to different parts of the field. Therefore, he cannot exclude the possibility that there is some back-door path that does not include the variable 'Plant yield previous year' and that is generating the association between the present plant yield and the amount of fertiliser added by the farmer. Remember that, to invoke such a possibility, one must be able to present some empirical evidence that such a back-door path might exist, but this would be easy to do. For instance, if the tractor slows down[29] as it begins to go up a slope (and therefore deposits more fertiliser), and if water (which is known to increase plant yield) tends to accumulate at the bottom of the slope, then we have a possible back-door path (fertiliser added←tractor slowed down←hill→water accumulation→plant yield).

The researcher knows that it is possible to randomly assign different levels of nitrate fertiliser to plots of ground in a way that is not caused by any attribute of these plots. He persuades the farmer not to add any fertiliser. The previous cause of the amount of fertiliser added has been erased in this new context, and so the arrow from 'Plant yield previous year' to 'Fertiliser added by farmer' is removed from the causal graph. Since the farmer has agreed not to add any fertiliser, the value of this variable is fixed at zero, and so all arrows coming out of this variable are also erased. The researcher decides to add nitrate fertiliser to different plots at either 0 or 20 kg/hectare based only on the value of randomly chosen numbers. Therefore, we add an arrow from 'Random numbers' to 'Researcher's actions' and an arrow from 'Researcher's actions' to 'Nitrogen added to soil'. Remember that an arrow signifies a *direct* cause – i.e. a causal effect that is not mediated through other variables in the causal explanation. Consequently, we cannot add an arrow from 'Researcher's actions' to 'Plant yield this year' unless we believe that the researcher's actions do cause a change in plant yield this year and that this cause is not completely mediated by some other set of variables in the causal system. The causal structure that exists after the experimental manipulation is shown in Figure 2.18.

Given this new causal scenario, we can now use d-separation to determine if there is a causal relationship between the amount of nitrate fertiliser added by the researcher and the plant yield that year. If one can trace a directed path beginning at 'Researcher's actions' and passing through 'Plant yield this year' by following the direction of the arrows then the two are not d-separated. This necessarily implies that there will be a

---

[29] Readers knowledgeable about tractors will have to assume that the governor of the tractor is defective.

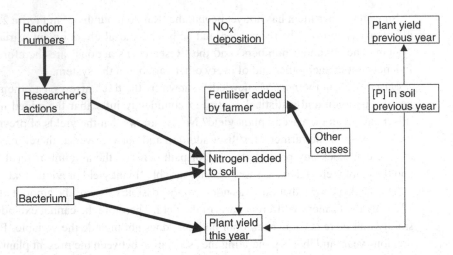

Figure 2.18 Experimental manipulation of the causal system shown in Figure 2.17 based on a randomised experiment

statistical association between the two variables. If no such directed path exists then the addition of nitrate fertiliser by the researcher does not cause a change in plant yield this year. In fact, these two variables are not d-separated in this causal graph, and so such a randomised experiment would detect an effect of fertiliser addition on plant yield. In Chapter 1 I said that if there is a statistical association between two variables, X and Y, then there can be only three elementary (but not mutually exclusive) causal explanations: either X causes Y (shown by a directed path leading from X and passing into Y), Y causes X (shown by a directed path leading from Y and passing into X) or there is some other variable (F) that is a cause of both X and Y (shown by a back-door path from F and into both X and Y). Because the researcher has agreed to act completely in accordance with the results of the randomisation process, we know that no arrows point into 'Researcher's actions' except the one coming from 'Random numbers'. The random numbers are not caused by any attribute of the system. Therefore, the researcher knows that there can be no back-door paths confounding his results because he knows that there are no arrows pointing into 'Researcher's actions' except for the one coming from 'Random numbers'. If there is a statistical association between 'Researcher's actions' and 'Plant yield this year' that cannot reasonably be attributed to random sampling fluctuations then the researcher knows that the association must be due to a directed path coming from 'Researcher's actions' and passing through 'Plant yield this year'. This is why such a randomised experiment, in conjunction with a way of calculating the probability of observing such a random event, can provide a strong inference concerning a causal effect. The reader should note that even the randomisation process might not allow the researcher to conclude that 'Nitrogen added to the soil' is a *direct* cause of increased plant yield. In Figure 2.18 the researcher has already concluded that there is a back-door path from these two variables emanating from the presence of the nitrogen-fixing bacterium, and so to make such a claim he would have to provide evidence beyond

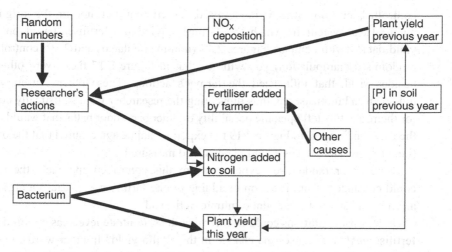

Figure 2.19 Experimental manipulation of the causal system shown in Figure 2.17 that is not based on a randomised experiment

a reasonable doubt that his actions did not somehow affect the abundance or activity of these bacteria.

Now, let's modify the causal scenario a bit. Imagine that the farmer has agreed to let the researcher conduct an experiment and promises not to add any fertiliser while the experiment is in progress, but insists that the researcher ensure that the parts of the field that had produced the lowest plant yield last year must absolutely receive more fertiliser this year. The researcher decides to allocate the fertiliser treatment in the following way: after choosing the random numbers as before, he also adds 5 kg/hectare to those plots whose previous yields were below the median value. Figure 2.19 shows this causal scenario. By doing so he is no longer conducting a true randomised experiment.

Now, using d-separation we see that there would be an association between 'Researcher's actions' and 'Plant yield this year' even if there was no causal effect of the amount of nitrate fertiliser added and the plant yield that follows. The reason is that there is now a back-door path linking the two variables through the common cause '[P] in soil previous year'. This path has been created by allowing 'Plant yield previous year' to be a cause of the researcher's actions. Yet all is not lost. He systematically assigned fertiliser levels based *only* on the yield data of the previous year plus the random numbers. This means that he knows that there are only two independent causes determining how much fertiliser each plot received. He also knows, because of d-separation, that any causal signal passing from any unknown variable into 'Researcher's actions' through 'Plant yield previous year' is blocked if he statistically controls for 'Plant yield previous year'. He can make this causal inference without knowing anything else about the causal system. Therefore, he knows that, once he statistically conditions on 'Plant yield previous year', any remaining statistical association, if it exists, must be due to a causal signal coming from 'Researcher's actions' and following a directed path into 'Plant yield this year'. This causal inference is just as solid as in the previous example, in which treatment allocation was due only to random numbers. What allows him to do this in this controlled, but not strictly randomised, experiment but not in the original non-manipulated system in which the farmer applied the fertiliser based on previous yield data? If you compare Figures 2.17 (non-manipulated) and 2.19 (controlled, non-randomised manipulation) you will see that in Figure 2.17 there were other causes, besides yield, that influenced the farmer's actions. These other causes were both unknown and unmeasured, thus preventing the researcher from statistically controlling for them, and this left open the possibility of other back-door paths that would confound the causal inference. In Figure 2.19 the experimental design ensured that the only cause (i.e. previous yields) was already known and measured.

Using either randomised experiments or this controlled approach, the researcher could conclude[30] that his action of adding nitrate fertiliser does cause a change in the alfalfa yield and in the amount of nitrate in the soil.

Under what conditions could he infer that the *soil* nitrate levels (as opposed to nitrate fertiliser *addition*) causes the change in the alfalfa yield? In other words, what would allow him to infer that the fertiliser addition increased soil nitrate concentration, which, in turn, increased the alfalfa yield? Although he was able to randomise and to exert experimental control over the amount of fertiliser added to the soil, this is not the same as randomly assigning values of soil nitrate to the plots, and he has not exerted *direct* experimental control over soil nitrate levels. Because of this he cannot unambiguously claim that the experiment has demonstrated that soil nitrate levels cause an increase in plant yield. In other words, there might be a back-door path from the fertiliser addition to each of soil nitrate and plant yield even though soil nitrate levels may have had no direct effect on plant yield. For instance, perhaps the fertiliser addition reduced the population level of some soil pathogen whose presence was reducing plant growth.

He can test the hypothesis that the association between soil nitrate levels and plant yield is due only to a back-door path emanating from the amount of added fertiliser by measuring soil nitrate levels and then statistically controlling for this variable. D-separation predicts that, if this new causal hypothesis is true, the effect of fertiliser addition will still exist. If the effect of fertiliser addition was due only to its effect on soil nitrate levels then d-separation predicts that the effect of fertiliser addition on plant yield will disappear once the soil nitrate level is statistically controlled. Since he knows, from previous biological knowledge, that there is at least one back-door path linking soil nitrate and plant yield (due to the effect of the nitrogen-fixing bacteria in the root nodules) then he can determine if there is some other common cause generating a back-door path if he can measure and then control for the amount of this bacterium.

2.17   A taste of things to come

Up to now we have been inferring the properties of the observational model (the joint probability distribution) given the causal model that generates it. Can we also do the

---

[30] Given the typical assumptions of the statistical test used, and assuming that he is not in the presence of an unusual event.

contrary? If we know the entire pattern of statistical independencies and conditional independencies in our observational model, can we specify the causal structure that must have generated it? No. It is possible for different causal structures to generate the same set of d-separation statements and, therefore, the same pattern of independencies. Nonetheless, it is possible to specify a *set* of causal models that all predict the same pattern of independencies that we find in the probability distribution; these are called *equivalent* models, and these are described in Chapter 8. By extension, we can exclude a vast group of causal models that could not have generated the observational data. There are two important consequences of this.

First, after proposing a causal model and finding that our observational data are consistent with it (i.e. that the data do not contradict any of the d-separation statements of our causal model), we can determine which other causal models would also be consistent with our data.[31] By definition, our data cannot distinguish between such equivalent causal models, and so we will have to devise other sorts of observations to differentiate between them.

Second, we can exploit the independencies in our observational data to generate such equivalent models even if we do not yet have a causal model that is consistent with our data. This leads to the topic of exploratory methods, which is discussed in Chapter 8. Such exploratory methods are very useful when theory is not sufficiently well developed to allow us to propose a causal explanation – a condition that occurs often in organismal biology.

However, before delving into these topics, we must first look at the mechanics of fitting such observational models, generating their correlational 'shadows' and comparing the observed shadows (the patterns of correlation and partial correlation) to the predicted shadows. This leads into the topic of path models and, more generally, structural equations. Chapters 3 to 7 deal with these topics.

---

[31] This statement must be tempered on account of practical problems involving statistical power.

3 Sewall Wright, path analysis and d-separation

3.1 A bit of history

The ideal method of science is the study of the direct influence of one condition on another in experiments in which all other possible causes of variation are eliminated. Unfortunately, causes of variation often seem to be beyond control. In the biological sciences, especially, one often has to deal with a group of characteristics or conditions which are correlated because of a complex of interacting, uncontrollable, and often obscure causes. The degree of correlation between two variables can be calculated with well-known methods, but when it is found it gives merely the resultant of all connecting paths of influence.

The present paper is an attempt to present a method of measuring the direct influence along each separate path in such a system and thus of finding the degree to which variation of a given effect is determined by each particular cause. The method depends on the combination of knowledge of the degrees of correlation among the variables in a system with such knowledge as may be possessed of the causal relations. In cases in which the causal relations are uncertain the method can be used to find the logical consequences of any particular hypothesis in regard to them.

So begins Sewall Wright's 1921 paper (Wright 1921), in which he describes his 'method of path coefficients'. In fact, he invented this method while still in graduate school (Provine 1986) and had even used it without presenting its formal description in a paper published the previous year (Wright 1920). The 1920 paper used his new method to describe and measure the direct and indirect causal relationships that he had proposed to explain the patterns of inheritance of different colour patterns in guinea pigs. The paper came complete with a path diagram – i.e. a causal graph – in which actual drawings of the colour patterns of guinea pig coats were used instead of variable names.

Wright was one of the most influential evolutionary biologists of the twentieth century, being one of the founders of population genetics and intimately involved in the modern synthesis of evolutionary theory and genetics. Despite these other impressive accomplishments, Wright viewed path analysis as one of his more important scientific contributions, and continued to publish on the subject right up to his death (Wright 1984). The method was described by his biographer (Provine 1986) as 'the quantitative backbone of his work in evolutionary theory'. His method of path coefficients is the intellectual predecessor of all the methods described in this book. It is therefore especially ironic that path analysis – the 'backbone' of his work in evolutionary theory – has been almost completely ignored by biologists.[1]

[1] This was written in the first edition. Path analysis and structural equation modelling are now more common.

This chapter has three goals. First, I want to explore why, despite such an illustrious family pedigree, path analysis and causal modelling have been largely ignored by biologists. To do this I will have to delve into the history of biometry at the turn of the twentieth century, but it is important to understand why path analysis was ignored in order to appreciate why its modern incarnation does not deserve such a fate. Next, I want to introduce a new inferential test that allows one to test the causal claims of the path model rather than only 'measuring the direct influence along each separate path in such a system'. The inferential method described in this chapter is not the first such test. Another inferential test was developed quite independently by sociologists in the early 1970s, based on a statistical technique called maximum-likelihood (ML) estimation. Since that method forms the basis of modern structural equation modelling, I will postpone its explanation until the next chapter. Finally, I will present some published biological examples of path analysis and apply the new inferential test to them.

3.2   Why Wright's method of path analysis was ignored

I suspect that scientists largely ignored Wright's work on path analysis for two reasons. First, it ran counter to the philosophical and methodological underpinnings of the two main contending schools of statistics at the turn of the twentieth century. Second, it was methodologically incomplete in comparison to R. A. Fisher's statistical methods (Fisher 1925), based on the analysis of variance combined with the randomised experiment, which had appeared at about the same time.

Francis Galton invented the method of correlation. Karl Pearson transformed correlation from a formula into a concept of great scientific importance and championed it as a replacement for the 'primitive' notion of causality. Despite Pearson's long-term programme to provide 'mathematical contributions to the theory of evolution' (Aldrich 1995), he had little training in biology, especially in its experimental form. He was educated as a mathematician and became interested in the philosophy of science early in his career (Norton 1975). Presumably his interest in heredity and genetics came from his interest in Galton's work on regression, which was itself applied to heredity and eugenics.[2] In 1892 Pearson published a book entitled *The Grammar of Science* (Pearson 1892). In the chapter entitled 'Cause and effect' he gave the following definition: 'Whenever a sequence of perceptions D, E, F, G is invariably preceded by the perception C ..., C is said to be the *cause* of D, E, F, G.' As will become apparent later, his use of the word 'perceptions', rather than 'events' or 'variables' or 'observations', was an important part of his phenomenalist philosophy of science. He viewed the

---

[2] Galton published his *Hereditary Genius* in 1869 (Galton 1869), in which he studied the 'natural ability' of men (women were presumably not worth discussing). He was interested in 'those qualities of intellect and disposition, which urge and qualify a man to perform acts that lead to reputation'. He concluded that '[those] men who achieve eminence, and those who are naturally capable, are, to a large extent, identical'. Lest we judge Galton and Pearson too harshly, remember that such views were considered almost self-evident at the time. Charles Darwin is reputed to have said of Galton's book: 'I do not think I ever in my life read anything more interesting and original . . . a memorable work' (Forrest 1974).

relatively new concept of correlation as having immense importance to science and the old notion of causality as so much metaphysical nonsense. In the third edition of his book (Pearson 1911) he even included a section entitled 'The category of association, as replacing causation'. He had this to say (Pearson 1911: 166):

The newer, and I think truer, view of the universe is that all existences are associated with a corresponding variation among the existences in a second class. Science has to measure the degree of stringency, or looseness of these concomitant variations. Absolute independence is the conceptual limit at one end to the looseness of the link, absolute dependence is the conceptual limit at the other end to the stringency of the link. The old view of cause and effect tried to subsume the universe under these two conceptual limits to experience – and it could only fail; things are not in our experience either independent or causative. All classes of phenomena are linked together, and the problem in each case is how close is the degree of association.

These words may seem curious to many readers because they express ideas that have mostly disappeared from modern biology. Nonetheless, these ideas dominated the philosophy of science at the beginning of the twentieth century and were at least partially accepted by such eminent scientists as Albert Einstein. Pearson was a convinced phenomenalist and logical positivist.[3] This view of science was expressed by people such as Gustav Kirchhoff, who held that all science can do is discover new connections between phenomena, not discover the 'underlying reasons'. Ernst Mach (Mach 1883), who dedicated one of his books to Pearson, viewed the only proper goal of science as providing economical descriptions of experience by describing a large number of diverse experiences in the form of mathematical formulae. To go beyond this and invoke unobserved entities such as 'atoms' or 'causes' or 'genes' was not science, and such terms had to be removed from its vocabulary. Accordingly, Mach (and Pearson) held that a mature science would express its conclusions as functional – i.e. mathematical – relationships that can summarise and predict direct experience, not as causal links that can explain phenomena (Passmore 1966).

Pearson had thought long and hard about the notion of causality, and he concluded, in accord with British empiricist tradition and the people cited above, that association was all that there was. Causality was an outdated and useless concept. The proper goal of science was simply to measure direct experiences (phenomena) and to economically describe them in the form of mathematical functions. If a scientist could predict the likely values of variable Y after observing the values of variable X then he would have done his job. The more simply and accurately he could do it, the better his science. Referring back to Chapter 2, Pearson did not view the equivalence operator of algebra (=) as an imperfect *translation* of a causal relationship because he did not recognise 'causality' as anything but correlation in the limit.[4] By the time that Wright published his method of path analysis, Pearson's British school of biometry was dominant. One of its fundamental tenets was that 'it is this conception of correlation between two occurrences embracing all relationships from absolute independence to complete

---

[3] It is more accurate to say that his ideas were a forerunner to logical positivism.

[4] And yet, citing David Hume, Pearson did accept that associations could be time-ordered from past to future. Nowhere in his writings have I found him express unease that such asymmetries could not be expressed by the equivalence operator.

dependence, which is the wider category by which we have to replace the old idea of causation' (Pearson 1911: 157).

Given these strong philosophical views, imagine what happened when Wright proposed using the biometrists' tools of correlation and regression ... to peek beneath direct observation and deduce systems of causation from systems of correlation! In such an intellectual atmosphere Wright's paper on path analysis was seen as a direct challenge to the biometrists. One has only to read the title ('Correlation and causation') and the introduction of Wright's (1921) paper, cited at the beginning of this chapter, to see how infuriating it must have seemed to the Pearson school.

The pagan had entered the temple, and, like the Macabees, someone had to purify it. The reply came the very next year (Niles 1922). Said Henry Niles: 'We therefore conclude that philosophically the basis of the method of path coefficients is faulty, while practically the results of applying it where it can be checked prove to be wholly unreliable.' Although he found fault in some of Wright's formulae (which were, in fact, correct) the bulk of Niles' scathing criticism was openly philosophical: '"Causation" has been popularly used to express the condition of association, when applied to natural phenomena. There is no philosophical basis for giving it a wider meaning than partial or absolute association. In no case has it been proved that there is an inherent necessity in the laws of nature. Causation is correlation...' (Niles 1922).

Any Mendelian geneticist during that time – of whom Wright was one – would have accepted as self-evident that a mere correlation between parent and offspring told nothing about the mechanisms of inheritance. Therefore, concluded these biologists, a series of correlations between traits of an organism told nothing of how these traits interacted biologically or evolutionarily.[5] The biometricians could never have disentangled the genetic rules determining colour inheritance in guinea pigs, which Wright was working on at the time, simply by using correlations or regressions. Even if distinguishing causation from correlation appeared philosophically 'faulty' to the biometricians, Wright and the other Mendelian geneticists were experimentalists for whom statements such as 'causation is correlation' would have seemed equally absurd. For Wright, his method of path analysis was not a statistical *test* based on standard formulae such as correlation or regression. Rather, his path coefficients were interpretative parameters for measuring direct and indirect causal effects based on a causal system that had already been determined. His method was a statistical translation, a mathematical analogue, of a biological system obeying asymmetric causal relationships.

As the fates would have it, path analysis soon found itself embroiled in a second heresy. Three years after Wright's 'Correlation and causation' paper, Fisher published his *Statistical Methods for Research Workers* (1925). Fisher certainly viewed correlation as distinct from causation. For him the distinction was so profound that he developed an entire theory of experimental design to separate the two. He viewed randomisation and experimental control as the only reliable way of obtaining causal knowledge. Later in

---

[5] Pearson was strongly opposed to Mendelism, and, according to Norton (1975), this opposition was based on his philosophy of science; Mendelians insisted on using unobserved entities ('genes') and forces ('causation').

his life Fisher even wrote an entire book criticising the research that identified tobacco smoking as a cause of cancer on the basis that such evidence was not based on randomised trials (Fisher 1959).[6] I have already described the assumptions linking causality and probability distributions, unstated by Fisher but needed to infer causation from a randomised experiment, as well as the limitations of these assumptions, when studying different attributes of organisms. Despite these limitations, Fisher's methods had one important advantage over Wright's path analysis: they allowed one to rigorously test causal hypotheses, while path analysis could only estimate the direct and indirect causal effects *assuming* that the causal relationships were correct.

Mulaik (1986) has described these two dominant schools of statistics in the twentieth century. His phenomenalist and empiricist school starts with Pearson. Examples of the statistical methods of this school were correlation, regression,[7] common-factor and principal component analyses. The purpose of these methods was primarily, as Mach directed, to provide an economical description of experience by describing a large number of diverse experiences in the form of mathematical formulae. The second school was the realist school, begun by Fisher. This second school emphasised the analysis of variance, experimental design based on the randomised experiment and the hypothetico-deductive method. These Fisherian methods were not designed to provide functional relationships but, rather, to ensure the conditions under which causal relationships could be reliably distinguished from non-causal relationships.

In hindsight, then, it seems that path analysis simply appeared at the wrong time. It did not fit into either of the two dominant schools of statistics and it contained elements that were objectionable to both. The phenomenalist school of Pearson disliked Wright's notion that one *should* distinguish 'causes' from correlations. The realist school of Fisher disliked Wright's notion that one *could* study causes by looking at correlations. Professional statisticians therefore ignored it. Biologists found Fisher's methods, complete with inferential tests of significance, more useful and conceptually easier to grasp, and so biologists ignored path analysis too. A statistical method viewed as central to the work of one of the most influential evolutionary biologists of the twentieth century was largely ignored by biologists.

3.3   D-sep tests

Wright's method of path analysis was so completely ignored by biologists that most biometry texts do not even mention it. Those that do (Li 1975; Sokal and Rohlf 1981) describe it as Wright originally presented it, without even mentioning that it was reformulated by others, primarily economists and social scientists, such that it permitted

---

[6] Fisher was a smoker. I wonder what he would have thought if, because of a random number, he had been assigned to the 'non-smoker' group in a clinical trial?

[7] Regression based on least squares had, of course, been developed well before Pearson by people such as Carl Friedrich Gauss and had been based on a more explicit causal assumption that the independent variable plus independent measurement errors were the causes of the dependent variable. This distinction lives on under the guise of type I and type II regression.

inferential tests of the causal hypothesis and allowed one to include unmeasured (or 'latent') variables. The main weakness of Wright's method – that it required one to assume the causal structure rather that being able to test it – had been corrected by 1970 (Jöreskog 1970), but biologists are mostly unaware of this.

Two different ways of testing causal models will be presented in this book. The most common method is called structural equation modelling and is based on maximum-likelihood techniques. This method will be described in Chapters 4 to 7, and it does have a number of advantages when testing models that include variables that cannot be directly observed and measured (so-called *latent* variables) and for which one must rely on observed indicator variables that contain measurement errors. SEM also has some statistical drawbacks. The inferential tests are asymptotic and can therefore require rather large sample sizes. The functional relationships must be linear. Data that are not multivariate normal are difficult to treat.

These drawbacks led me to develop an alternative set of methods that can be used for small sample sizes, non-normally distributed data or non-linear functional relationships (Shipley 2000). Since these methods are derived directly from the notion of d-separation that was described in Chapter 2, I will call these *d-sep* tests. The main disadvantage of d-sep tests is that they are not applicable to causal models that include latent (unmeasured) variables.

The link between causal conditional independence and probabilistic independence, given by d-separation, suggests an intuitive way of testing a causal model: simply list all the d-separation statements that are implied by the causal model and then test each of them using an appropriate test of conditional independence. There are a number of problems with this naïve approach. First, even models with a small number of variables can include a large number of d-separation statements. Second, we need some way of combining all these tests of independence into a single composite test. For instance, if we had a model that implied 100 independent d-separation statements and tested each independently at the traditional 5 per cent significance level then we would expect, on average, five of these tests to reach significance simply due to random sampling fluctuations. Even worse, the d-separation statements in a causal model are almost never completely independent, and so we would not even know what the true overall significance level would be. Each of these problems can be solved.

3.4   Independence of d-separation statements

Given an acyclic[8] causal graph, we can use the d-separation criterion to predict a set of conditional probabilistic independencies that must be true if the causal model is true. However, many of these d-separation statements can be themselves predicted from other d-separation statements and are therefore not independent. Happily, Pearl (1988)

---

[8] This restriction will be partly removed later. Remember that d-separation also implies probabilistic independence in cyclic causal models in which all variables are discrete and in cyclic causal models in which functional relationships are linear.

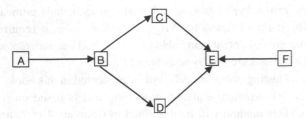

Figure 3.1 A directed acyclic graph involving six variables

describes a simple method of obtaining the minimum number of d-separation statements needed to completely specify the causal graph and proves that this minimum list of d-separation statements is sufficient to predict the entire set of d-separation statements. This minimum set of d-separation statements is called a *basis set*.[9] The basis set is not unique. This method is illustrated with Figure 3.1.

To obtain the basis set, the first step is to list each unique pair of non-adjacent vertices – that is, list each pair of variables in the causal model that do not have an arrow between them. So, in Figure 3.1 the list is {(A,C), (A,D), (A,E), (A,F), (B,E), (B,F), (C,D), (C,F), (D,F)}. Pearl's (1988) basis set is given by d-separation statements consisting of each such pair of vertices conditioned on the parents of the vertex having higher causal order. The number of pairs of variables that don't have an arrow between them is always equal to the total number of pairs minus the number of arrows in the causal graph. In general, if there are V variables and A arrows in the causal graph, the number of elements in the basis set will be

$$\frac{V!}{2(V-2)!} - A$$

Unfortunately, the conditional independencies derived from such a basis set are not necessarily mutually independent in finite samples (Shipley 2000). A basis set that does have this property[10] is given by the set of unique pairs of non-adjacent vertices, of which each pair is conditioned on the set of causal parents of both (Shipley 2000). Remember that an exogenous variable has no parents, so the set of 'parents' of such a variable is empty (such an empty set is written $\{\phi\}$). The second step in getting the basis set that will be used in the inferential test, described next, is to list all causal parents of each vertex in the pair. Using Figure 3.1 and the notation for d-separation introduced in Chapter 2,[11] Table 3.1 summarises the d-separation statements that make up the basis set.

Each of the d-separation statements in Table 3.1 predicts a (conditional) probabilistic independence. How you test each predicted conditional independence depends on the

---

[9] Let **S** be the set of d-separation facts (and therefore the set of conditional independence relationships) that are implied by a directed acyclic graph. A basis set **B** for **S** is a set of d-separation facts that implies, using the laws of probability, all other elements of S, but no proper subset of **B** sustains such implications.

[10] I call this the 'union' basis set. I describe a function from the GGM library of R (basiSet) in a few more pages.

[11] In other words, $X \perp\!\!\!\perp Y|Q$ means that vertex X is d-separated from vertex Y, given the set of vertices **Q**.

Table 3.1 A basis set for the DAG shown in Figure 3.1 along with the implied d-separation statements

Non-adjacent variables	Parent variables of either non-adjacent variable	D-separation statement
A, C	B	$A \parallel C \vert B$
A, D	B	$A \parallel D \vert B$
A, E	C, D, F	$A \parallel E \vert CDF$
A, F	None	$A \parallel F$
B, E	A, C, D, F	$B \parallel E \vert ACDF$
B, F	A	$B \parallel F \vert A$
C, D	B	$C \parallel D \vert B$
C, F	B	$C \parallel F \vert B$
D, F	B	$D \parallel F \vert B$

nature of the variables, and so different d-separation statements in your basis set could be tested with different statistical tests of (conditional) independence. For instance, if the two variables involved in the independence statement are normally and linearly distributed, you could test the hypothesis that the Pearson partial correlation coefficient is zero. Other tests of conditional independence are described below. At this point, assume that you have used tests of independence that are appropriate for the variables involved in each d-separation statement and that you have obtained the exact probability level assuming such independence. By 'exact' probability levels, I mean that you cannot simply look at a statistical table and find that the probability is ≤ 0.05; rather, you must obtain the actual probability level – say, p – 0.036.

Because the conditional independence tests implied by the basis set are mutually independent, we can obtain a composite probability for the entire set using Fisher's test. Since this test seems not to have a name, I have called it Fisher's C (for 'combined') test. If there are a total of k independence tests in the basis set, and p_i is the exact probability of the ith test assuming independence, then the test statistic is $C = -2 \sum_{i=1}^{k} Ln(p_i)$. If all k independence relationships are true then this statistic will follow a chi-squared distribution with 2k degrees of freedom. This is not an asymptotic test unless you use asymptotic tests for some of the individual independence hypotheses. Furthermore, you can use different statistical tests for different individual independence hypotheses. In this sense, it is a very general test.

3.5 Testing for probabilistic independence

In this section, I want to be more explicit concerning what 'independence' and 'conditional independence' mean and the different ways that one can test such hypotheses given empirical data. Let's first start with the simplest case: that of unconditional independence.

The difference between the value of a random value X_i and its expected value μ_X is $(X_i-\mu_X)$. Since these differences can be both negative or positive, and we want to know simply the deviation around the expected value, not the direction of the deviation, we can take the square of the difference: $(X_i-\mu_X)^2$. The expected value of this squared difference[12] is the variance:

$$E[(X_i - \mu_X)^2] = E[(X_i - \mu_X)(X_i - \mu_X)]$$

The covariance is simply a generalisation of the variance. If we have two different random variables (X, Y) measured on the same observational units then the covariance between these two variables is defined as $E[(X_i - \mu_X)(Y_i - \mu_Y)]$. If X and Y behave independently of each other then large positive deviations of X from its mean (μ_X) will be just as likely to be paired with large or small, negative or positive, deviations of Y from its mean (μ_Y). These will cancel each other out in the long run (remember, we are envisaging a complete statistical population when we talk about 'expectation') and the expected value of the product of these two deviations, $E[(X_i - \mu_X)(Y_i - \mu_Y)]$, will be zero. Therefore, the probabilistic independence of X and Y implies a population zero covariance.[13] If X and Y tend to behave similarly, increasing or decreasing together, then large positive values of X will often be paired with large positive values of Y and large negative values of X will often be paired with large negative values of Y. In such cases, the covariance will be large and positive. If X and Y tend to behave in opposite ways, the covariance between them will be negative.

A Pearson correlation coefficient is simply a standardised covariance. Neither a variance nor a covariance has any upper or lower bounds. Changing the units of measurement (say, from metres to millimetres) will change both the variance and the covariance. If we divide the covariance between two variables by the product of their variances (taking the square root of this product in order to ensure that the range goes from +1 to −1) then we obtain a Pearson correlation coefficient. Box 3.1 summarises these points.

The formulae in Box 3.1 are valid so long as both X and Y are random variables. If we want to conduct an inferential test of independence using these formulae, we have to pay attention to the probability distributions of X and Y and the form of the relationship between them in case they are not independent. Different assumptions concerning these points require different statistical methods.

Case 1: X and Y are both normally distributed and any relationship between them is linear.

Tests of the independence of X and Y involving this set of assumptions are treated in any introductory statistics book. First, one can transform the Pearson correlation coefficient so that it follows Student's t-distribution. If X and Y, sampled randomly and measured on n units, are independent (so, the null hypothesis is that $\rho = 0$) then the

---

[12] The formula to estimate this in a sample is given in Box 3.1.

[13] But not the converse! One can have a zero covariance among variables that are still dependent if the relationship is non-linear.

Box 3.1 Variances, covariances and correlations

Population variance (sigma^2, σ^2) of a random variable X:

$$E[(X - \mu_X)^2]$$

Variance (s^2) of a random variable X from a sample of size n:

$$\frac{\sum_i (X_i - \bar{X})^2}{n - 1}$$

Population covariance (sigma$_{XY}$, σ_{XY}) between two random variables X, Y:

$$E[(X - \mu_X)(Y - \mu_Y)]$$

Covariance (s$_{XY}$) between two random variables X, Y from a sample of size n:

$$\frac{\sum_i (X_i - \bar{X})(Y_i - \bar{Y})}{n - 1}$$

Population Pearson correlation (rho$_{XY}$, ρ_{XY}) between two random variables, X, Y:

$$\frac{E[(X - \mu_X)(Y - \mu_Y)]}{\sqrt{E[(X - \mu_X)^2]E[(Y - \mu_Y)^2]}} = \frac{\sigma_{XY}}{\sqrt{\sigma_X^2 \sigma_Y^2}}$$

Pearson correlation coefficient (r$_{XY}$) between two random variables, X, Y from a sample of size n:

$$\frac{s_{XY}}{\sqrt{s_X^2 s_Y^2}}$$

Variances and covariances can be obtained via the cov() function in R:

```
cov(x, y = NULL, use = "everything",
method = c("pearson", "kendall", "spearman"))
```

The closely related function cor() gives the correlations; specifiying the method to be 'pearson' gives the Pearson correlations:

```
cor(x, y = NULL, use = "everything",
method = c("pearson", "kendall", "spearman"))
```

---

following transformation will follow a Student's t-distribution[14] with n–2 degrees of freedom:

$$t_r = \frac{r\sqrt{n - 2}}{\sqrt{1 - r^2}}$$

---

[14] For partial correlations, described below, one simply replaces r with the value of the partial correlation coefficient, and the numerator (n–2) becomes (n–2–p), where p is the number of conditioning variables.

This test is exact. So long as you have at least three independent observations then you can test for the independence of X and Y.[15] The R function cor.test, with method = 'pearson', performs this test.

It is also possible to transform a Pearson correlation coefficient so that it asymptotically follows a standard normal distribution – i.e. a normal distribution with a mean of zero and a variance of 1. For sample sizes of at least 50 (and approximately even for sample sizes as low as 25) one can use Fisher's Z-transform:

$$z = 0.5\sqrt{n-3}\ln\left(\frac{1+r}{1-r}\right)$$

If X and Y are independent then the probability of z can be obtained from a standard normal distribution. Finally, one can use Hotelling's (1953) transformation,[16] which is acceptable for sample sizes as low as 10:

$$z = \sqrt{(n-1)}\left[0.5\ln\left(\frac{1+r}{1-r}\right) - \frac{1.5\ln\left(\frac{1+r}{1-r}\right)+r}{4(n-1)}\right]$$

Case 2: X and Y are continuous but not normally distributed and any relationship between them is only monotonic.

If X or Y are not normally distributed and any relationship between them is not linear but is monotonic[17] then we can use Spearman's correlation coefficient. Although there exist statistical tables giving probability levels for Spearman's correlation coefficient, one can use exactly the same formulae as for Pearson's correlation coefficient so long as the sample size is greater than 10 (Sokal and Rohlf 1981).

The first step is to convert X and Y to their ranks. In other words, sort each value of X from smallest to largest and replace the actual values of X by its order in the rank; the smallest number becomes 1, the second smallest number becomes 2, and so on. Do the same thing for Y. Now that you have converted each X and each Y to its rank, you can simply put these numbers into the formula for a Pearson's correlation coefficient and test as before.

One complication is when there are ties. Spearman's coefficient assumes that the underlying values of X and Y are continuous, not discrete. Given such an assumption, equal values of X (or Y) will occur only because of limitations in measurement. To correct for such ties, first sort the values ignoring ties, and then replace the ranks of tied values by the mean rank of these tied values. Box 3.2 gives an example of the calculation of a Spearman rank correlation coefficient.

---

[15] Of course, with so few observations you would have so little statistical power that only very strong associations would be detected.

[16] Both Fisher's and Hotelling's transformations can be used to test null hypotheses in which ρ equals a value different from zero. This useful property allows one to compute confidence intervals around the Pearson correlation coefficient.

[17] A non-monotonic relationship is one in which X increases with increasing Y over part of the range and decreases with increasing Y over another part of the range. If you think that a graph of X and Y has hills and valleys then the relationship is non-monotonic.

Box 3.2 Spearman's rank correlation coefficient

Here are 10 simulated pairs of values and the accompanying scatterplot (Figure 3.2). The X values were drawn from a uniform distribution and rounded to the nearest unit. The Y values were drawn from the following equation, $Y_i = X_i^{0.2} + \beta(5, 1)$, where the random component is drawn from a beta distribution with shape parameters of 5 and 1.

Values of X, Y and their ranks

X	Y	Rank X	Rank Y	Rank X	Rank Y
2	2.08	1	3	1	3
3	2.02	2	2	2.5	2
15	2.68	10	10	10	10
10	2.47	8	9	8	9
5	2.21	5	4	5	4
12	2.23	9	5	9	5
3	1.86	3	1	2.5	1
4	2.25	4	6	4	6
9	2.31	6	8	6.5	8
9	2.28	7	7	6.5	7

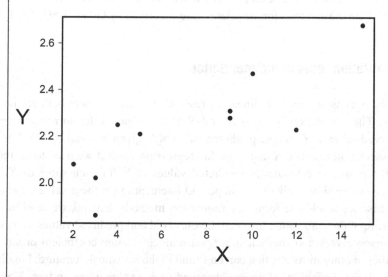

Figure 3.2 A scatterplot of randomly generated pairs of values from a bivariate non-normal distribution and possessing a non-linear monotonic relationship

In the above table, X and Y are the original values. Columns 3 and 4 of the table are the ranks of X and Y before correcting for ties (the underlined values). Columns 5 and 6 are the ranks after correcting for the two pairs of ties values of X (there were two values of 3 and two values of 9). To calculate the Spearman rank correlation coefficient of X and Y, simply use the values in columns 5 and 6 and enter them into the formula for the Pearson's correlation coefficient. In the above example the Spearman rank correlation coefficient is 0.726. Assuming that X and Y are independent in the statistical population, we can convert this to a standard normal variate using Hotelling's z-transform, giving a value of 2.47. This value has a probability under the null hypothesis of 0.014.

The R function cor.test, with method = 'spearman', performs this test.

Case 3: X and Y are continuous and any relationship between them is not even monotonic.

This case applies when the relationship between X and Y might have a very complicated form, with X and Y being positively related in some parts of the range and negatively related in other parts, and therefore when neither a Pearson nor a Spearman correlation can be applied. This situation requires more computationally demanding methods, including form-free regression and permutation tests. Each of these topics is dealt with much more fully in other publications but will be intuitively introduced here because these notions are needed for the analogous case in conditional independence. Form-free regression is a vast topic, which includes kernel smoothers, cubic-spline smoothers (Wahba 1991) and local (loess) smoothers (Cleveland and Devlin 1988; Cleveland, Devlin and Grosse 1988; Cleveland, Grosse and Shyu 1992). Collectively, these methods form the basis of generalised additive models[18] (Hastie and Tibshirani 1990). Permutation tests for association are described by Good (1993; 1994).

3.6   Permutation tests of independence

To begin, consider a simple linear regression of Y on X, where both are random variables. The correlation between X and Y is the same as the correlation between the observed value of Y and the predicted value of Y given X – that is, E[Y|X]. To test for an association between X and Y in this regression context we need to do three things. First, we have to estimate the predicted values of Y for each value of X. For linear regression we simply obtain the slope and intercept to get these values, and in the general case we would use form-free regression methods. Second, we need to calculate a measure of the association between the observed and predicted values of Y; we can use a Pearson correlation coefficient, a Spearman correlation coefficient or any of a large number of other measures that can be found in the statistical literature. Finally, we need to know the probability of having observed such a value when, in fact, X and Y really are independent. This is where a permutation test comes in handy.

Remember the definition of probabilistic independence given in Chapter 2. We know that, if X and Y are independent, the probability of observing any particular value of Y is the same regardless of whether we know the value of X. In other words, any value of

---

[18] Note that smooth.spline() and loess() implement cubic-spline smoothers and loess smoothers in R, respectively . The mgcv package of R includes many functions for generalised additive models.

X is just as likely to be paired with any other value of Y as with the particular Y that we happen to observe. The permutation test works by making this true in our data. After calculating our measure of association in our data, we randomly rearrange the values of X and/or Y using a random number generator. In this new randomly mixed 'data set' the values of X and Y really are independent, because we forced them to be so; we have literally forced our null hypothesis of independence to be true, and the value of the association between X and Y is due only to chance. We do this a very large number of times until we have generated an empirical frequency distribution of our measure of association.[19] The exact number of times that we randomly permute our data will depend on the true probability level of our actual data and the accuracy that we want to obtain in our probability estimate. Manly (1997) shows how to determine this number, but it is typically between 1,000 and 10,000 times. On modern computers this will appear instantaneously unless the intermediate calculations are intensive. The last step is to count the proportion of times that we observe at least as large a value of association within the permuted data sets, or its absolute value for a two-tailed test, as we actually observed in our original data. Box 3.3 gives an example of this permutation procedure.

3.7 Form-free regression

The first graph in Box 3.3 (Figure 3.3(a)) shows a highly non-linear relationship between X and Y, and it is unlikely that we would be able to deduce the actual function that generated these data.[20] On the other hand, if we concentrate on smaller and smaller sections of the graph (Figures 3.3(b) and (c)), the relationship becomes simpler and simpler. The basic insight of form-free regression methods is that even complicated functions can be quite well approximated by simple linear, quadratic or cubic functions in the neighbourhood of a given value of X. Within such a neighbourhood, shown by the boxes in the graphs of Box 3.3, we can use these simpler functions to calculate the expected value of Y at that particular value of X. We then go on to the next value of X, move the neighbourhood so that it is centred around this new value of X and calculate the expected value of the new Y, and so on. In this way, we do not actually estimate a parametric function predicting Y over the entire range of X, but we do get very good estimates of the predicted values of Y given each unique value of X. To obtain the predicted values of Y given X, we use weighted regression (linear, quadratic or cubic) where each (X, Y) pair in the data set is weighted according to its distance from the value of X around which the neighbourhood is centred. In local, or loess,[21] regression the neighbourhood size can be chosen according to different criteria, such as minimising the residual sum of squares, and the weights are chosen on the basis of the tricube weight

---

[19] For small samples one can generate all unique permutations of the data. The use of random permutations, described here, is generally applicable, and the estimated probabilities converge on the true probabilities as the number of random permutations increases.

[20] The actual function was: $Y = X\sin(X) + \varepsilon$, where the error term comes from a unit normal distribution.

[21] The word 'loess' comes from the geological term 'loess', which is a deposit of fine clay or silt along a river valley. I suppose that this evokes the image of a very wavy surface that traces the form of the underlying geological formation. At least some statisticians have a sense of the poetic.

Box 3.3 Loess regression

The following three graphs (Figure 3.3) show a simulated data set generated from a complicated non-linear function (the solid line of Figure 3.3(a)) along with a loess regression (the broken line) using a local quadratic fit and a neighbourhood size of one-half the range of X. Figure 3.3(b) shows the same complicated non-linear function in the range of 1 to 3 of the X values and Figure 3.3(c) shows this in the range of 1.5 to 2.5 of the X values.

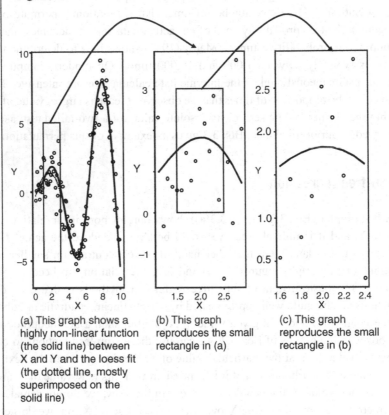

(a) This graph shows a highly non-linear function (the solid line) between X and Y and the loess fit (the dotted line, mostly superimposed on the solid line)

(b) This graph reproduces the small rectangle in (a)

(c) This graph reproduces the small rectangle in (b)

Figure 3.3 A simulated data set

The loess regression (the broken line in Figure 3.3(a)) doesn't actually give a parametric function linking Y to X, but it does give the predicted value of Y for each unique value of X – that is, it gives the sample estimate of $E[Y|X]$; the solid and broken lines in the figure completely overlap except in the range of $X = 2$. To estimate a permutation probability of the non-linear correlation of X and Y, we can first calculate the Pearson correlation coefficient between the observed Y values (the circles in the figure) and the predicted values of Y given X (the loess estimates). In this example, $r = 0.956$. If we don't want to assume any particular probability distribution for the residuals then we can generate a permutation frequency distribution for the correlation coefficient. To do this, we randomly permute the order of the observed Y values (or the predicted values; it doesn't matter which) to get a 'new'

set of Y* values and recalculate the Pearson correlation coefficient between Y* and
E[Y*|X]. The following histogram (Figure 3.4) shows the relative frequency of the
Pearson correlation coefficient in 5,000 such permutations; the arrow indicates the
value of the observed Pearson correlation coefficient. None of the 5,000 permutation data sets had a Pearson correlation coefficient whose absolute value was at least
0.956. Since the residuals were actually generated from a unit normal distribution,
we can calculate the probability of observing a value of 0.956 with 101 observations.
It is approximately 1×10^{39}.

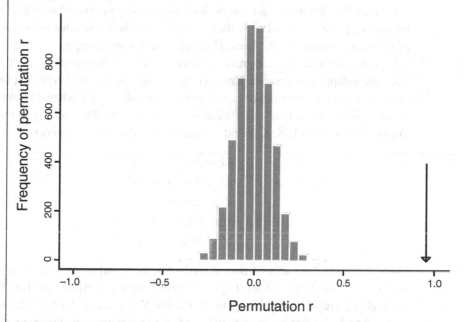

Figure 3.4 The frequency distribution of the Pearson correlation coefficient in 5,000 random
permutations of the simulated data set involving the observed Y values and the predicted
loess values

Note: The arrow shows the observed Pearson correlation in the original simulated data set.

function. Shipley and Hunt (1996) describe this in more detail in the context of plant
growth rates.

3.8   Conditional independence

So far we have been talking about unconditional independence – that is, the independence of two variables without regard to the behaviour of any other variables. Such
unconditional independence is implied by two variables in a causal graph that are d-
separated without conditioning on any other variable. D-separation upon conditioning implies *conditional* independence. The notion of conditional independence seems
paradoxical to many people. How can two variables be dependent, even highly correlated, and still be independent upon conditioning on some other set of variables?

Consider the following causal graph: ε1→X←Z→Y←ε2. Does it seem equally para-doxical if I say that X and Y will behave similarly due to the common causal effect of Z, but that they will no longer behave similarly if I prevent Z from changing? If Z doesn't change then the only changes in X and Y will come from the changes in ε1 and ε2, and these two variables are d-separated and therefore unconditionally independent. A moment's reflection will convince you that if Z is allowed to change (vary) then both X and Y will change as well in a systematic fashion, since they are both responding to Z. If the variables in the causal graph are random then the correlation between X and Y will be attributable to the fact that both share common variance due to Z. If we restrict the variance in Z more and more then X and Y will share a smaller and smaller amount of common variance. In the limit, if we prevent Z from changing at all then X and Y will no longer share any common variance; the only variation in X and Y will come from the independent error variables ε1 and ε2, and so X and Y will then be indepen-dent. In such a case we would be comparing values of X and Y when Z is constant. This is the intuitive meaning of conditional independence. To illustrate, I generated 10,000 independent sets of ε1, X, Z, Y and ε2 according to the following generating equations:

$$\varepsilon 1 = N(0, 1 - 0.9^2)$$
$$\varepsilon 2 = N(0, 1 - 0.9^2)$$
$$Z = N(0, 1)$$
$$Y = 0.9Z + \varepsilon 1$$
$$X = 0.9Z + \varepsilon 2$$

Since X, Y and Z are all unit normal variables, the population correlations are $\rho_{X,Z} = 0.9$, $\rho_{Y,Z} = 0.9$ and $\rho_{X,Y} = 0.81$. Figure 3.5 shows three scatterplots. Notice that X and Y are highly correlated even though neither X nor Y is a cause of the other. Figure 3.5(a) shows the relationship between X and Y when no restrictions are placed on the variance of Z. The sample correlation between X and Y in this graph is 0.8016, compared to the population value of 0.81. Figure 3.5(b) plots only those values of X and Y for which the value of Z is between −2 and 2, which means that the variance of Z is restricted by just a little bit. The sample correlation between X and Y has been decreased slightly to 0.7591. Figure 3.5(c) plots those values of X and Y for which the value of Z is between −0.5 and 0.5, thus restricting the variance of Z much more. The sample correlation between X and Y is now only 0.2294. Clearly, the degree of association between X and Y is decreasing as Z is prevented more and more from varying.

If we calculate the correlation between X and Y as we restrict the variation in Z more and more, we can get an idea of what happens to the correlation between X and Y in the limit when the variance of Z is zero. This limit is the correlation between X and Y when Z is fixed (or 'conditioned') to a constant value; this is called the *partial* correlation between X and Y, conditional on Z, and it is written $\rho_{XY.Z}$ or $\rho_{XY|Z}$. Figure 3.6 plots the sample correlation between X and Y as Z is progressively restricted in its variance.

As expected, as the range of Z around its mean (zero) becomes smaller and smaller, the correlation between X and Y also becomes smaller and approaches zero. Given the causal graph that governed these data we know that X and Y are not unconditionally

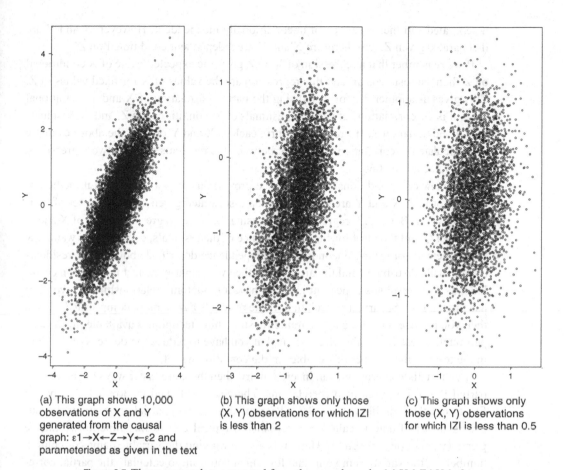

(a) This graph shows 10,000 observations of X and Y generated from the causal graph: ε1→X←Z→Y←ε2 and parameterised as given in the text

(b) This graph shows only those (X, Y) observations for which |Z| is less than 2

(c) This graph shows only those (X, Y) observations for which |Z| is less than 0.5

Figure 3.5 Three scatterplots generated from the causal graph ε1→X←Z→Y←ε2

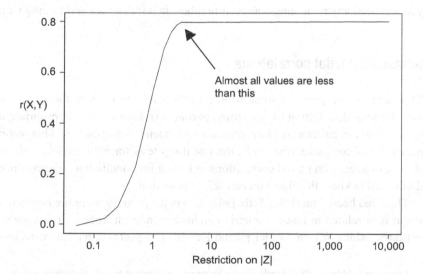

Figure 3.6 The Pearson correlation coefficient between X and Y in the data shown in Figure 3.5 when the absolute value of Z is restricted to various degrees

Note: The limiting value of the correlation coefficient when |Z| is restricted to a constant value is the partial correlation between X and Y.

d-separated and therefore are not unconditionally independent. However, X and Y are d-separated given Z, and therefore X and Y are independent conditional on Z.

If we remember that a regression of X on Z gives the expected value of X conditional on Z then the residuals around this regression are the values of X for fixed values of Z. This gives us another way of visualising the partial correlation of X and Y conditional on Z: it is the correlation between the residuals of X, conditional on Z, and the residuals of Y, conditional on Z. If I regress, in turn, each of X and Y on Z in the above example and calculate the correlation coefficient between the residuals of these two regressions, I get a value of −0.0060.

This view of a conditional independence provides us with a very general method of testing for it. If X and Y are predicted to be d-separated given some other set of variables $Q = \{A, B, C, \ldots\}$, regress (perhaps using form-free regression) each of X and Y on the set Q and then test for the independence of the residuals, using, if you want, any of the methods of testing unconditional independence described above. If the residuals are normally distributed and linearly related then you can use the test for Pearson correlations. If the residuals appear, at most, to have a monotonic relationship then you can use the test for a Spearman correlation. If the residuals have a more complicated pattern then you can use one of the non-parametric smoothing techniques available, followed a permutation test. The only difference is that you have to reduce the degrees of freedom in the tests by the number of variables in the conditioning set.

If your statistical program can invert a matrix then there are faster ways of calculating partial Pearson or Spearman correlations. These are explained in Box 3.4. The pcor() function in the ggm library calculates Pearson or Spearman partial correlations. If your matrix or data frame is called 'my.dat' then a typical call to this function would be pcor(u,cov(my.data)). Here, u is a vector giving the variable names or column numbers of the variables in your data for which you want to calculate the partial correlation. The first two names or numbers in u index the variables whose partial correlation you want and any remaining names or numbers in u index the conditioning variables.

3.9   Spearman partial correlations

This next section presents some Monte Carlo results to explore the degree to which the sampling distribution of Spearman partial correlations, after appropriate transformation, follows either a standard normal or a Student's t distribution. This section is not necessary to understand the application of d-sep tests for path models, only to justify the use of Spearman partial correlations in testing for conditional independence. If you don't need to know this then you can skip this section.

There has been remarkably little published in the primary literature concerning inferential tests related to non-parametric conditional independence.[22] It is known that the expected values of a first-order partial Kendall or Spearman partial correlations need

---

[22] Kendall and Gibbons (1990) briefly discuss Spearman and Kendall partial correlations and provide a table of significance values for first-order Kendall partial correlations for small sample sizes.

Box 3.4 Calculating partial covariances and correlations

Given a sample covariance matrix $\mathbf{S}$, the inverse of this matrix is called the *concentration* matrix, $\mathbf{C}$. The negative of the off-diagonal elements c_{ij} give the partial covariance between variables i and j, conditional on (holding constant) all the other variables included in the matrix. This gives an easy way of estimating partial covariances and partial correlations of any order. To get the partial covariance between variables X and Y conditional on a set of other variables $\mathbf{Q}$, simply create a covariance matrix in which the only variables are X, Y and the remaining variables in $\mathbf{Q}$. Invert the matrix, and then this partial covariance is the negative of the element in the row pertaining to X and the column pertaining to Y – namely $-c_{XY}$. The partial correlation between X and Y is given by

$$r_{X,Y|Q} = \frac{-c_{XY}}{\sqrt{c_{XX} \cdot c_{YY}}}$$

The partial correlation between two variables conditioned on n other variables is said to be a *partial correlation of order n*. The unconditional correlation coefficient is simply a partial correlation of order 0. Some texts give recursion formulae for partial correlations of various orders, although partials of higher orders are very tedious to calculate by such means. For instance, the formula for a partial correlation of order 1 between X and Y, conditional on Z, is

$$\rho_{X,Y|Z} = \frac{\rho_{XY} - \rho_{XZ}\rho_{YZ}}{\sqrt{\left(1 - \rho_{XZ}^2\right)\left(1 - \rho_{YZ}^2\right)}}$$

As an example, consider the following causal graph: W→X→Z→Y. 100 independent (W,X,Y,Z) observations were generated according to structural equations with all path coefficients equal to 0.5 and the variances of all four variables equal to 1.0. Here is the sample covariance matrix:

	W	X	Y	Z
W	1.43347870	−0.75265627	−0.06269845	0.10179918
X	−0.75265627	1.52762094	−0.53911722	−0.03777874
Y	−0.06269845	−0.53911722	1.71116716	−0.90033856
Z	0.10179918	−0.03777874	−0.90033856	1.73196991

The inverse of the matrix (rounded to the nearest 100th) obtained by extracting only the elements of the covariance matrix pertaining to W, X and Y is

	W	X	Y
W	1.43	−0.75	−0.01
X	−0.75	1.53	−0.56
Y	−0.01	−0.56	1.24

The partial correlation between W and Y, conditional on X, is

$$r_{WY|X} = \frac{(-1) - 0.01}{\sqrt{1.43 \cdot 1.24}} = 0.0075$$

The same method can be used to obtain partial Spearman partial correlations, by simply ranking the variables as described in Box 3.2 and then proceeding in the same way as for Pearson partial correlations.

Table 3.2 Results of a Monte Carlo study of the distribution of z-transformed Spearman partial correlations

Distribution of ε_i	Sample size	Order of partial	Linear/ non-linear	Mean of z	Standard deviation of z	Two-tailed 95% quantile	Theoretical probability
Normal	25	1	L	0.08	1.03	2.04	0.04
Normal	50	1	L	0.08	0.97	2.01	0.05
Normal	400	1	L	0.08	1.04	2.16	0.03
Normal	50	2	L	0.03	0.99	1.86	0.06
Normal	50	3	L	−0.07	1.00	1.85	0.06
Gamma(1)	25	3	L	0.01	1.05	2.09	0.04
Gamma(1)	50	3	L	−0.02	0.96	1.82	0.07
Gamma(1)	50	3	NL	0.03	0.96	2.02	0.04
Gamma(5)	50	3	NL	−0.07	0.99	1.93	0.05
Beta(1,1)	50	3	L	−0.02	0.99	2.00	0.05
Beta(1,1)	50	3	NL	0.03	1.02	2.08	0.04
Beta(1,5)	50	3	NL	0.03	1.02	2.08	0.04
Beta(5,1)	50	3	NL	−0.05	0.99	1.78	0.07
Beta(5,1)	400	3	NL	0.00	1.02	2.01	0.04
Binomial	50	3	NL	0.01	0.99	1.95	0.05

Notes: Four different distributional types were simulated for the random components. The sample size was the number of observations per simulated data set. Linear (L) and non-linear (NL) functional relationships were used. The empirical mean, the standard deviation and the two-tailed 95 per cent limits of 1,000 simulated data sets are shown.

not be strictly zero even when two variables are conditionally independent given the third (Shirahata 1980; Korn 1984). On the other hand, Conover and Iman (1981) recommend the use of partial Spearman correlations for most practical cases in which the relationships between the variables are at least monotonic. A Spearman partial correlation is simply a Pearson partial correlation applied to the ranks of the variables in question. Therefore, the conditional independence of non-normally distributed variables with non-linear, but monotonic, functional relationships between the variables can be tested with Spearman's partial rank correlation coefficient simply by ranking each variable (and correcting for ties, as described in Box 3.2) and then applying the same inferential tests as for Pearson partial correlations. For instance, if one accepts Conover and Iman's (1981) recommendations then a Spearman partial rank correlation will be approximately distributed as a standard normal variate when z-transformed.

How robust is this recommendation? To explore this question, Table 3.2 presents the results of some Monte Carlo simulations to determine the effects of sample size, the distributional form of the variables and the effect of non-linearity on the sampling distribution of the z-transformed Spearman partial correlation coefficient. The random components of the generating equations (ε_i) were drawn from four different probability distributions: normal, gamma, beta or binomial. I chose the shape parameters of the gamma and beta distributions to produce different degrees of skew and kurtosis.

Gamma($\lambda = 1$) is a negative exponential distribution. Gamma($\lambda = 5$) is an asymmetric distribution with a long right tail. Beta(1,1) is a uniform distribution, Beta(1,5) is a highly asymmetric distribution with a long right tail and Beta(5,1) is a highly asymmetric distribution with a long left tail. The final (discrete) probability distribution was symmetric with an expected value of 2 and had ordered states of X = 0, 1, 2, 3 or 4; these were generated from a binomial distribution of the form $C(5,X)0.5^X0.5^{1-X}$. Random numbers were generated using the random number generators given by Press et al. (1986). The generating equations were of the form

$$X_1 = \varepsilon_1$$
$$X_i = \alpha_i X_{(i-1)}^{\beta_i} + \varepsilon_i; \quad i > 1$$

These generating equations are based on a causal chain ($X_1 \rightarrow X_2 \rightarrow X_3 \rightarrow \ldots$) with sufficient variables (3, 4 or 5) to produce zero partial associations of orders 1 to 3. When β_i equals 1.0 the relationships between the variables are linear and when β_i is different from 1.0 then the relationships between the variables are non-linear but monotonic. The results in Table 3.2 are based on models with $\beta_i = 1$ (linear) and 0.5 (non-linear) but other values give similar results. All the simulation results in Table 3.2 are based on 1,000 independent simulated data sets. In interpreting Table 3.2, remember that the z-transformed Spearman partial correlations should be approximately distributed as a standard normal variate whose population mean is zero, whose population standard deviation is 1.0 and whose two-tailed 95 per cent limit is 1.96.

Generally, the sampling distribution of the z-transformed Spearman rank partial correlations is a very good approximation of a standard normal distribution. In fact, the only significant deviation from a standard normal distribution (based on a Kolmogorov–Smirnov test) was observed for the ranks of normally distributed variables, for which one would not normally use a Spearman partial correlation. The empirical standard deviations were always close to 1.0 and the empirical means only once differed significantly, though very slightly, from zero at high levels of replication. Approximate 95 per cent confidence intervals for the empirical 0.05 significance level (i.e. the two-tailed 95 per cent quantiles), based on 1,000 simulations, are 0.037 to 0.064 (Manly 1997).

The results of this simulation study support the recommendations of Conover and Iman (1981). These results are also consistent with the theoretical values given by Korn (1984) for the special case of a Spearman first-order partial based on trivariate normal and trivariate lognormal distributions, where the limiting values of the Spearman partial correlation are less than, or equal to, an absolute value of 0.012, thus giving an expected absolute z-score of ≤ 0.024. Korn (1984) gives a pathological example in which the above procedure will not work even after ranking the data because there is a non-monotonic relationship between the variables; he recommends that one first check[23] to see if the relationship between the ranks are approximately linear before using Spearman partial correlations.

---

[23] This can be done by simply plotting the scatterplots of the ranked data.

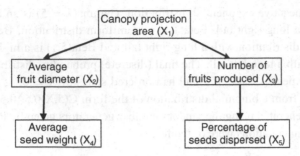

Figure 3.7 Proposed causal relationships between five variables related to seed dispersal in St Lucie cherry

3.10   Seed production in St Lucie cherry

St Lucie cherry (*Prunus mahaleb*) is a small species of tree that is found in the Mediterranean region and that relies on birds for the dispersal of its seeds. As in most plants, seedlings from seeds that are dispersed some distance from the adult are more likely to survive, since they will not be shaded by their own parent or eaten by granivores that are attracted to the parent tree. For species whose seeds can survive the passage through the digestive tract of the dispersing animal, it is also evolutionarily and ecologically advantageous for the fruit to be eaten by the animal, since the seed will be deposited with its own supply of fertiliser. Not all frugivores of St Lucie cherry are useful fruit dispersers. Some birds just consume the pulp, either leaving the naked seed attached to the tree or simply dropping the seed to the ground directly beneath the parent. In order to estimate selection gradients, Jordano (1995) measured six traits of 60 individuals of this species: the canopy projection area (a measure of photosynthetic biomass), average fruit diameter, the number of fruits produced, average seed weight, the number of fruits consumed by birds and the percentage of these consumed fruits that were properly dispersed away from the parent by passage through the gut. Based on five of these variables for which I had data (I was missing the total number of fruits consumed by birds) I proposed the path model shown in Figure 3.7 (Shipley 1997), using the exploratory path models described in Chapter 8.

We can use this model to illustrate the d-sep test. The first step is to obtain the d-separation statements in the basis set that are implied by the causal graph in Figure 3.7. There are six such statements, since there are five variables and four arrows. Table 3.3 lists these d-separation statements. You can also obtain this basis set using the basiSet() function of the ggm library of R. To do this you must first enter the causal graph shown in Figure 3.7 using the DAG() function that was described in Chapter 2.

```
Figure3.7←DAG(X2~X1,X3~X1,X4~X2,X5~X3)
basiSet(Figure3.7)
```

We next have to decide how to test the independencies that are implied by these six d-separation statements. The original data showed heterogeneity of variance, as often

Table 3.3 The d-separation statements in the basis set of the causal graph shown in Figure 3.7

D-separation statement	Pearson partial correlations		Spearman partial correlations	
	Estimate	Probability assuming independence	Estimate	Probability assuming independence
$X_4 \parallel X_1 \mid X_2$	−0.066	0.617	−0.063	0.635
$X_4 \parallel X_3 \mid X_2 X_1$	0.142	0.289	0.144	0.279
$X_4 \parallel X_5 \mid X_2 X_3$	0.004	0.976	0.075	0.574
$X_2 \parallel X_3 \mid X_1$	0.021	0.873	0.059	0.655
$X_2 \parallel X_5 \mid X_3 X_1$	−0.155	0.244	−0.160	0.229
$X_1 \parallel X_5 \mid X_3$	0.076	0.565	0.102	0.443

Note: Also shown are the Pearson and Spearman partial correlations that are implied by the d-separation statements. The probabilities, assuming that the population partial correlations are zero, are listed as well.

happens with size-related variables, but transforming each variable to its natural logarithm stabilises the variance. Figure 3.8 shows the scatterplot matrix of these ln-transformed data.

Since the relationships appear to be linear and histograms of each variable did not show any obvious deviations from normality, we can test the predicted independencies using Pearson partial correlations. The results[24] are shown in Table 3.3. Fisher's C statistic is 7.73, with 12 degrees of freedom, for an overall probability of 0.806. The difference between the observed and predicted (partial) correlations would occur in about 80 per cent of data sets (in the long run) even if the data really were produced by the causal structure in Figure 3.7. This doesn't mean that the data really were produced by such a causal structure but it does mean that we have no reason to reject it on the basis of the statistical test. If we want to reject it anyway then we will need to produce reasonable doubt. Perhaps the assumption of normality, upon which the test of the Pearson partial correlations is based, was producing incorrect probability estimates. Table 3.3 also lists the Spearman partial correlations. The overall probability of the model ($\chi^2 = 9.99$, 12 df), based on the individual probability levels of these Spearman partial correlations, was 0.616. On the other hand, there are equivalent models that also produce non-significant probability estimates (Shipley 1997), and if any of these equally well-fitting alternative models do not contradict what is known of the biology of these

[24] For the case of normally distributed variables and linear relationships between them, you could use another function, shipley.test(), from the ggm library. (I would apologise for such an egotistical name but I did not create this library!) This function takes three arguments: amat, S, n. The first argument is a square Boolean matrix normally created using the DAG function, The second argument is the sample covariance matrix of the variables, obtained via the cov() function. The third argument is the sample size of the data set from which the covariance matrix was calculated. If you calculate your covariance matrix from the ranks of your data then you will get a result based on Spearman partial correlations.

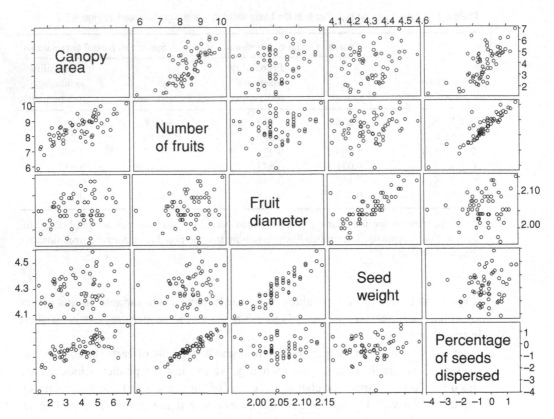

Figure 3.8 Scatterplot matrix of the empirical observations (all variables transformed to their natural logarithms)

trees then they might constitute reasonable doubt.[25] Equivalent models are explained in Chapter 8.

The original data used by Jordano (1995) was fitted to a latent variable model using maximum likelihood methods.[26] Neither the model chi-square statistic nor the model degrees of freedom were given. It is therefore not possible to judge the fit of that original model,[27] but it is possible to extract those d-separation statements involving only the measured variables available to me from the original latent-variable model. Jordano's published model implies four d-separation statements in the basis set that can be tested: {(canopy projection area $\parallel$ average fruit diameter), (canopy projection area $\parallel$ average seed weight), (number of fruits produced $\parallel$ average fruit diameter), (number of fruits produced $\parallel$ average seed weight)}. Combining the null probabilities using Fisher's C test, based on Pearson correlations, gives a probability of 0.005 ($\chi^2 = 21.85$, 8 df).

---

[25] The model was actually developed using the exploratory methods of Chapter 8. This, too, should give us reason to question the model until independent data can be tested against it. At this point, all we can reasonably say is that the data are consistent with the model and so deserve further study.

[26] These methods are described in Chapters 4 to 7.

[27] One measured variable (the total number of seeds dispersed) was not provided to me, so I can't fit his original model.

Using Spearman correlations the probability is 0.019 ($\chi^2 = 18.24$). These low probabilities, based on a subset of the original measured variables, provide reasonable doubt concerning Jordano's (1995) model.

3.11   Generalising the d-sep test

These are the steps involved in a d-sep test.

(1) Write down your causal hypothesis in the form of a DAG.
(2) Use the basiSet() function to get the set of d-sep statements forming your basis set, and thus the series of null hypotheses of (conditional) independence implied by them.
(3) Conduct the series of statistical tests of (conditional) independence specified in step (2) using whatever statistical tests are appropriate for each null hypothesis.
(4) Combine the null probabilities, obtained from step (3), using Fisher's C test.
(5) Reject your causal hypothesis if the null probability from Fisher's C test is below your chosen significance level; otherwise, conclude that your data are consistent with your causal hypothesis.

Stated thus, it is clear that the d-sep test is less a statistical 'test' than a recipe for creating your own test based on the specific properties of your data. So long as you can (a) specify your causal hypothesis in the form of a DAG, (b) obtain direct measures of each variable in this DAG and (c) perform an appropriate test of (conditional) independence for each d-sep statement then you can create your very own d-sep test.

Once the d-sep test tells you that the data are consistent with the causal process represented in the DAG, you can calculate the path coefficients and related statistics. The details of how to do this will vary depending on the nature of your data but the first step will be the same: you fit a series of structured equations by exactly following the causal structure specified by your DAG. In other words, you identify each effect variable and, for each one, you fit the data to a model in which this effect variable is predicted jointly by its causal parents and by no other variable.

In this section I want to explain how you can use the full power of linear models, generalised linear models, mixed models, non-linear models and generalised additive models to carry out step (3) above. I do not explain how to conduct these various analyses here since each type could (and does) fill an entire book. If you are reading this book then you probably know at least how to fit and interpret a linear model (simple and multiple regression, ANOVA, ANCOVA,[28] etc.).

Consider any d-sep statement of the form $X \perp\!\!\!\perp Y | \mathbf{Q}$, where $\mathbf{Q}$ is the set of conditioning variables. We can translate this d-sep statement into a statistical model either as $Y \sim f(X) + f(\mathbf{Q}) + \varepsilon$ or as $X \sim f(Y) + f(\mathbf{Q}) + \varepsilon$. The notation $f(\cdot)$ simply means 'some function of' whatever variables are inside the brackets and ε is a random variable following some probability distribution. Depending on the details of the data this

---

[28] Analysis of covariance; a general linear model that blends ANOVA and regression.

function may be linear or non-linear. Depending on the details of the data the random variable could be normally distributed or follow some other probability distribution, it could be a scalar or a vector of values, and so on. Depending on these differences you would then choose a statistical model appropriate for these details. Should variable X or Y be the dependent variable (i.e. the variable whose probability distribution we need to know)? It doesn't matter for what follows, and so you can choose the one that is easier to model. Once you decide the nature of your dependent variable (i.e. what probability distribution you should assume for ε), the nature of the predictor variables (i.e. whether they are continuous, discrete counts or ordered/unordered factors) and the form of the functions linking the dependent and predictor variables, you will have decided which type of statistical model to use.

The basis set of your DAG, obtained from the basiSet() function, consists of a series of d-sep claims, each having the general form $X \perp\!\!\!\perp Y | Q$. For each of these d-sep claims you will choose an appropriate statistical model based on the considerations discussed above. In order to test each d-sep claim, which is a null hypothesis stating that variables X and Y are independent conditional on the set Q of other variables, you will then obtain the probability that the partial slope of X in this model is zero in the statistical population. Why? Because a partial slope of X that is zero is equivalent to saying that X and Y are independent conditional on the variables in Q, which is the null hypothesis specified by the d-sep claim. As an added complication, the order in which the independent variables (i.e. X and Q) are added to the model can change the null hypothesis that is associated with a zero partial slope in some types of models, and the null hypothesis that we want to test is that X is independent of Y after taking into account the variables in Q. Therefore, a good rule of thumb is to always enter the Q variables into the model before X. I have explained how to do this (Shipley 2009) in the context of nested data having both normally distributed and binomially distributed variables, along with an example. However, you can adapt this to many different contexts as long as your causal model is a DAG and as long as you can test the (conditional) independence claims implied by your basis set with an appropriate statistical test that provides null probabilities.

An example

The leaves of most flowering plants are photosynthetic organs. Since carbon fixation is so central to the survival of plants, one might expect there to be a tight integration of leaf form and physiology to provide for this necessary function. However, land plants face a dilemma. They need to keep their tissues turgid, but these humid tissues find themselves surrounded by air or soil that is not saturated with water. The leaves (and other tissues) are protected by a cuticle to prevent dehydration. Unfortunately, this severely restricts not only the diffusion of water vapour but also other gases, especially the carbon dioxide that is required for photosynthesis, from diffusing into the leaves. The production of stomates is the evolutionary solution to this problem. Stomates are small openings on the surface of the leaves through which gases can diffuse, and the size of the stomatal openings is controlled by guard cells.

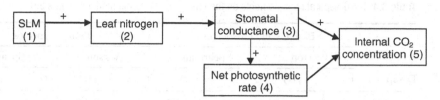

Figure 3.9 Proposed causal relationships between five variables related to interspecific leaf morphology and gas exchange

As soon as the stomates begin to open, carbon dioxide begins to diffuse from the outside air into the intercellular spaces of the leaf through a process of passive diffusion. Since the leaf is photosynthesising, carbon dioxide is being removed from the intercellular spaces, creating a diffusion gradient. However, the air inside the leaf is always saturated with water vapour. As soon as the stomates begin to open, this water vapour also begins to diffuse out of the leaf, since the outside air is not saturated with water. In essence, the leaf has to accept a loss (water) in order to effect a gain (carbon). Cowan and Farquhar (1977) have proposed a theoretical model of stomatal regulation to predict how the leaf should control its stomates in order to maximise carbon gain relative to water loss. The basic insight of this model is that the leaf should restrict carbon fixation below the maximum level, because when the internal CO_2 level in the leaf reaches a certain level the main carboxylating enzyme (ribulose-1,5-bisphosphate carboxylase/oxygenase: RuBisCO) becomes saturated, and further increases in carbon fixation require the regeneration of adenosine triphosphate (ATP) from the light reaction of photosynthesis. The second stage results in a greatly reduced rate of increase of carbon fixation per increase in the internal CO_2 concentration, but the rate of water loss continues at its former rate. Thus, Cowan and Farquhar's principal insight is that the leaf should maintain the intercellular CO_2 concentration at the break point between RuBisCO limitation and ribulose-1,5-bisphosphate (RuP_2) regeneration limitation so that the carboxylating capacity and the capacity to regenerate RuBisCO are co-limiting.

Based on these theoretical notions, Martin Lechowicz and I (Shipley and Lechowicz 2000) have proposed a path model based on five variables:

(1) specific leaf mass (SLM: leaf dry mass divided by leaf area, g/m^2);
(2) leaf organic nitrogen concentration ($mmol/m^2$);
(3) stomatal conductance to water ($mmol/m^2/s$);
(4) net photosynthetic rate ($\mu mol/m^2/s$); and
(5) internal CO_2 concentration ($\mu l/l$).

The proposed model is shown in Figure 3.9. Our data were the mean values from 40 herbaceous species typical of wetland environments.

There are five outliers in the data in relation to the internal CO_2 concentration. These are 'C_4' species. The other 35 species are C_3 species. C_4 species have an additional metabolic pathway in which atmospheric carbon is first fixed by phosphoenolpyruvate (PEP) carboxylase in the mesophyll cells to form malate or aspartate. This molecule, a 4-carbon acid, is then transferred into bundle-sheath cells deeper in the leaf. Here these

Table 3.4 The d-separation statements on the basis implied by the model in Figure 3.9

| | Both C_3 and C_4 species | | | | Only C_3 species | | | |
| | Pearson | | Spearman | | Pearson | | Spearman | |
D-sep	r	p(r)	r	p(r)	r	p(r)	r	p(r)
1 ∥ 3|2	−0.286	0.0777	−0.234	0.1523	−0.298	0.0871	−0.226	0.1986
1 ∥ 4|3	0.165	0.3163	0.217	0.1841	0.109	0.5392	0.188	0.2860
1 ∥ 5|3,4	0.035	0.8328	0.099	0.5560	0.043	0.8139	0.215	0.2303
2 ∥ 4|1,3	−0.092	0.5837	−0.069	0.6809	0.160	0.3743	0.156	0.3870
2 ∥ 5|1,3,4	0.262	0.1169	0.058	0.7327	−0.079	0.6678	−0.006	0.9758
Fisher's C	$\chi^2 = 13.15$, 10 df,		$\chi^2 = 9.713$, 10 df,		$\chi^2 = 9.301$, 10 df,		$\chi^2 = 10.621$, 10 df,	
	$p = 0.216$		$p = 0.466$		$p = 0.503$		$p = 0.388$	

Notes: Also shown are the Pearson and Spearman partial correlations and their two-tailed probabilities. Results are shown for the full data set of 40 species and for the 35 species of C_3 species only. Numbers refer to the variables shown in Figure 3.9.

C_4 acids are decarboxylated, and the freed carbon dioxide enters the normal Calvin cycle of the dark reaction of photosynthesis. An advantage of C_4 photosynthesis is that plants exhibiting it are able to absorb CO_2 strongly from a lower concentration of CO_2 within the leaf. They can do this without RuBisCO acting as an oxygenase, rather than a carboxylase, under conditions of low CO_2 and high O_2. This means that C_4 plants do not exhibit the wasteful process of photorespiration under conditions of high illumination and low water availability. Because of this, they are able to maintain high rates of photosynthesis even when the stomates are nearly closed. The basis set implied by the model in Figure 3.9, along with the relevant statistics, is summarised in Table 3.4.

There is no strong evidence for any deviation of the data from the predicted correlational shadow, as given by the d-separation statements. However, a reasonable alternative model would be that the leaf nitrogen content, which is primarily due to enzymes related to photosynthesis, directly causes the net photosynthetic rate. In other words, what if Cowan and Farquhar's (1977) model of stomatal regulation is wrong, and the leaf is regulating its stomates to maximise the net rate of CO_2 fixation independently of water loss? In this case, the observed rate of stomatal conductance would be a consequence of the net photosynthetic rate rather than its cause and the net photosynthetic rate would be directly caused by leaf nitrogen content. We can test this alternative model too, and Table 3.5 summarises the results.

This alternative model is clearly rejected when both the C_3 and C_4 species are analysed together, since there are only about two out of 10,000 changes of observing such a large difference by chance. This lack of fit is coming from the predicted independence between leaf nitrogen level (2) and stomatal conductance (3), conditioned jointly on specific leaf mass (1) and net photosynthetic rate (4). This, of course, is the critical distinction between the path model in Figure 3.9 and the alternative model. When looking only at the C_3 species, the alternative model does not have a large degree of lack of fit, though the critical prediction still shows a reasonably large lack of fit ($r_{2,3|1,4} = 0.371$, $p = 0.0338$) and is always poorer than that provided by the structure shown in Figure 3.9.

Table 3.5 The d-separation statements implied by an alternative model

| | Both C$_3$ and C$_4$ species | | | | Only C$_3$ species | | | |
| | Pearson | | Spearman | | Pearson | | Spearman | |
D-sep	r	p(r)	r	p(r)	r	p(r)	r	p(r)
$1 \parallel 3\vert2$	−0.286	0.0777	−0.234	0.1523	−0.298	0.0871	−0.226	0.1986
$1 \parallel 3\vert4$	0.286	0.0777	0.279	0.0853	0.221	0.2092	0.1723	0.3298
$1 \parallel 5\vert3,4$	0.035	0.8328	0.099	0.5560	0.043	0.8139	0.215	0.2303
$2 \parallel 3\vert1,4$	0.599	7×10^{-5}	0.569	2×10^{-4}	0.371	0.0338	0.339	0.0541
$2 \parallel 5\vert1,3,4$	0.262	0.1169	0.058	0.7327	−0.079	0.6678	−0.006	0.9758
Fisher's C:	$\chi^2 = 33.96$, 10 df, p = 0.0002		$\chi^2 = 27.59$, 10 df, p = 0.0021		$\chi^2 = 16.00$, 10 df, p = 0.1000		$\chi^2 = 14.27$, 10 df, p = 0.161	

Notes: In this alternative the model in Figure 3.9 is changed to make leaf nitrogen cause the net photosynthetic rate, which then causes the observed rate of stomatal conductance, along with the Pearson and Spearman partial correlations and their two-tailed probabilities. Results are shown for the full data set of 40 species and for the 35 species of C$_3$ species only.

Because of such results, and other reasons described in the original reference, I prefer the causal structure shown in Figure 3.9. However, such a conclusion must remain tentative. After all, the conclusion is based on only 40 species, and a larger sample size might detect some more subtle lack of fit that was too small to be found in the present data set.

Given the model in Figure 3.9, and given that we have not been able to reject it, we can now fit the path equations. Although Sewall Wright's original method was based on standardised variables, I prefer to use the original variables, because the variables each have well-established units of measurement. The least-squares regression equations, using only the C$_3$ species, are shown below. The residual variation is indicated by $N(0,\sigma)$.

Ln(% nitrogen) = 0.78+0.90Ln(SLM)+$N(0,0.243)$, R = 0.85
Ln(conductance)= −6.60+1.15Ln(% nitrogen)+$N(0,0.56)$, R = 0.69
Ln(photo) = 3.08+0.55Ln(conductance)+$N(0,0.31)$, R = 0.81
Ln(CO$_2$ internal) = 6.42+0.14Ln(conductance)−0.1Ln(photo)+$N(0,0.04)$, R = 0.77

Each of the slopes is significant at a level below 10^{-4} and the sign of each is in the predicted direction. With these path equations we can begin to simulate how the entire suite of leaf traits would change if we change the specific leaf mass (the exogenous variable in this model) or if we observe species with different specific leaf masses. We get the functional relationships by back-converting the variables in the equations from their logarithms. Of course, each of these variables may also change with changing environmental conditions. By including these environmental variables we could generate the response surfaces across which the suite of leaf traits would move as the environment changes.

A suggestion when proposing your own causal models

The following is a common experience among educators; I call it the 'real-world para-dox'. The students arrive on their first day to class with a pre-existing, 'naïve' under-standing of some phenomenon and how to deal with it. The teacher then spends several weeks in a classroom setting in which he or she teaches the theoretical concepts of a discipline and how to apply these theoretical concepts in practice. These theoretical concepts, and their application, differ from the original, 'naïve' understanding of the stu-dents. The students demonstrate, through their course work, that they understand these new concepts and how to apply them. After the course is finished and the final mark is posted, the students then return to the 'real world' and are presented with the same type of problem that they have already mastered in the classroom. They then proceed to ignore their classroom training and fall back on their 'naïve' methods. This occurs because people often associate ideas and methods with the context in which they were obtained. The new theoretical ideas and methods were obtained in a very particular and structured context (a classroom with a professor looking on), which is a context that is very different from the 'real' world. Because the original naïve concepts were obtained in the real world, these are the default ones that the students use. The same thing might happen to you when you finish reading this book.

I have noticed that beginners to causal modelling, even after having the notions pre-sented so far in this book explained to them, often make a common mistake when it comes time to leave the classroom and do causal modelling on their own data. They will show me a DAG with lots of arrows and, when I ask why they placed an arrow going from variable x to variable y, they will respond that this is because previous studies have shown that x and y are correlated. This is, of course, the wrong answer! An arrow between x and y (x→y) in a DAG means that changing x would cause a change in y even if we could experimentally hold constant all other variables in our DAG. It is irrel-evant that we cannot actually carry out such a manipulation. The wrong answer of our student occurs because their previous real-world experience of data analysis involved purely observational models involving associations between variables.

To help you avoid this mistake, I suggest that you take some time before reading further and close this book. Go to wherever you do your 'real' work and place in front of you a picture of your research organism or site and then begin writing down a DAG representing the causal structure linking the variables that interest you. For each pair of variables (x, y), say out loud: 'I don't care if x and y are associated. If I could physically hold constant all other variables in my model except for x and y, but not any variables outside my model, and then if I could manipulate x, would the value of variable y change or not?' If the answer is 'Yes' then add an arrow going from x to y. If the answer is 'No' then don't add an arrow. It doesn't matter if this experimental manipulation can be performed in practice. The point is to equate an arrow in a DAG with a hypothetical controlled experiment, not with a statistical association.

4 Path analysis and maximum likelihood

James Burke (1996), in his fascinating book *The Pinball Effect*, demonstrates the curious and unexpected paths of influence leading to most scientific discoveries. People often speak of the 'marriage of ideas'. If this is true then the most prolific intellectual offspring come not from the arranged marriages preferred by research administrators but from chance meetings, and even illicit unions. The popular view of scientific discoveries as being linear causal chains from idea to solution is profoundly wrong; a better image would be a tangled web with many dead ends and broken strands. If most present knowledge depends on unlikely chains of events and personalities, what paths of discovery have been deflected because the right people did not come together at the right time? Which historical developments in science have been changed because two people, each with one half of the solution, were prevented from communicating due to linguistic or disciplinary boundaries? The second stage in the development of modern structural equation modelling is a case study in such historical contingencies and interdisciplinary incomprehension.

During the First World War, and in connection with the American war effort, Sewall Wright was on a committee allocating pork production to various states based on the availability of corn.[1] He was confronted with a problem that had a familiar feel. Given a whole series of variables related to corn availability and pork production, how do all these variables interact to determine the relationship between supply and demand, and the fluctuations between these two? It occurred to him that his new method of path analysis might help. He calculated the correlation coefficients between each pair of variables for five years, giving 510 separate correlations. After much trial and error he developed a model involving only four variables (corn price, summer hog price, winter hog price and hog breeding) and only fourteen paths that still gave a 'good match' between observed and predicted correlations. He described his results in a manuscript that was submitted as a bulletin of the US Bureau of Animal Industry. It was promptly rejected, perhaps because officials at the Bureau of Agricultural Economics considered it as an intrusion onto their turf. Happily for Wright, he had also shown it to the son of the secretary of agriculture (Henry A. Wallace), who was interested in animal breeding and quantitative modelling. Wallace, using his political influence, intervened to have the manuscript published as a United States Department of Agriculture (USDA) bulletin (Wright 1925).

---

[1] This next section is based on Wright's biography (Provine 1986).

Although economists later developed methods that were very similar to path analysis, Wright's foray into economics does not seem to have been very influential. During the Second World War he presented a seminar on path analysis to the Cowles Commission, in which economists were developing methods that were the forerunner of structural equation modelling. Neither Wright nor the economists recognised the link between the two approaches or the usefulness of such a marriage (Epstein 1987). Nonetheless, some economists were independently trying to express causal processes in functional form[2] (Haavelmo 1943). In economics, constraints on the covariance matrix (for example, zero partial correlations due to d-separation) were called *over-identifying constraints*. Since most work in this area was in parameter estimation, not theory testing, such constraints were mostly avoided because they made consistent estimation difficult.

In the 1950s the political scientist Herbert Simon began to derive the causal claims of a statistical model.[3] This led some social scientists to think about expressing causal processes as statistical models that implied certain structural, or *over-identifying*, constraints. One such person was Hubert Blalock, who began deriving over-identifying constraints, in the form of zero partial correlations, that were implied by the structure of the causal process (Blalock 1961; 1964). Wright's method of path analysis had been largely rediscovered by social scientists, with the important difference that the emphasis shifted from being an *a posteriori* description of an assumed causal process – as Wright viewed his method – to being a (tentative) test of an hypothesised causal process. The late 1960s and early 1970s saw many applications of path analysis in sociology, political science and related social science disciplines.

The most important next step was the work of people such as Karl Jöreskog (1967; 1969; 1970; 1973) and Ward Keesling (1972), who developed ways of combining confirmatory factor analysis (see Chapter 8) and path analysis using maximum-likelihood estimation techniques. The advance was not simply in using a new method of estimating the path coefficients. More importantly, the use of maximum likelihood allowed the resulting series of equations describing the hypothesised causal process (a series of *structural equations*) to be tested against data in order to see if the over-identifying constraints (the zero partial correlation coefficients and other types of constraints) agreed with the observations. This advance solved the main weakness of Wright's original method of path analysis, since one did not simply have to *assume* the causal structure, as Wright did. Now one could test the statistical consequences of the causal structure and therefore potentially falsify the hypothesised causal structure.[4] Unfortunately, by the 1970s most biologists had forgotten about Wright's method of path analysis, and disciplinary boundaries prevented the new SEM approach from penetrating into biology.

---

[2] Some economists referred to Wright's work in passing (Goldberger 1972; Griliches 1974), but only for historical completeness.

[3] Summarised by Simon (1977).

[4] The logical and axiomatic relationships between probability distributions and causal properties had not yet been developed. This led to much confusion concerning the causal interpretation of structural equation models (Pearl 1997). One reason why I discuss these points in detail is to prevent the same sterile debates from recurring between biologists.

Wright's method was essentially the application of multiple regression based on standardised variables in the order specified by the path diagram (the causal graph). This, along with ANOVA and most other familiar statistical methods, consists of modelling the individual observations. In other words, the path coefficients were obtained using least-square techniques by minimising the squared differences between the observed and predicted values of the individual observations, as is usual in multiple regression. Structural equation models, of which modern path analysis is a special case,[5] concentrate instead on the pattern of covariation between the variables and minimise the difference between the observed and predicted pattern of covariation among them. The basic steps are as follows.

(1) Specify the hypothesised causal structure of the relationships between the variables in the form of a causal diagram (often a DAG).
(2) Translate the causal model into an observational model. Write down the set of linear equations that follow this structure and specify which parameters (slopes, variances, covariances) are to be estimated from the data (i.e. that are *free*) and which are *fixed* (i.e. are not to be changed to accommodate the data) based on the causal hypothesis.
(3) Derive the predicted variance and the covariance between each pair of variables in the model using covariance algebra. Covariance algebra gives the rules of path analysis that Wright had already derived.
(4) Estimate these free parameters using maximum likelihood or related methods, while respecting the values of the fixed parameters. This estimation is done by minimising the difference between the observed covariances of the variables in the data and the covariances of the variables that are predicted by the causal model.
(5) Calculate the probability of having observed the measured minimum difference between the observed and predicted covariances, assuming that the observed and predicted covariances are identical except for random sampling variation.
(6) If the calculated probability that the remaining differences between observed and predicted covariances due only to sampling variation is sufficiently small (say, below 0.05), one concludes that the observed data were not generated by the causal process specified by the hypothesis, and the proposed model is rejected. If, on the contrary, the probability is sufficiently large (say, above 0.05) then one concludes that the data are consistent with such a causal process.

4.1   Testing path models using maximum likelihood

Step 1: translate the hypothetical causal system into a path diagram

This first step should be almost second nature by now, but there are a few notational conventions that must be introduced. Path diagrams contain three different types of variables. Variables that have been directly observed and measured are enclosed in squares; these variables are called *manifest* variables in SEM terminology. Variables that are

---

[5] There are a number of different names for this general class of models: Structural equation models, LISREL models, covariance structure models.

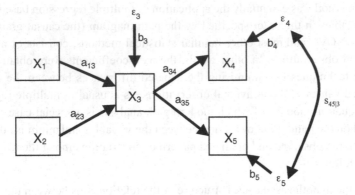

Figure 4.1 A path model involving five variables
Note: $S_{45|3}$ is a free covariance between ε_4 and ε_5.

hypothesised to have a causal role in the model but that have not been directly observed or measured are enclosed in circles; these variables are called *latent* variables in SEM jargon.[6] The third type of variable is the residual *error* variable,[7] and it is not enclosed at all. This type of variable represents all other unmodelled causes of the variable into which it points. It is also generally defined as a normally distributed random variable with a zero mean and a variance of 1, although it is possible to model it with a different variance. A second common classification is between a variable that has no causal parents in the model, called *exogenous*, and a variable that is caused by some other variable in the model, called *endogenous*.[8] Finally, there are two types of arrows. A straight arrow indicates a causal relationship between two variables, just as it does in the directed graphs of previous chapters. A curved, double-headed arrow indicates an unknown causal relationship linking the two variables. This means that there is a free covariance in the structural equations. These conventions are shown in Figure 4.1.

Step 2: translate the causal model into an observational model in the form of a set of structural equations

As the arrows in Figure 4.1 suggest, the hypothesised relationships are asymmetric causal ones. When we translate the causal model into mathematical equations we obtain an observational (statistical) model. Because we are now dealing with a statistical model, we must make assumptions concerning the form of the functional relationships and the sampling distribution of the random variables. Contrary to the path diagram, this new model is not strictly a causal model, because it is expressed in the language of

---

[6] By convention, a path model is simply a structural equation model that does not involve unmeasured, or latent, variables.

[7] A common error is to assume that the residual error in a structural equation model is the same as the residual error in a regression model. The two are not necessarily the same. In a regression model the residuals are always uncorrelated (or orthogonal) with the predictors. The residuals in a structural equation model need not be uncorrelated with the predictors or with each other.

[8] Yes, I know; I didn't invent these terms. I will try to limit the jargon to a minimum, but you will have to be aware of these terms in order to read the literature.

algebra using the equivalence operator ($=$). It is an imperfect translation of the causal model, and we must not forget this and begin manipulating these algebraic equations in ways contrary to the original asymmetric causal relations expressed in the path diagram. In almost all structural equation models the relationships are assumed to be additively linear. In most structural equation models the random variables are assumed to be multivariate normal.[9] 'Multivariate normal' means that the variables follow a multivariate normal distribution, which is a somewhat stronger assumption that assuming simply that each variable is normally distributed. Different ways of assessing this assumption, and the degree of non-normality that can be tolerated, will be described in Chapter 6.

If your causal model is sufficiently detailed that you are willing to hypothesise the numerical values of some parameters (path coefficients, variances or covariances) then you can include this information in the model by specifying the parameter to be fixed. If you are not able or willing to make such an assumption (except, of course, that the parameter is not zero) then the parameter is estimated from the data and is therefore *free*. Specifying that a variable is not a direct cause of another (i.e. that there is not an arrow from the one to the other in the path diagram) is the same as specifying that the path coefficient of this 'missing' arrow is fixed at zero. Each parameter that is fixed adds one degree of freedom to the inferential test.

In such models we are usually interested in the relationships between the variables, not the mean values of the variables themselves.[10] For this reason, all variables are 'centred' by subtracting the mean value of each variable from each observation. For instance, if the mean of X_1 in Figure 4.1 was 6, then we would replace each value of X_1 by $(X_1 - 6)$. This trick ensures that the mean of each transformed variable is zero and therefore that the intercepts are zero, which frees us from having to estimate them. Assuming that all our variables are already centred, these are the structural equations corresponding to Figure 4.1, where $Cov(X_1, X_2)$ means 'the population covariance between X_1 and X_2':

$$X_1 = N(0, \sigma_1)$$
$$X_2 = N(0, \sigma_2)$$
$$X_3 = a_{13}X_1 + a_{23}X_2 + b_3\varepsilon_3$$
$$X_4 = a_{34}X_3 + b_4\varepsilon_4$$
$$X_5 = a_{35}X_3 + b_5\varepsilon_5$$
$$Cov(X_1, X_2) = Cov(X_1, \varepsilon_3) = Cov(X_1, \varepsilon_4) = Cov(X_1, \varepsilon_5) = Cov(X_2, \varepsilon_3)$$
$$= Cov(X_2, \varepsilon_4) = Cov(X_2, \varepsilon_5) = Cov(\varepsilon_3, \varepsilon_4) = Cov(\varepsilon_3, \varepsilon_5) = 0$$
$$Cov(\varepsilon_4, \varepsilon_5) = \sigma_{45}$$

This process of converting your causal diagram into a series of linear equations containing free and fixed parameters is a critical step in the lavaan package of R that we will

---

[9] More specifically, the endogenous variables are assumed to be multivariate normal. The exogenous variables do not have to be normally distributed, in the same way that the independent variables in a regression do not have any distributional assumptions.

[10] Means can also be modelled, but this requires a little bit more work.

use. Notice that some parameters (σ_i, a_{ij}) in the above equations do not have numerical values and therefore have to be estimated; before looking at the data they are 'free' to take on any value.

Let's go through these equations more slowly to understand exactly how the causal model in Figure 4.1 has been represented in equation form. First, variables X_1 and X_2 are exogenous in the model; we don't know, or are not interested in explicitly modelling, the causal parents of these two variables. In the equations I have specified that X_1 and X_2 are each normally distributed random variables whose mean is zero and whose standard deviation is unknown. Therefore, these two standard deviations are free and must be estimated from the data.[11] Next, X_3 is written as a linear function of both X_1 and X_2 in accordance with Figure 4.1. Since I don't know the numerical strength of the direct causal effects of these two variables, the path coefficients (a_{13} and a_{23}) are also free and must be estimated from the data. If my causal hypothesis had been sufficiently well developed that I could specify what the values of these path coefficients were then I would have entered the predicted values rather than having to estimate them from the data. In addition, the combined direct effect of the other unknown causes of X_3 are not known either, and so b_3, the path coefficient from the error variable (ε_3), is also free and must be estimated. Remember that all the error variables (ε) are unit normal variables – i.e. with a zero mean and a standard deviation of 1. Multiplying a unit normal variable by a constant (b_3 in this case) makes its variance equal to the constant. Therefore, the part of the variance of X_3 that is not accounted for by X_1 and X_2 is b_3. In this particular equation the residual is exactly analogous to the residuals of a multiple regression, since it is made to be uncorrelated with either X_1 or X_2, but this is not always the case. Next, each of X_4 and X_5 are also written as linear functions of X_3 with the accompanying free path coefficients.

Since there are five variables in the model, there are 10 different pairs of variables, and therefore 10 different covariances between the unique pairs of variables. Since X_1 and X_2 are causally independent, the covariance between these two variables must be zero (remember d-separation). X_1, X_2 and the unknown other causes of X_3 (i.e. ε_3) are also independent of each other and of the unknown other causes of X_4 and X_5, and so each of these pairs of covariances must also be zero. Finally, the causal model in Figure 4.1 states that there is some causal influence linking X_4 and X_5 but the researcher does not know what it is. Perhaps X_4 causes X_5? Perhaps X_5 causes X_4? Perhaps there is a reciprocal causal relationship? Perhaps there is some unknown common cause of both X_4 and X_5? Adding a free covariance (the translation of a curved double-headed arrow) is an admission of ignorance as to the causal origin of the covariance. Each of the above causal relationships would generate a non-zero covariance between X_4 and X_5 even after controlling for X_3. Therefore, we allow ε_4 and ε_5 to have a non-zero covariance, and the numerical value of this non-zero covariance must also be estimated from the data.

---

[11] Fixing the error variance at 1.0 and freely estimating the path coefficient associated with the error variable, or fixing the path coefficient to 1.0 and freely estimating the error variance, are two equivalent ways of doing the same thing.

Box 4.1 Basic rules of covariance algebra

The notation E(X) means the expected value of X. Therefore, the population covariance between two variables – symbolised here as $Cov(X_1, X_2)$ – is defined as

$$Cov(X_1, X_2) = E[(X_1 - E(X_1))(X_2 - E(X_2))] = E(X_1, X_2) - E(X_1)E(X_2).$$

If the variables are centred about their expected values, this reduces to

$$Cov(X_1, X_2) = E(X_1, X_2).$$

Since a variance is simply the covariance of a variable with itself, we can write the population variance (Var) as

$$Var(X_1) = Cov(X_1, X_1).$$

If k is a constant and X_1, X_2, X_3 are random variables then we can also state the following useful rules:

(1) $Cov(k, X_1) = 0$
(2) $Cov(kX_1, X_2) = kCov(X_1, X_2)$
(3) $Cov(k_1X_1, k_2X_2) = k_1k_2Cov(X_1, X_2)$
(4) $Cov(X_1 + X_2, X_3) = Cov(X_1, X_3) + Cov(X_2, X_3)$

This completes the best translation of the causal model into the observational (statistical) model that can be obtained consistent with the statistical assumptions that are needed to estimate the free parameters. It will be important to evaluate these assumptions when judging whether the results of the analysis can be trusted, as is true of any statistical method.

Step 3: derive the predicted variance and the covariance between each pair of variables in the model using covariance algebra

The set of structural equations allows us to derive the predicted values for the covariances between each pair of variables. Since a Pearson correlation coefficient is simply a covariance that has been standardised, one can also derive the predicted values for the correlations between each pair of variables. This step uses the rules of path analysis that were derived originally by Sewall Wright. Box 4.1 summarises a few basic rules of covariance algebra that will be useful in discussing this section.

From Chapter 2 you know that two vertices in the path diagram that are d-separated correspond to two random variables that are independently distributed, meaning that the population covariance between them must be zero. If the two vertices are not d-separated then the corresponding random variables are not independently distributed, and so (given the assumption of linearity made by SEM) the covariance between them cannot be zero. This justifies the list of zero covariances in the structural equations given above. For those vertices that are not unconditionally d-separated (and are therefore correlated in some way), we can use the rules of covariance algebra to obtain formulae

---

Box 4.2 The Bentler–Weeks model

Definitions: let the endogenous (i.e. dependent) variables in the model be written in a column vector called η and let the exogenous (i.e. independent) variables (including the error variables) be written in a column vector called ε. Let the coefficients of the effects of dependent causes to dependent effects be a matrix called β (rows are dependent effects and columns are dependent causes) and let the coefficients of the effects of independent causes to dependent effects be a matrix called γ (rows are dependent effects and columns are independent causes). Then the system of structural equations can be written as $\eta = \beta\eta + \gamma\varepsilon$.

For instance, the path model in Figure 4.1 would be written

$$\begin{bmatrix} x_3 \\ x_4 \\ x_5 \end{bmatrix} = \begin{bmatrix} 0 & 0 & 0 \\ a_{34} & 0 & 0 \\ a_{35} & 0 & 0 \end{bmatrix} \begin{bmatrix} x_3 \\ x_4 \\ x_5 \end{bmatrix} + \begin{bmatrix} a_{13} & a_{23} & b_3 & 0 & 0 \\ 0 & 0 & 0 & b_4 & 0 \\ 0 & 0 & 0 & 0 & b_5 \end{bmatrix} \begin{bmatrix} x_1 \\ x_2 \\ \varepsilon_3 \\ \varepsilon_4 \\ \varepsilon_5 \end{bmatrix}.$$

In reduced form the equation is $\eta = (\mathbf{I} - \beta)^{-1}\gamma\varepsilon$.

Predicted covariances between exogenous variables are $E\lfloor\varepsilon\varepsilon'\rfloor = \varsigma$.

Predicted covariances between endogenous and exogenous variables:

$$E[\eta\varepsilon'] = (\mathbf{I} - \beta)^{-1}\gamma\varsigma$$

Predicted covariances between endogenous variables:

$$E[\eta\eta'] = (\mathbf{I} - \beta)^{-1}\gamma\varsigma\gamma'(\mathbf{I} - \beta)^{-1'}$$

---

giving their covariances. Take, for instance, variables X_1 and X_3. Since X_3 is a linear function of X_1, X_2 and ε_3, we can write

$$Cov(X_1, X_3) = Cov(X_1, (a_{13}X_1 + a_{23}X_2 + b_3\varepsilon_3))$$
$$= a_{13}Cov(X_1, X_1) + a_{23}Cov(X_1, X_2) + b_3Cov(X_1, \varepsilon_3).$$

Looking at the path diagram and applying the d-separation operation, we see that X_1 is independent of both X_2 and ε_3 and therefore the population covariances involving X_1 and these two variables are zero. Therefore, the population covariance between X_1 and X_3 is simply $a_{13}Cov(X_1, X_1)$ or $a_{13}Var(X_1)$. In this way we can obtain formulae for the expected value for each pair of variables in the model.

This must seem like a lot of work. Don't worry; most SEM programs, including the lavaan package of R, do all this work for you. The important point at this stage is that you have an intuitive understanding of why we can express the covariances between each pair of variables as a function of path coefficients, variances and covariances. For those who are used to working with matrix algebra, Box 4.2 gives a more formal derivation of the predicted covariance matrix based on the Bentler–Weeks model (see Bentler 1995 for a concise description).

If we go back to the analogy of correlations being the shadows that are cast by causal processes, the predicted covariance matrix is a description of the 'shape', but not the 'size', of the shadow cast by the hypothesised causal process shown in Figure 4.1. Imagine that we were describing the shadow cast by a solid square whose size was unknown to us (i.e. the length of whose sides are *free* parameters). We would describe the shadow as having four equal sides of unknown length (the first constraint) with four sides that meet in such a way that they make four corners having 90 degree angles (the second constraint). The general *shape* of the shadow is fixed (a square) but the numerical *values* (the lengths of the sides) are free parameters and can be estimated by measuring the real shadow.

Step 4: estimate the free parameters by minimising the difference between the observed and predicted variances and covariances

The hypothesised object was the solid square, and from this we have predicted the shape of the shadow that it would cast. Is our hypothesis correct? To decide, we superimpose our hypothesised square shadow on top of the actual shadow and increase or decrease the length of the sides of our hypothesised shadow until our hypothesised square is as close to the observed shadow as possible while respecting the constraints (i.e. we can't change the angles of the sides or make some sides longer than others). Once we have superimposed the two shadows as close as we can, we then measure the remaining lack of fit. This is the same basic logic used to fit and test a structural equation model. We first choose values for the free parameters in our predicted covariance matrix that make it as numerically close as possible to the observed covariance matrix while respecting the constraints applied to the predicted covariance matrix. How this is done depends on the assumptions that have been made concerning the distributional form of the random variables; in SEM the usual assumption is that the random variables follow a multivariate normal distribution. Then we see how much difference remains between the observed and predicted covariance matrices.

The general strategy for obtaining the best values for the free parameters is easy enough to grasp: choose values of the free parameters that make the numerical values of the predicted covariance matrix as close as possible to the actual covariances measured in the data. This is usually done using a method called *maximum-likelihood estimation* (Fisher 1950). Scott Eliason (1993) and Kenneth Bollen (1989) describe the mechanics of this technique, and Box 4.3 gives a brief introduction for those who are interested in this topic. In essence, the numerical algorithm used to maximise the likelihood is a bit like playing the child's game of 20 questions (Is it alive? Is it a mammal? Is it a carnivore? Does it live in Africa? And so on . . .). Box 4.3 gives a more precise definition of the likelihood function, but this function can be intuitively understood to measure the discrepancy between the observed data and the sort of data that would have been observed had the free parameter been equal to our chosen value. We start with an initial guess of the values of the free parameters and calculate the likelihood of the data given the current parameter values. We then see if we can modify our guess of the values of the free parameters in such a way as to improve the likelihood. We continue with this

Box 4.3 Maximum-likelihood estimation

The probability of occurrence of a random variable, x_i, is given by a probability function (for discrete variables) or a probability density function (for continuous variables). For instance, the probability density function of a univariate normal random variable is

$$f(x; \mu, \sigma) = \frac{1}{\sqrt{2\pi\sigma^2}} e^{\frac{-(x-\mu)^2}{2\sigma^2}}$$

The notation means that x is the random variable and μ and σ are population parameters that are fixed. Now, if we take a series of N independent observations of the random variable, then the joint probability density function for these N observations is $f(x_1; \mu, \sigma)f(x_2; \mu, \sigma)f(x_3; \mu, \sigma)\ldots f(x_N; \mu, \sigma)$.

The objective of maximising the likelihood of a parameter is to find a value of this parameter that maximises this joint probability density function – in other words, finding a value for the parameter (e.g. μ) that maximises the likelihood of having observed the series of observations. This objective turns the probability density function on its head. Now the observed values (x_i) are fixed and we view the population parameters as variables. We are envisaging a whole series of different normal distributions and we want to choose the most likely one given our data. So, the likelihood function of the univariate normal distribution is

$$L(\mu, \sigma; x) = \frac{1}{\sqrt{2\pi\sigma^2}} e^{\frac{-(x-\mu)^2}{2\sigma^2}}$$

and the joint likelihood function of the entire set of data is $L(\mu, \sigma; x_1)L(\mu, \sigma; x_2)L(\mu, \sigma; x_3)\ldots L(\mu, \sigma; x_N)$.

The natural logarithm of a series of positive numbers is an increasing function of these numbers. Because it is difficult to maximise a product but easier to maximise a sum, we use the logarithm of the likelihood function. For instance, imagine that we have observed eight values (1.20, 0.08, 0.34, 0.57, 0.46, 0.48, 0.56, 1.01) from a normal distribution whose population variance (σ^2) is 1 and we want to find the maximum-likelihood value for the population mean (μ). Figure 4.2 shows a graph of the log-likelihood function over the range of $\mu = -4$ to 4.

We see that this function is maximal at around 0.58. This is the sample mean of our eight values, showing that the standard formula for the sample mean – an unbiased estimator – is also a maximum-likelihood estimator. Maximum-likelihood estimates are not always unbiased in small samples (for instance, the maximum-likelihood estimate of the variance is not) but they are consistent, meaning that such estimates converge on the true value as the sample size increases. In other words, maximum-likelihood estimates are asymptotic estimates.

In general, the maximum (or minimum) of the likelihood function occurs when its first derivative is zero. To see if one has found a maximum one then checks to see if the second derivative is negative. In SEM, the usual assumption is that the data are multivariate normal. In other words, we assume that the probability density function

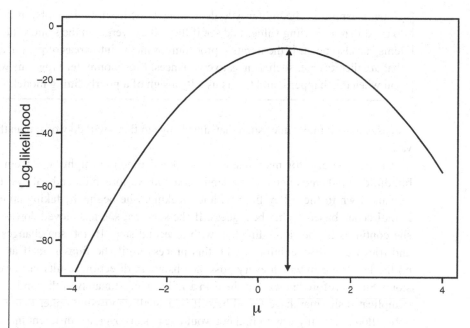

Figure 4.2 The log-likelihood function for the population mean (μ)

is multivariate normal. The likelihood function for this distribution is

$$L(\mu, \Sigma; x) = \frac{1}{(2\pi)^{n/2}|\Sigma|^{1/2}} e^{-[x-\mu]\Sigma^{-1}[x-\mu]}$$

In SEM we usually centre the variables about their means, so the only parameters whose likelihood functions we need to estimate are those in the population covariance matrix Σ. These parameters are the free parameters that we have already derived. With this much more complicated function we are not able to derive the maximum-likelihood estimates directly, and so we have to use numerical methods that involve iteration.

For instance, in the predicted covariance matrix for the causal graph shown in Figure 4.1 we have free parameters for the variances of the two independent observed variables, the three error variables, the one free covariance between X_4 and X_5, and the four path coefficients. Let's group all these free parameters together in a vector called θ. Now, if we take a first guess at the values of these free parameters, we can calculate the predicted covariance matrix based on these initial values; let's call the predicted covariance matrix that results from this guess $\Sigma_{(1)}(\theta)$ to emphasise that this matrix will change if we change our values in θ. We now calculate the value of the log-likelihood function. Next, we change our initial estimates of the free parameters and recalculate the predicted covariance matrix, $\Sigma_{(2)}(\theta)$, in such a way as to increase the log-likelihood. We continue in this way until we can't increase the value of the log-likelihood anymore.

Problems can occur if the log-likelihood function contains 'potholes' or local maxima. If this happens then the iterative procedure can get 'trapped' without finding the global maximum. The only way to determine if one has found a global maximum is to try different starting values and see if they all converge on the same values. Problems can also occur if the iterative procedure wanders into areas of parameter space that are illegitimate, such as negative variances. Most computer programs will warn you when this happens, and this is usually a sign of a poorly fitting model.

process until we find values such that any change to them will do worse than the present values.

Another analogy for maximising a likelihood function might be a woman who is blindfolded and finds herself in a landscape with various hills and valleys. Her job is to walk down to the valley floor without peeking. She begins by taking an initial step in a direction based on her best guess. If she sees that she has moved down-slope then she continues in the same direction with a second step. If not, she changes direction and tries again. She continues with this process until she finds herself at a position on the landscape in which every possible change in direction results in movement up-slope. She therefore knows that she is in a valley. Unfortunately, if the landscape is very complicated she may have found herself in a small depression rather than on the true valley floor. The only way to find out would be to start over at a different initial position and see if she again ends up in the same place.

Let S be the observed covariance matrix, involving p dependent (endogenous) and q independent (exogenous) variables. Let Σ be the maximum-likelihood estimate of the model covariance matrix. Since these maximum-likelihood estimates depend on the values of the free parameters, which we group together in a vector θ, we will write the population covariance matrix as $\Sigma(\theta)$. The maximum likelihood fitting function, F_{ML}, that compares the difference between the observed and predicted covariance matrices is

$$F_{ML} = \log|\Sigma(\theta)| + trace(S\Sigma^{-1}(\theta)) - \log|S| - (p+q)$$

This function has three important properties. First, the values of the free parameters, θ, that minimise it are also the values that make the predicted covariance matrix as similar as possible to the observed covariance matrix while respecting the constraints implied by the causal model. Second, the values of θ that *minimise* this function are the same values that *maximise* the multivariate normal-likelihood function, and so such values of the free parameters that define the population covariance matrix (θ) are called *maximum-likelihood estimates*. Third, and most importantly, if the observed data (and therefore S) really were generated by the causal process that the structural equations are modelling then the only remaining differences between $\Sigma(\theta)$ and S at the minimum of F_{ML} will be due to normally distributed random sampling variation. Given these assumptions, $(N-1)F_{ML}$ is asymptotically distributed as a chi-square distribution.

I said that one probable reason why biologists did not accept Wright's method of path analysis was that his original method could derive the logical consequences of a causal model but could not test it. The method described above, developed by Jöreskog (1970), was the first to solve this important shortcoming of path analysis.

Step 5: calculate the probability of having observed the measured minimum difference, assuming that the observed and predicted covariances are identical except for random sampling variation

The central chi-square distribution has only one parameter: the degrees of freedom. In testing a structural equation model we are comparing the fit between the observed and predicted elements of the covariance matrix. If we have v variables then there will be v^2 elements in the covariance matrix. Since this matrix is symmetric about its diagonal, some of these elements are redundant. The number of unique elements is $v(v + 1)/2$. If we were to compare the observed and predicted values of these unique elements using $(N - 1)F_{ML}$ and all the predicted values were obtained independently of the observed values, this would define the degrees of freedom for the chi-squared test. However, we have had to use our data to estimate the free parameters that partly determine the predicted covariance matrix. Each free parameter that we have to estimate 'uses up' one degree of freedom. The degrees of freedom available to test the model are

$$\frac{v(v + 1)}{2} - (p + q)$$

As before, q is the number of free variances of exogenous variables (including the error variables) in the model and p is the number of free path coefficients in the model. I say 'free' because it may sometimes be possible to specify the value of a variance or path coefficient based on theory or prior experience and therefore constrain the model to have the specified value no matter what the data say.

We therefore specify, as the null hypothesis, that there is no difference between the observed and predicted covariance matrices except what would be expected given the random sampling variation of N independent observations all taken from the same multivariate normal distribution. Given this hypothesis, the following statistic (the *maximum-likelihood chi-squared statistic*) will asymptotically follow a central chi-square distribution with the degrees of freedom given above:

$$(N - 1)F_{ML} \xrightarrow{N \to \infty} \chi^2_{\frac{v(v-1)}{2}-(p+q)}$$

In practice one uses a computer program to do all these calculations.

Step 6: if the calculated probability is sufficiently small (say, below 0.05) then one concludes that the model was wrong, but if the probability is sufficiently large (say, above 0.05) then one concludes that the data are consistent with such a causal process

At first blush this step appears much easier to understand than the previous ones. In fact, it is the step that causes the greatest confusion. The previous steps are more mathematically involved but they are largely automated, and so the user does not need more than an intuitive grasp of what is happening. This last step requires the user to interpret the meaning of the resulting probability for the biological model. This interpretation can often lead to confusion.

In most of the statistical tests used by biologists, the biologically interesting hypothesis is the alternative hypothesis; the null hypothesis functions as a straw man that is erected only to see if we have sufficiently strong evidence to knock it down. This is useful because it forces us to have strong evidence (evidence beyond reasonable doubt) before we can accept the biologically interesting alternative hypothesis. In SEM, on the other hand, models are constructed on the basis of biological arguments in such a way as to reflect what we hypothesise to be correct. In other words, our model and the resulting predicted covariance matrix embody what we view to be biologically interesting. The null hypothesis, not the alternative, is therefore the biologically interesting hypothesis. A probability below the chosen significance level means that the predicted model is wrong and should be rejected (i.e. the null hypothesis should be rejected). Although the flipping of the null and alternative hypotheses might seem strange, it is exactly the same logic as testing the null hypothesis that the slope of a simple linear regression equals, say, 0.75. Notice that we are reversing the burden of proof: we are requiring strong evidence, evidence beyond reasonable doubt, before we are willing to reject our preferred hypothesis. This leads naturally to the temptation to conclude that the predicted model is correct simply because we have not obtained strong evidence to the contrary! In fact, all that we can conclude is that we have no good evidence to reject our model and that the data are consistent with it. The degree to which we have good evidence in favour of our model will depend on how well we can exclude other models that are also consistent with the data. This leads naturally to the subject of equivalent models (Chapter 8).

At this stage, some numerical examples will help. I will generate 100 independent 'observations' following the causal graph shown in Figure 4.1. Here are the generating equations; these are the same as those shown previously except that the free parameters have been replaced by actual values:

$$X_1 = N(0, 1)$$
$$X_2 = N(0, 1)$$
$$X_3 = 0.5X_1 + 0.5X_2 + 0.5\varepsilon_3$$
$$X_4 = 0.5X_3 + 0.707\varepsilon_4$$
$$X_5 = 0.5X_3 + 0.707\varepsilon_5$$
$$Cov(X_1, X_2) = Cov(X_1, \varepsilon_3) = Cov(X_1, \varepsilon_4) = Cov(X_1, \varepsilon_5) = Cov(X_2, \varepsilon_3)$$
$$= Cov(X_2, \varepsilon_4) = Cov(X_2, \varepsilon_5) = Cov(\varepsilon_3, \varepsilon_4) = Cov(\varepsilon_3, \varepsilon_5) = 0$$
$$Cov(\varepsilon_4, \varepsilon_5) = 0.5$$

First, we look at the observed covariance matrix obtained from these 100 observations. This matrix is the observational 'shadow' that was cast by the causal process shown in Figure 4.1 and quantified by the above equations. Table 4.1 shows this covariance matrix.

The first step is to specify the hypothesised causal model. Imagine that we actually had two different competing models and wished to test between them. The first model is the model shown in Figure 4.1. This is the correct model that generated these data, although the model contains free parameters that have not been specified by our theory.

Table 4.1 The observed unique variances and covariances between variables X_1 to X_5 from 100 simulated observations based on the causal process shown in Figure 4.1

X_1	X_2	X_3	X_4	X_5
X_1 0.931				
X_2 0.171	1.094			
X_3 0.630	0.762	1.350		
X_4 0.384	0.368	0.743	1.265	
X_5 0.324	0.385	0.611	0.624	0.949

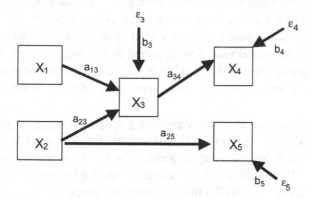

Figure 4.3 An alternative path model involving five variables

The second causal model, albeit one that is incorrect, is shown in Figure 4.3. The next step is to translate our two hypothesised causal graphs into structural equations. The translation of our first model has already been given. The translation of this second (incorrect) model is the following:

$$X_1 = N(0, \sigma_1)$$
$$X_2 = N(0, \sigma_2)$$
$$X_3 = a_{13}X_1 + a_{23}X_2 + b_3\varepsilon_3$$
$$X_4 = a_{34}X_3 + b_4\varepsilon_4$$
$$X_5 = a_{25}X_2 + b_5\varepsilon_5$$
$$Cov(X_1, X_2) = Cov(X_1, \varepsilon_3) = Cov(X_1, \varepsilon_4) = Cov(X_1, \varepsilon_5) = Cov(X_2, \varepsilon_3)$$
$$= Cov(X_2, \varepsilon_4) = Cov(X_2, \varepsilon_5) = Cov(\varepsilon_3, \varepsilon_4) = Cov(\varepsilon_3, \varepsilon_5) = 0$$
$$Cov(\varepsilon_4, \varepsilon_5) = 0$$

Note the differences between these structural equations and the ones derived from the correct model. First, the population covariance between the residual errors of X_4 and X_5 (i.e. $Cov(\varepsilon_4, \varepsilon_5)$) is zero in this incorrect model. Second, X_5 is hypothesised to be directly caused by X_2 rather than being indirectly caused by both X_1 and X_2 through their effects on X_3.

We next have to obtain the maximum-likelihood estimates of the free parameters of each model. To do this we have to provide starting values for the iterative process.

In the path models discussed in this chapter the choice of starting values for the free parameters is usually not critical, and so I will use the default values of 1 for all of them. Remember that the fitting of these free parameters is an iterative process in which the estimates that are obtained during each iteration are changed in such a way as to reduce the maximum-likelihood fitting function, as described in Box 4.3. At the very first iteration, when all free parameters are equal to 1.0, both the correct model and the incorrect model produce a predicted covariance matrix that poorly fits the observed values; the maximum-likelihood fitting function is 0.45707 for the correct model and 0.74821 for the incorrect model. The correct model took five iterations to converge on the maximum-likelihood estimates, giving a final value of 0.04890 for the maximum-likelihood fitting function. Since this value, multiplied by 99 (i.e. $N - 1$), is the maximum-likelihood chi-square statistic, the final value of the chi-square statistic was 4.8411. The incorrect model took four iterations to converge on the maximum-likelihood estimates, giving a final value of 0.39369 for the maximum-likelihood fitting function. Therefore, the final value of the chi-square statistic for this incorrect model was 38.98.

To see if these chi-square statistics are significantly different from what one would expect given a correct model, we next need to determine the degrees of freedom. In both models we had five measured variables, giving a total of 15 unique variances and covariances – i.e. $v(v+1)/2$, or $5(6)/2$. In the correct model we had to estimate 10 free parameters: the variances of X_1 and X_2, the three error variances, the four path coefficients and one free covariance (between X_4 and X_5). The correct model therefore had $15 - 10 = 5$ degrees of freedom. In the incorrect model we had to estimate nine free parameters: the variances of X_1 and X_2, the three error variances, four path coefficients but no free covariance. The incorrect model therefore had $15 - 9 = 6$ degrees of freedom. SEM programs do all these calculations for you.

Remember what we are testing. *If* the hypothesised model is correct *then* the maximum-likelihood chi-square statistic will follow a chi-square distribution. If the predicted and observed covariance matrices were identical then the maximum-likelihood chi-square statistic would be zero. The further the predicted covariance matrix deviates from the observed covariance matrix, the larger the maximum-likelihood chi-square statistic will be. Of course, even if our causal model were correct we would not expect the two matrices to be identical, because of sampling variation; the predicted covariance matrix contains the predicted population values but the observed matrix is from a random sample of 100 observations. However, if the only differences were due to random sampling fluctuations then the maximum-likelihood chi-square statistic will closely follow (for large samples) a theoretical chi-square distribution with the appropriate degrees of freedom. To evaluate our two models, we have only to hypothesise that each is the true model and then calculate the probability, based on this null hypothesis, of observing at least as large a difference between the observed and predicted covariance matrices as measured by our statistic.

First, let's look at the results for the correct model. The probability of observing a chi-square value of at least 4.8411 with five degrees of freedom is 0.44. In other words, there is a probability of 0.44 of seeing such a result even if our null hypothesis were correct.

Table 4.2 The observed covariance matrix for the 100 independent observations

	X_1	X_2	X_3	X_4	X_5
Observed covariance matrix					
X_1	0.931				
X_2	0.171	1.094			
X_3	0.630	0.762	1.350		
X_4	0.384	0.368	0.743	1.265	
X_5	0.324	0.385	0.611	0.624	0.949
Predicted values using the correct model					
X_1	0.931				
X_2	0.000	1.094			
X_3	0.526	0.665	1.233		
X_4	0.289	0.366	0.678	1.229	
X_5	0.238	0.301	0.557	0.594	0.925
Predicted values using the incorrect model					
X_1	0.931				
X_2	0.000	1.094			
X_3	0.526	0.666	1.234		
X_4	0.290	0.366	0.679	1.230	
X_5	0.000	0.385	0.234	0.129	0.949

Note: Also shown are the predicted maximum-likelihood covariance matrices based on the correct model (Figure 4.1) and the incorrect model (Figure 4.3).

In fact, our null hypothesis is correct, since we generated our data to agree with it. The result is telling us what we know to be true: the data are perfectly consistent with the model given normally distributed sampling variation. On the other hand, the maximum-likelihood chi-square statistic for the incorrect model was 38.98 with six degrees of freedom. The probability of observing such a large difference between the observed and predicted covariance matrices, assuming that the data were actually generated according to the incorrect model, is 7.2×10^{-7}. We either have to accept that an extremely rare event has occurred (one chance in about 1.5 million times) or reject the hypothesis that our data were generated according to the incorrect model. Again, the result is telling us what we know to be true: the data are not consistent with the model.

Compare the two predicted covariance matrices with the observed covariance matrix to see where the differences lie (Table 4.2). The biggest differences involve X_5. First, the predicted covariance between X_1 and X_5 is zero in the incorrect model while the observed value is 0.324. This is because X_1 is d-separated from X_5 in the incorrect model. The fitting procedure had to respect this constraint when fitting the incorrect model and so constrained this predicted covariance to be zero. The correct model allows X_1 to be an indirect cause of X_5 through its effect on X_3. The fitting procedure had to respect the constraint that the partial covariance between X_1 and X_5 be zero when controlling for X_3 but, since this constraint actually existed in the generating process, such a constraint did not distort the estimates. In the same way, the incorrect model required that the partial covariance between X_3 and X_5 as well as the partial covariance between X_4 and X_5 be zero when controlling for X_2. Since neither of these constraints actually existed in the correct causal process, the fitting procedure was forced to distort the estimates in order to meet these incorrect constraints.

Let's look next at the maximum-likelihood estimates for the free parameters in the two different models. Here again are the true population values used to generate the data:

$$X_1 = N(0, 1)$$
$$X_2 = N(0, 1)$$
$$X_3 = 0.5X_1 + 0.5X_2 + 0.5\varepsilon_3$$
$$X_4 = 0.5X_3 + 0.707\varepsilon_4$$
$$X_5 = 0.5X_3 + 0.707\varepsilon_5$$
$$Cov(X_1, X_2) = Cov(X_1, \varepsilon_3) = Cov(X_1, \varepsilon_4) = Cov(X_1, \varepsilon_5) = Cov(X_2, \varepsilon_3)$$
$$= Cov(X_2, \varepsilon_4) = Cov(X_2, \varepsilon_5) = Cov(\varepsilon_3, \varepsilon_4) = Cov(\varepsilon_3, \varepsilon_5) = 0$$
$$Cov(\varepsilon_4, \varepsilon_5) = 0.5$$

Here are the maximum-likelihood estimates with their asymptotic standard errors in parentheses, based on the true model:

```
X₁ = 0.931 N(0,1)
     (0.132)
X₂ = 1.094 N(0,1)
     (0.156)
X₃ = 0.565 X₁ + 0.608 X₂ + 0.531 E₃
     (0.076)     (0.070)     (0.075)
X₄ = 0.550 X₃ + 0.856 E₄
     (0.084)     (0.122)
X₅ = 0.452 X₃ + 0.673 E₅
     (0.074)     (0.096)
Covariance(X₄,X₅) = 0.287
                    (0.082)
```

Notice that each estimate is close to the population value. The standard errors are asymptotic, not exact, but with 100 observations these are quite close to the actual sample standard errors, and so two times each value defines an approximate 95 per cent confidence interval. For instance, the path coefficient from X_1 to X_3 is 0.565 with a standard error of 0.076, so an approximate 95 per cent confidence interval would be $0.565 \pm 2(0.076)$, or between 0.413 and 0.717; the true population value was 0.5. We could obtain the maximum-likelihood estimates for the incorrect model as well, but since we already know that the data are very unlikely to have been generated by this incorrect model at least some of the estimates will be incorrect.

We can place the estimates of the free parameters of the correct model directly on the path diagram (Figure 4.4). I prefer this because the path diagram makes it explicit that these estimates are based on a causal model with asymmetric relationships. The estimates shown in Figure 4.4 are not the ones Sewall Wright would have used. First, his

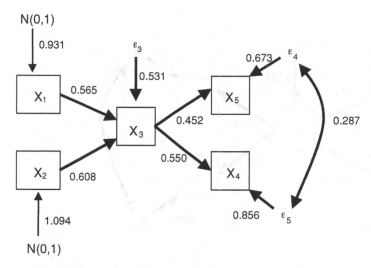

Figure 4.4 Fully parameterised path model of Figure 4.1 with centred, but not standardised, variables
Note: The numerical values are the maximum-likelihood values of the free parameters based on centred, but not standardised, variables.

estimates were not based on maximum-likelihood methods but, rather, on least-squares methods. Second, he used standardised variables so that the decomposition of direct and indirect effects were based on correlations rather than covariances. If the causal model is correct then the maximum-likelihood and least-squares estimates will be the same[12] since least-squares (partial) regression coefficients are also maximum-likelihood estimates, but if the causal model is wrong then the two types of estimates will differ. The standardised estimates are easily obtained by first standardising the variables to zero mean and unit variance. In fact, most SEM programs, including the lavaan package of R, will output these standardised estimates. For instance, if you have saved your output from the sem function in an object called 'fm' then the standardised estimates of the parameters can be obtained using the standardizedSolution() function. Figure 4.5 shows the path diagram for the correct model based on standardised variables.

4.2   Decomposing effects in path diagrams

One important use of path diagrams is to 'decompose' an association between variables into different types of causal relationships. In fact, this was the main goal of Wright's original method of path coefficients. Remembering the notions of causal graphs that were introduced in Chapter 2, we can differentiate between types of effects: direct causal effects, indirect causal effects, effects due to shared causal ancestors and unknown causal relationships.[13] One way of visualising this classification of associations is shown in Figure 4.6.

---

[12] This assumes that the data are multivariate normal.
[13] Of course, there can always be associations due to random sampling fluctuations.

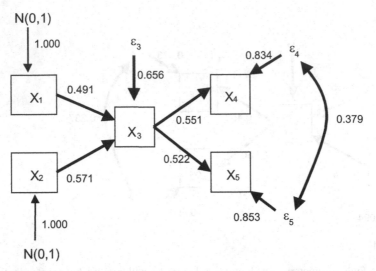

Figure 4.5 Fully parameterised path model of Figure 4.1 with centred and standardised variables
Notes: The numerical values are the maximum-likelihood values of the free parameters based on centred and standardised variables. The units are therefore standard deviations from the mean.

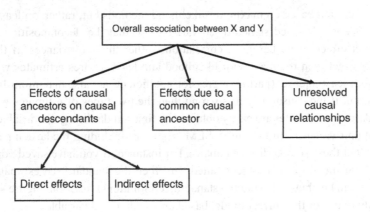

Figure 4.6 A classification of associations (for example, correlations or covariances)

This decomposition of a statistical association into different types of causal relationships is based on the fundamental association linking causality with probability distributions, as described in Chapters 1 and 2. The overall association between two variables is simply the overall correlation or covariance between them. This overall association can be generated by a number of different causal relationships at the same time. Since the consequences of interventions or manipulations will depend critically on these different types of relationships, it is important to be able to distinguish and quantify them.

If you can trace a path on the causal graph from a causal ancestor to a descendant by following the direction of the arrows then this path defines an hypothesised causal effect from the ancestor to its descendant. These effects can be of two different types. A

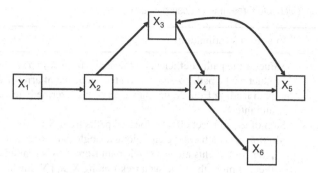

Figure 4.7 A path diagram used to illustrate the decomposition of associations

direct effect is an effect of the ancestor on its descendant that is not transmitted through any other variable in the model; of necessity, this means that the relationship is one of parent and child. In other words, it is the effect that would occur if all other variables in the model did not change.[14] The magnitude of this direct effect is measured by the path coefficient on the arrow going from the parent to the child. The units of this effect are the same as those used to measure the variables. If the variables are not standardised then the path coefficient measures the number of unit changes in the child per unit change of the parent. For instance, if X is measured in grams, Y is measured in millimetres, there is an arrow from X to Y ($X \rightarrow Y$) and the path coefficient for this arrow equals 0.6 then this means that a 1 gram change in X will provoke a 0.6 millimetre change in Y once X is d-separated from all other causes in the model. It is also the quantitative effect of X on Y when all other variables are held constant. If the variables are standardised then the units are standard deviations from the mean. As usual, these points are much easier to grasp when looking at a causal graph. Consider Figure 4.7.

In Figure 4.7 there are six different direct effects. There are always as many direct effects as there are one-headed arrows in the path diagram. If we were to fit this model to data then the path coefficient from X_1 to X_2 would measure the direct effect of X_1 on X_2. If the three other variables were held constant then this direct effect would quantify by how much X_2 would change given a one- unit change in X_1. However, since X_2 has no other causal ancestors, this direct effect would also quantify by how much X_2 would change given a one-unit change in X_1 even if the other variables were not held constant.

Indirect effects are the effects of a causal ancestor on its descendant that are completely transmitted through some other variable. This intervening variable is sometimes called a *mediator* of the causal effect. For instance, the effect of X_1 on X_3 along the path $X_1 \rightarrow X_2 \rightarrow X_3$ is an indirect effect of X_1 that is mediated by X_2. To quantify this effect one *multiplies* the path coefficients along this path. This indirect effect measures by how much X_3 would change following a change in X_1 if all causal parents of X_3 except for X_2 were held constant. In general, an indirect effect measures by how much the effect variable would change following a change in the indirect cause when this effect is transmitted only along the path in question. It is possible for the same causal variable to exert

[14] This may also be true even if other variables do change, as described below.

Table 4.3 Rules for estimating decomposition into three different causal sources

Effect involving variables X and Y	Rule for its estimation
Direct effect	Value of the path coefficient on the arrow from X to Y.
Indirect effect along a single path	Product of the path coefficients on the sequence of arrows along the path leading from X, through at least one intermediate variable, and into Y.
Overall indirect effect along all paths	Sum of the indirect effects along all paths from X to Y.
Effect due to common causal ancestor (Z) of both X and Y	Multiply the path coefficients along a single path from Z to Y and the path coefficients along a single path from Z to X (called a *trek*). If there is more than one such trek linking X and Y due to common causes, sum these together.
Effect due to unresolved causal relationship	Path coefficient on the double-headed arrow between X and Y.
Effect due to all common causal ancestors of both X and Y	Sum together the effects due to each common causal ancestor of both X and Y.
Overall effect	Sum together the direct effect, the total indirect effects, the total effects due to common causal ancestors and any remaining unresolved causal relationship between X and Y. This will equal the covariance or correlation (if using standardised variables) between X and Y.

Notes: Given two variables (X and Y) in a path model, the overall covariance or correlation (if using standardised variables) between them can be decomposed into three different causal sources. Shown are the rules of the estimation of each source.

both a direct and an indirect effect on the same descendant. An example of this is the effect that X_2 has on X_4 in Figure 4.7; X_2 had a direct effect on X_4 since X_2 is the causal parent of X_4, but X_2 also exerts an indirect effect on X_4 through its effect on X_3.

Both direct and indirect effects involve variables in which one is a causal ancestor of the other. In these cases there is a directed path from one variable to the other. The third way in which an association can be decomposed in a path diagram is when the association between the two variables is due to another variable that is a causal ancestor of both. In Figure 4.7 the association between X_5 and X_6 is due to the effect of X_4 (their common ancestor) on both. To quantify this effect one *multiplies* the path coefficients along the path[15] from X_4 to X_5 and along the path from X_4 to X_6. Such effects do not measure any causal effect of one variable on the other and represent what Pearson would have called a 'spurious' association.

Finally, path diagrams can include unresolved causal relationships between variables; these are shown by double-headed arrows. Including such an effect in the model is an admission of ignorance; we do not know which is the cause, which is the effect, or if the association is due to a common cause that is not included in the model. Such unresolved effects are quantified simply by the covariance between the two variables.[16] In tracing indirect effects along paths that include such double-headed arrows one can go in either

[15] The two paths together ($X_5 \leftarrow X_4 \rightarrow X_6$) are sometimes called a *trek*.
[16] Or a correlation coefficient if the variables are standardised, since a correlation is simply a standardised covariance.

Table 4.4 Decomposition of Figure 4.7

Variable pair	Direct	Indirect	Common causal ancestor	Unresolved causal relationship
X_1, X_2	$X_1 \rightarrow X_2$	None	None	None
X_1, X_3	None	(1) $X_1 \rightarrow X_2 \rightarrow X_3$ (2) $X_1 \rightarrow X_2 \rightarrow X_4 \rightarrow X_5 \leftarrow\rightarrow X_3$	None	None
X_1, X_4	None	(1) $X_1 \rightarrow X_2 \rightarrow X_3 \rightarrow X_4$ (2) $X_1 \rightarrow X_2 \rightarrow X_4$	None	None
X_1, X_5	None	(1) $X_1 \rightarrow X_2 \rightarrow X_3 \rightarrow X_4 \rightarrow X_5$ (2) $X_1 \rightarrow X_2 \rightarrow X_3 \leftarrow\rightarrow X_5$ (3) $X_1 \rightarrow X_2 \rightarrow X_4 \rightarrow X_5$	None	None
X_1, X_6	None	(1) $X_1 \rightarrow X_2 \rightarrow X_3 \rightarrow X_4 \rightarrow X_6$ (2) $X_1 \rightarrow X_2 \rightarrow X_4 \rightarrow X_6$	None	None
X_2, X_3	$X_2 \rightarrow X_3$	(1) $X_2 \rightarrow X_4 \rightarrow X_5 \leftarrow\rightarrow X_3$	None	None
X_2, X_4	$X_2 \rightarrow X_4$	(1) $X_2 \rightarrow X_3 \rightarrow X_4$	None	None
X_2, X_5	None	(1) $X_2 \rightarrow X_3 \leftarrow\rightarrow X_5$ (2) $X_2 \rightarrow X_4 \rightarrow X_5$ (3) $X_2 \rightarrow X_3 \rightarrow X_4 \rightarrow X_5$	None	none
X_2, X_6	None	(1) $X_2 \rightarrow X_3 \rightarrow X_4 \rightarrow X_6$ (2) $X_2 \rightarrow X_4 \rightarrow X_6$	None	None
X_3, X_4	$X_3 \rightarrow X_4$	None	(1) $X_3 \leftarrow X_2 \rightarrow X_4$	None
X_3, X_5	None	(1) $X_3 \rightarrow X_4 \rightarrow X_5$	None	$X_3 \leftarrow\rightarrow X_5$
X_3, X_6	None	(1) $X_3 \rightarrow X_4 \rightarrow X_6$	(1) $X_3 \leftarrow X_2 \rightarrow X_4 \rightarrow X_6$	None
X_4, X_5	$X_4 \rightarrow X_5$	None	(1) $X_4 \leftarrow X_3 \leftarrow\rightarrow X_5$	None
X_4, X_6	$X_4 \rightarrow X_6$	None	None	None
X_5, X_6	None	$X_5 \leftarrow\rightarrow X_3 \rightarrow X_4 \rightarrow X_6$	(1) $X_5 \leftarrow X_4 \rightarrow X_6$	none

Note: Decomposition of the total association between each pair of variables in Figure 4.7 into direct effects, indirect effects, effects due to common causal ancestors and pure unresolved causal effects.

direction but can traverse the double-headed arrow only once. Table 4.3 summarises the rules for decomposing the overall covariance or correlation between two variables in the path model and Table 4.4 lists the decomposition of Figure 4.7.

4.3 Multiple regression expressed as a path model

Since path analysis looks rather similar to multiple regression, let's look at how to represent a multiple regression as a path model. A multiple-regression equation uses a series of predictor variables (say, X_1, X_2 and X_3) to predict, or account for, the observed variation in the dependent variable Y. The predictor variables are often called the 'independent' variables, but this term can be misleading since they do not have to be independent of one another at all. Except when these predictor variables are measured in controlled experiments, they are often not independent of one another. Figure 4.8 shows such a multiple regression in the form of a path model.

It is clear from the path diagram that the partial-regression coefficients that are estimated with multiple regression are the direct effects of each predictor on the

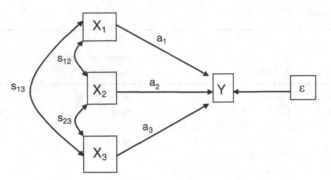

Figure 4.8 A multiple regression of X_1, X_2 and X_3 on Y, expressed as a path diagram

dependent variable. The indirect effects, which are simply the unresolved causal relationships between the predictors, are ignored. If the free covariances (the s_{ij} in Figure 4.8) really are zero, the direct effects will also be the overall effects, but the regression equation will not tell you this.[17] Furthermore, the model in Figure 4.8 cannot be tested as a causal claim. There are only four observed variables in this model, and therefore there are $4(5)/2 = 10$ unique elements in the covariance matrix. There are also 10 free parameters that have to be estimated (the three path coefficients, the three free covariances, the error variance and the variances of the three predictor variables). In other words, we have used up all our degrees of freedom in estimating our free parameters and have none left over to test the causal implications of the model.[18] This is true of every regression model. If you use the inferential test described in Chapter 3 you will find that no variable is d-separated from any other variable, either unconditionally or after conditioning on any set of other observed variables. The regression model places no statistical constraints with which to test the causal implications of the model.

Multiple regression can certainly be used to decide if the path coefficients (i.e. the partial-regression coefficients) are different from zero. Multiple regression can help us to decide if the error (or residual) variance is less than the total variance of Y in the statistical population (this is the F ratio). Multiple regression cannot help us to decide if the causal assumptions of the model are correct; it cannot tell us if the predictor variables are causes of Y. This is because multiple regression makes no testable predictions about the causal structure generating your data. Multiple regression can allow us to predict but not to explain. Statistics texts are quite correct when they say that one cannot draw causal conclusions from regression. The causal conclusions must come from somewhere else. The best way would be to conduct a controlled randomised experiment in which the values of the X variables are randomly assigned to the experimental units, since we would then have good reason to assume that the free covariances between them really are zero. If this is not possible then we have to construct our models, and collect our observations, in such a way that we can constrain the patterns of covariation based on our causal hypothesis and then test these constraints.

---

[17] If these covariances are not zero then one can run into problems of collinearity.

[18] When we get to the topic of identification, we will see that multiple regression is an example of a just-identified model.

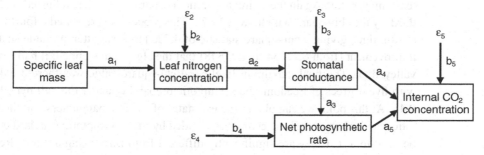

Figure 4.9 The proposed path model relating leaf morphology and leaf gas exchange
Note: The letters with subscripts show the free parameters whose maximum-likelihood estimates must be obtained.

4.4   Maximum-likelihood estimation of the gas exchange model

In Chapter 3 we looked at the model of Shipley and Lechowicz involving specific leaf mass, leaf nitrogen concentration, stomatal conductance, net photosynthetic rate and the internal concentration of CO_2. Let's fit and test these same data (log-transformed) to the proposed path model using maximum-likelihood methods. Remember that five of the 40 species were actually C_4 species and that these were clear outliers in the data set. Because we require approximate multivariate normality, we cannot include these five species in the data set. The analysis will be restricted to the remaining 35 species. Since the resulting chi-square statistic and the standard errors of the free parameters are only asymptotically correct, we can expect that the estimated standard errors are somewhat narrower than they should be and the probability value of the chi-square statistic will not be exact.[19] This model is reproduced in Figure 4.9.

The first step is to specify the structural equations and to indicate which parameters are free. There are five free path coefficients (a_1 to a_5) and five free variances (the variance of specific leaf mass and the four error variances ε_2 to ε_5). Since there are five measured variables there will be five degrees of freedom. The free parameters are shown in Figure 4.9. Next, I have to specify initial values for these free parameters. In my experience one rarely has problems with convergence of the maximum-likelihood estimates when there are no latent variables in the model, unless parts of the model are under-identified, and so I will make all free parameters equal to 1 except a_5, which I set at -1. I do this because I expect increasing photosynthetic rates to reduce the internal CO_2 concentration. One can sometimes have problems with convergence if some variances are much (i.e. orders of magnitude) larger than others, but I know that this is not the case with these variables.

The value of the chi-square statistic based on the initial values of the free parameters was 150.96 – obviously a very poor fit. The numerical algorithm searched for changes in these initial values that would improve the fit while respecting the constraints and came up with a second set of values. The value of the chi-square statistic based on this second set of values of the free parameters was 65.96 – obviously still a poor fit, but at

---

[19]   Chapter 6 describes the effects of sample size on the maximum-likelihood chi-square statistic.

least much better. Again the estimates of the free parameters were adjusted, and after the third try the chi-square statistic was 19.72. This process was repeated a fourth, and then a fifth, time, giving a chi-square statistic of 4.72. The sixth attempt made such a small improvement (from 4.71954 to 4.71648) that the algorithm stopped; it had reached the valley floor. The final maximum-likelihood chi-square value was therefore 4.72, and, with five degrees of freedom, the asymptotic probability under the null hypothesis was 0.45. At this point we can obtain the estimates of the free parameters and their asymptotic standard errors. These estimates, divided by their asymptotic standard errors, can be used to test if they are significantly different from zero, using a z-test. Remember, because we have only 35 observations, such a z-test will be somewhat liberal, since the real standard errors will be a bit larger that their asymptotic estimates. Here are the maximum-likelihood estimates. The asymptotic standard errors are given in round brackets and the z-value (whose absolute value will be less that 1.96 95 per cent of the time) is given in square brackets.

$$Ln(SLM) = N \begin{pmatrix} 0.183 \\ 0, (0.044) \\ [4.123] \end{pmatrix}$$

$$Ln(Nit) = \begin{matrix} 0.898Ln(SLM) \\ (0.096) \\ [9.338] \end{matrix} + N \begin{pmatrix} 0.057 \\ 0, (0.014) \\ [4.123] \end{pmatrix}$$

$$Ln(Cond) = \begin{matrix} 1.146Ln(Nit) \\ (0.206) \\ [5.525] \end{matrix} + N \begin{pmatrix} 0.300 \\ 0, (0.070) \\ [4.123] \end{pmatrix}$$

$$Ln(Photo) = \begin{matrix} 0.548Ln(Cond) \\ (0.069) \\ [7.977] \end{matrix} + N \begin{pmatrix} 0.091 \\ 0, (0.022) \\ [4.123] \end{pmatrix}$$

$$Ln(CO_2) = \begin{matrix} -0.162Ln(Photo) \\ (0.020) \\ [-8.133] \end{matrix} + \begin{matrix} 0.142Ln(Cond) \\ (0.013) \\ [10.526] \end{matrix} + N \begin{pmatrix} 0.001 \\ 0, (0.0002) \\ [4.123] \end{pmatrix}$$

The probability associated with the maximum-likelihood chi-square statistic (0.45) tells us that the data are consistent with the constraints that our model has placed on them. Given the small sample size we know that this probability estimate is not exact but is far from being significant. We already knew this based on the d-sep test (Chapter 3), whose probability estimates are exact. Although I will discuss tests for multivariate normality in Chapter 6, one such test is due to Mardia (1970; 1974). The normalised version of Mardia's coefficient is asymptotically distributed as a standard normal variate, though it requires very large sample sizes. At least it gives a rough guide to deviations from multivariate normality, and the value of this coefficient in the present data set, 0.174, is much smaller than the 1.96 needed for significant non-normality at the 0.05 level. These points tell us that the causal structure that was proposed is consistent with the data. We conclude that there is no good reason to reject the model, and so we can go on to look at the maximum-likelihood estimates for the free parameters.

The first thing to notice is that the z-scores associated with each of the 10 free parameters are all much larger than the 1.96 value that indicates significance at the 0.05 level. In other words, each of the free parameters is significantly different from zero. Since d-separation predicts that each of the five free path coefficients must be different from zero, this is good news. A path coefficient that is not significantly different from zero doesn't necessarily mean that it really is zero. It is always possible that the true value is close enough to zero that we can't detect the difference; this is the well-known problem of statistical power. Nonetheless, a path coefficient that is very close to zero requires us to be able to provide reasonable doubt that it is not really zero as our model requires. The five variances (for specific leaf mass and the four error variables) are also significantly different from zero. The fact that the four error variances are not zero simply means that there are other unknown causes contributing to the variance of each of the dependent variables. However, we have no strong reason to suppose that there are unknown variables that are common causes of two or more measured variables, since, if there were, the covariances between the error variables would not all be zero, as specified by the model, and the model X^2 value would be large. On the other hand, since we have a small data set, we have little statistical power to detect common unknown causes that are numerically weak.

We can obtain the sort of path model that Wright would have produced simply by first standardising each variable by subtracting its mean and dividing by its standard deviation. In this way, the variance of each variable is always 1.0. The residual variance of each variable is, as always, the variance of the error variable (ε, or the path coefficient from the error variable to the measured variable if the variance of the error variable is fixed at unity). This means that the explained variance is simply $1-\text{Var}(\varepsilon)$, and the square root of this gives the Pearson correlation coefficient for the (multiple) regression (i.e. R). Here are the standardised equations:

$$Ln(Nitrogen) = 0.848Ln(SLM) + 0.530\varepsilon_1 \quad R = 0.845$$

$$Ln(Conductance) = 0.688Ln(Nitrogen) + 0.726\varepsilon_2 \quad R = 0.688$$

$$Ln(Photo) = 0.807Ln(Conductance) + 0.590\varepsilon_3 \quad R = 0.807$$

$$Ln(CO_2) = -1.144Ln(Photo) + 1.480Ln(Conductance) + 0.484\varepsilon_4 \quad R = 0.875$$

Now that we have the maximum-likelihood estimates for the free parameters, we can estimate the effect of the variables on each other along different paths of influence. These effects are the amount by which the variable at the end of the path will change (in natural logarithmic units, since they are transformed) after a one-unit change in the variable at the beginning of the path, when all variables not involved in the path are held constant. Table 4.5 summarises these effects based on the non-standardised variables.

Using the values in Table 4.5 we can see how the overall effects are decomposed into direct effects, indirect effects and the effects of common causal ancestors. For instance, the overall effect of a 1 Ln-unit increase in stomatal conductance on the internal CO_2 concentration was to increase it by only 0.053 Ln-units. However, the direct effect was 0.142 units. The small overall effect was due to the fact that stomatal conductance also had an indirect negative effect on the internal CO_2 concentration. Increasing stomatal conductance increased the net photosynthetic rate by 0.548 units and increasing

Table 4.5 Effects of the variables on each other along different paths of influence

Path	Effect along path
SLM→Nitrogen	0.898
SLM→Nitrogen→Conductance	1.029
SLM→Nitrogen→Conductance→Photosynthesis	0.564
SLM→Nitrogen→Conductance→Internal CO_2	0.146
SLM→Nitrogen→Conductance→Photosynthesis→Internal CO_2	−0.091
Nitrogen→Conductance	1.146
Nitrogen→Conductance→Photosynthesis	0.628
Nitrogen→Conductance→Internal CO_2	0.167
Nitrogen→Conductance→Photosynthesis→Internal CO_2	−0.102
Conductance→Internal CO_2	0.142
Conductance→Photosynthesis	0.548
Conductance→Photosynthesis→Internal CO_2	−0.089
Photosynthesis→Internal CO_2	−0.162

Note: Predicted changes, in units of natural logarithms, of the variable at the end of the path after a one-unit change in the variable at the beginning of the path, when holding constant all variables not involved in the path.

photosynthetic rate decreased the internal CO_2 concentration by 0.162. The indirect effect of stomatal conductance on internal CO_2 concentration through the mediating effect of net photosynthesis was $0.548 \times (-0.162) = -0.089$. The overall effect was therefore $0.142 - 0.089 = 0.053$ Ln-units.

The importance of decomposing effects can be seen more clearly when considering the standardised path coefficients. The overall effect (thus the overall predicted correlation) between stomatal conductance and internal CO_2 was 0.557. This was due to a direct effect of 1.48 plus an indirect effect of $0.807 \times -1.144 = -0.923$. The overall effect (thus the overall predicted correlation) between net photosynthetic rate and internal CO_2 was only $1.48 - 0.923 = 0.557$, which is not even significantly different from zero at the 0.05 level. Clearly, it would be biologically absurd to suggest that the photosynthetic rate, which is removing carbon dioxide from the internal air spaces of the leaf, does not affect the internal concentration of CO_2. The apparent contradiction is resolved by decomposing this overall effect. The direct standardised effect of net photosynthetic rate on internal CO_2 was -1.144. However, both net photosynthetic rate and the internal CO_2 concentration have a common causal ancestor: the stomatal conductance. The standardised effect of the path net photosynthesis←stomatal conductance→internal CO_2 was $0.807 \times 1.480 = 1.194$. The overall correlation between net photosynthesis and internal CO_2 was therefore $-1.144 + 1.194 = 0.050$. In other words, the direct effect and the effect of the common causal ancestor almost cancelled each other out. This is just what the physiological model of stomatal regulation of Cowan and Farquhar (1977) would predict (see Chapter 3).

4.5   Using lavaan to fit path models

In this section we see how to specify a simple path model in R using the lavaan package and its main workhorse function: 'sem'. The simplest way to obtain the lavaan package is by clicking on 'Packages' from the main R console, then 'Install packages', and, finally (after lavaan has been installed), by loading it. Rather than overwhelming you with too much detail, I will limit myself to the details and information needed to specify and fit the sort of simple path models that we have seen so far in this chapter. We will deal with additional complications as we come across more complicated models in subsequent chapters. We first generate a 'data set' from a known causal structure, called 'sim.data', and then fit an hypothesised causal model to these data. You can (and will) do this using real data, but simulated data have an important advantage over real data when learning about a method or about the quirks and default assumptions of R functions: we already know the 'true' answer and so will immediately notice if the output tells us differently. We will generate our data following the DAG in Figure 3.1, in which (a) there are linear relationships between direct cause–effect pairs and (b) each variable is a standard normal variate (i.e. with a zero mean and a unit standard deviation).

First we generate 100 independent 'observations' of the two exogenous observed variables (A and F) as standard normal variates using the 'rnorm' function from the base library. This function takes three arguments: (a) the number of random values wanted, (b) the population mean and (c) the population standard deviation of the normal distribution from which these random values will be generated. The default is a standard normal probability distribution and so the default mean is zero and the default standard deviation is 1.

```
# generate data for the two exogenous observed standard unit
normal variables
var.A← rnorm(100)
var.F←rnorm(100)
```

Now we generate 100 independent 'observations' of the four remaining exogenous variables. In order to ensure that these endogenous variables are also standard normal variates we need to choose standard deviations of the residuals so that the total variance is equal to unity. We do this by taking advantage of the fact that the total explained variance of an endogenous variable is equal to the sum of the squares of the path coefficients linking it to its causal parents. Subtracting this explained variance from unity gives the unexplained (i.e. residual) variance; the square root of this unexplained variance is its standard deviation.

```
# generate data for the four endogenous observed standard
unit normal variables
var.B←0.5*var.A+rnorm(100,0,sqrt(1-0.5^2))
var.C← -0.5*var.B+rnorm(100,0,sqrt(1-0.5^2))
var.D← +0.5*var.B+rnorm(100,0,sqrt(1-0.5^2))
var.E← 0.5*var.C+ 0.5*var.D -0.5*var.F +rnorm(100,0,sqrt(1-
3*0.5^2))
sim.data←data.frame(A = var.A,B = var.B,C = var.C,D =
var.D,E = var.E,F = var.F)
```

The true values of all the path coefficients are 0.5 except for the path coefficient for the pair C←B, which is –0.5.

Since this is your first exposure to the lavaan package we will begin with the most basic notions needed to specify the model. You do this by using the tilde (~) operator. You probably already know this operator, since it is used in many other R functions, where it means 'is a function of'. For instance, lm(y~x) means 'variable y is a (linear) function of x'. In lavaan, you should instead read this operator to mean 'is caused by'. Thus, in lavan, y~x should be read 'y is caused by x'. We therefore translate the DAG from Figure 3.1 using this operator, enclosing the lines that define our model in quotes and saving it as an R object called 'true.model'. Beware: the variable names that you use in this step have to agree exactly with the names found in the data set to which you will fit this model, though you can have variable names in your data frame that are not used in the model specification.

```
true.model1←"
B~A
C~B
D~B
E~C+D+F
"
```

To fit this hypothesised model to the data, we use the 'sem' function in the lavaan library. The result will be saved in an R object called 'fit.true'. Again, we'll start with the simplest form of this function, which takes two arguments: the object containing our hypothesised model (true.model1) and the data frame containing our observations (sim.data).

```
fit.true←sem(true.model1,data = sim.data)
```

Here is part of the output:

```
lavaan (0.5-15) converged normally after 18 iterations
    Number of observations                        100
    Estimator                                      ML
    Minimum Function Test Statistic            10.092
    Degrees of freedom                              8
    P-value (Chi-square)                        0.259
```

The first line of the output is important, and tells us that the maximum-likelihood estimation procedure converged and did not encounter any problems; if there were convergence problems then an error message would be given. The second line tells us how many observational units were actually used in this fitting procedure; it used all 100 lines in our data set, but this will not necessarily be the case if there are missing values. Options when there are missing values will be discussed later. The third line tells us which method ('Estimator') was used to fit the model to the data. In this case maximum likelihood ('ML') was used because it is the default method in the 'sem' function. The maximum-likelihood chi-square statistic is given in the next line ('Minimum

Function Test Statistic'), whose value is 10.092. We know that it is the maximum-likelihood chi-square statistic because this is the statistic associated with the maximum-likelihood estimator. If we change the estimation method then the test statistic reported in this line will also differ; these different methods are described in Chapter 6. The final two lines give the degrees of freedom associated with the test statistic (8) and the probability of observing at least this level of difference between the observed and predicted covariance matrices (0.259) (a) assuming that the data were generated according to the causal model specified in the 'true.model' object and (b) on the basis the distributional assumptions implicit in our method of calculating the maximum likelihood.

At this point, if you have mastered the ideas in this chapter, you will see a problem. The model in Figure 3.1 has six observed variables (v) and 12 free parameters (six non-zero path coefficients, two exogenous variances and four residual variances) and no free covariances. Therefore, the degrees of freedom should be $v(v+1)/2-12 = 9$. However, the output says that we have eight degrees of freedom, and that means that lavaan actually estimated 10 free parameters instead of nine. Why? We can see which free parameters have been included in the model using the 'parTable' function. The triplet 'lhs op rhs' in the header gives the parameter in question (i.e. 'left-hand side' 'operator' 'right-hand side') and the 'user' column tells us if the parameter was explicitly included in the model specification by us (1) or if it was implicitly included as a default (0):

```
parTable(fit.true)
    id lhs op rhs user group free ustart exo label eq.id unco
1    1   B  ~   A    1     1    1     NA   0           0    1
2    2   C  ~   B    1     1    2     NA   0           0    2
3    3   D  ~   B    1     1    3     NA   0           0    3
4    4   E  ~   C    1     1    4     NA   0           0    4
5    5   E  ~   D    1     1    5     NA   0           0    5
6    6   E  ~   F    1     1    6     NA   0           0    6
7    7   B ~~   B    0     1    7     NA   0           0    7
8    8   C ~~   C    0     1    8     NA   0           0    8
9    9   D ~~   D    0     1    9     NA   0           0    9
10  10   E ~~   E    0     1   10     NA   0           0   10
11  11   A ~~   A    0     1    0     NA   1           0    0
12  12   A ~~   F    0     1    0     NA   1           0    0
13  13   F ~~   F    0     1    0     NA   1           0    0
```

Notice that the six free path coefficients are listed and were also explicitly included by us (the first six lines). Following this are six free variances (two for the two exogenous variables A and F and four for the four residual variances associated with our four endogenous variables). We didn't actually specify these free variances when setting up our model (we could have), but these variances are estimated by default unless we tell lavaan not to do it (more on how to do this below). You should always be as explicit as possible when specifying your model, rather than trusting lavaan to make the right default choices for you. In this case the default choice of freely estimating all variances is reasonable, because we rarely have sufficient theoretical knowledge to specify a particular fixed value for such variances, and so we would almost always want to freely estimate them. However, there is a free parameter listed on the 12th line of the output that was estimated even though we didn't explicitly ask for it and even though it contradicts a causal claim of Figure 3.1: the covariance between the two exogenous variables A and F (i.e. A~~F). The double tilde (~~) is a new operator that you haven't seen yet, and it represents a variance or covariance. Thus, A~~A means 'the covariance between A and itself' (i.e. the variance of A) and A~~F means 'the covariance between A and F'.[20] By default, lavaan assumes that you will always allow free covariances between each unique pair of exogenous variables, and so proceeds to estimate it, thus using up one degree of freedom. I don't like this choice for a default, because adding a free covariance is usually an explicit part of our causal hypothesis. A free covariance between A and F means that there is some causal process linking the two even though we aren't in a position to completely specify it. A fixed (zero) covariance means that these two exogenous variables are independent. Both possibilities are causal claims, and so I urge you to always explicitly specify the nature of such covariances. I assume that the default choice of freeing covariances between exogenous variables derives from the philosophical position that 'everything is always correlated with everything'. Since exogenous variables are, by definition, not explicitly caused by any other variable in the model, this philosophical position means that they must still be associated, and so we must estimate this. I recommend instead that you think explicitly about this rather than assuming it. You should include such free covariances only if your causal understanding of the process allows that there might be some causal process generating a correlation between two exogenous variables but doesn't allow you to explicitly model this relationship. If your causal hypothesis explicitly requires the two exogenous variables to be independent then explicitly add a zero covariance.

To do this we first add an argument in the 'sem' function to tell lavaan to estimate the variances for the exogenous variables (fixed.x = FALSE) rather than fixing these exogenous variables at their sample values, which is the default (i.e. fixed.x = TRUE). Then we tell lavaan to estimate the variances of these exogenous variables while ensuring that the covariances between them is zero. This is done by explicitly fixing the covariance between these estimated exogenous variances at zero. We do this by adding the line 'A~~0*F' in our model specification, meaning that the covariance parameter (A~~F) must remain fixed at zero during the process of likelihood maximisation. Incidentally, we could fix any parameter to any value using this general syntax, though if we fix a parameter to an impossible value, such as a negative variance, we will generate error messages. Thus:

```
true.model2←"
B~A
C~B
D~B
E~C+D+F
# here is where the covariance is fixed at zero.
```

---

[20] Remember that a variance is simply the covariance of a variable with itself.

```
A~~0*F
"
fit.true2←sem(true.model2,data = sim.data,fixed.x = FALSE)
fit.true2
lavaan (0.5-15) converged normally after 18 iterations
    Number of observations                          100
    Estimator                                        ML
    Minimum Function Test Statistic              11.414
    Degrees of freedom                                9
    P-value (Chi-square)                          0.248
```

This result now has nine degrees of freedom[21] and so agrees with the causal hypothesis specified in Figure 3.1. We now want to look at the parameter estimates (the path coefficients etc.) using the 'summary' function. I haven't reproduced the full output here, only the relevant part:

```
summary(fit.true2)
                      Estimate Std.err Z-value P(>|z|)
Regressions:
  B ~
    A                   0.574   0.105   5.466   0.000
  C ~
    B                  -0.561   0.096  -5.844   0.000
  D ~
    B                   0.492   0.078   6.292   0.000
  E ~
    C                   0.493   0.046  10.753   0.000
    D                   0.391   0.055   7.079   0.000
    F                  -0.566   0.059  -9.659   0.000
Covariances:
  A ~~
    F                   0.000
Variances:
    B                   0.794   0.112
    C                   0.951   0.135
    D                   0.632   0.089
    E                   0.249   0.035
    A                   0.719   0.102
    F                   0.725   0.102
```

The first part of the output gives the estimates and standard errors of the six path coefficients. The associated z-scores and probability estimates are based on the implicit null hypothesis that these path coefficients are zero in the statistical population – i.e. that they don't exist. Since the causal model requires that they do exist, these null probabilities

---

[21] Because the previously free covariance between A and F, which used up one degree of freedom, is now fixed and so does not use the information in our data to determine its value.

should be low. If some of these path coefficients are not significantly different from zero then either they are too small to detect given the sample size of your data (i.e. not enough statistical power) or else your causal model is wrong. We next see that the covariance between A and F (i.e. A~~F) is zero but with no associated standard error. This is because this covariance wasn't estimated from the data; rather, we forced this value in our model definition. Finally, we see the estimates of the residual variances for the four endogenous variables and of the variances of the two exogenous variables, along with the standard errors of these estimates. Lavaan doesn't print out the z-statistics and associated null probabilities of such null hypotheses because they are rarely of interest,[22] but there is nothing stopping you from computing these if you want. In particular, testing the null hypothesis that a residual variance is zero in the statistical population is the same thing as testing that your model has identified all the causal parents of that particular variable and so there is no remaining residual variance except for sampling fluctuations.

The 'summary' function is convenient but doesn't provide all the information contained in the object produced by the 'sem' function. This further information can be obtained by other extractor functions. For instance, what is the proportion of the variance of each endogenous variable that is captured by the path model – i.e. the R^2? We can get this information as follows:

```
inspect(fit.true2,"rsquare")
          B          C          D          E
0.2300349  0.2545846  0.2836265  0.6949891
```

This tells us that our path model has accounted for 23 per cent, 25 per cent, 28 percent and 69 per cent of the total variance of our four endogenous variables, respectively. These amounts are close to the population values, since our data really were generated by this causal model.

The 'parameterEstimates' function gives the same information on parameter estimates as found in the 'summary' function but also gives confidence intervals for these estimates (using various bootstrap methods if requested) and standardised values of the parameter estimates – i.e. estimates calculated on variables that have been transformed to have zero means and unit standard deviations. The extractor function 'standardizedSolution' is similar to the more general 'parameterEstimates' function but gives only standardised values. The general form of this extractor function is:

```
parameterEstimates(object, ci = TRUE, level = 0.95,
boot.ci.type = "perc", standardized = FALSE, fmi = "default")
```

Remembering that this method revolves around the comparison of an observed covariance matrix with the matrix predicted by the model, you might want to know the values in this model predicted covariance matrix. You can obtain them via the

---

[22] Is variable E completely determined by its causal parents (C, D and F)? If so, its residual variance is zero in the population. Thus $z = (0.216-0)/0.031 = 6.97$ and the associated null probability, calculated in R, is `2(1-pnorm(6.97))`, or 3.2e–12.

'fitted' function. Even more useful is the difference between the observed and predicted covariance matrices. This is obtained via the 'residuals' extractor function. If the variables are measured in different units then such residuals are difficult to interpret, and it is better to look at the residuals based on standardised values.

You will learn more about using lavaan when you get to the chapters dealing with latent variables, alternative 'fit' measures and information theory measures of model fit via the 'AIC' and 'BIC' functions, multiple groups or hierarchically structured data. There is also a 'cheat-sheet' of lavaan functions at the end of this book. However, two further details require our attention: how to choose and specify starting values for parameter estimation and how to name parameters.

As you begin using more complicated models you will probably encounter a frustrating problem in which your model either fails to converge or else reports convergence errors. Unless you have a profound knowledge of the code buried in lavaan then its error messages will probably resemble something written by the Oracle of Apollo at Delphi.[23] Sometimes such errors are due to your specification of the model (more on this in a later chapter), in which case you must go back and modify it. However, in some cases the problem is not with your model structure but, rather, in the starting values used during the process of likelihood maximisation. If poor starting values are used then the iterative maximum-likelihood procedure can wander into forbidden areas of parameter space, or else begin so far away from those values that it actually maximises the likelihood that the iterative process never converges. Remembering our analogy between maximum-likelihood estimation and a blind woman trying to find a valley floor in a landscape, the more rugged and heterogeneous the landscape the more difficult this becomes and the more important it becomes for the blind woman to start her search relatively close to the valley. During the process of maximising the likelihood the procedure might wander into an area of parameter space that contains negative values for some variances; this is like our blind woman wandering into a swamp and getting stuck. You will see this when you look at the parameter estimates.

Alternatively, the maximum-likelihood procedure might terminate and report that convergence has not occurred; this is like our blind woman starting so far away from the valley that she gives up before she gets to the end. The default number of iterations in the 'sem' function is 150. You can change this number to, say, 200 by including the following argument in the 'sem' function: `control = list(iter.max = 200);` and there are other control parameters that you can modify as well. Check the documentation for the 'nlminb' function, since this is the default function controlling the maximisation of the likelihood function (you could also choose a different one). However, convergence problems that are not related to the causal structure of the model can usually be solved by choosing better starting values – i.e. placing our blindfolded woman close enough to the valley floor that she won't wander into a swamp or have to walk too far. The syntax for specifying a particular starting value for a parameter in the model definition is 'start()*'. I have modified the model specification from 'true.model2' and called it 'true.model3':

---

[23] Sacrificing a goat doesn't help.

```
true.model3←"
#causal relationships
B~start(0.5)*A
C~start(-0.5)*B
D~start(0.5)*B
E~start(0.5)*C+start(0.5)*D+start(-0.5)*F
# fixed covariance
A~~0*F
# exogenous variances
A~~start(1)*A
F~~start(1)*F
# endogenous residual variances
B~~start(0.7)*B
C~~start(0.7)*C
D~~start(0.7)*D
E~~start(0.3)*E
"
```

Fitting 'true.model3' results in exactly the same parameter estimates as before, and it took one fewer iteration to converge.

For relatively simple models, such as those considered in this chapter, which don't involve latent variables or grouped data, you should only rarely experience convergence problems, but this will not be the case when you begin using more complicated models. There are no foolproof methods of solving such problems but there are some good rules of thumb.

(1) Try to keep the range of your different variables to within the same order of magnitude. For instance, if you have one variable, measured in centimetres, that varies from 200 to 2,000 and another variable, measured in kilograms, that varies from 3 to 10 then convert the first variable to metres (2 to 20).

(2) Propose starting values that are at least within the same order of magnitude as the likely maximum-likelihood value and that have the same sign.

(3) Propose starting values of exogenous variances that are equal to their sample variances.

(4) Take an educated guess about the proportion of the variance of each endogenous variable that might remain unexplained by their causal parents and then propose starting values for these endogenous variances equal to the sample variances of these variables multiplied by your guess. For instance, in 'true.model3' I proposed a starting value of 0.7 for the residual variance associated with variable B because the sample variance of this variable was 0.98, and I figured that A might explain about 30 per cent of the variance of B, thus

$$(1 - 0.3)*0.98 \approx 0.7$$

In my experience, poor starting variables are particularly troublesome for free variances and free covariances.

All parameters in lavaan have a name. By default, the name of each parameter consists of three parts. The first part is the name of the variable that appears on the left-hand side of the formula. The middle part is the operator used in the formula. The last part is the name of the variable that appears on the right-hand part of the formula. For example, the first line in the model definition of object 'true.model3' is B~start(0.5)*A, corresponding to the directed edge A→B in Figure 3.1. Therefore, the name of the parameter (i.e. path coefficient) is B~A. The last line in object 'true.model3' is E~~start(0.3)*E, corresponding to the residual error variance for variable E. Therefore, the name of this parameter is E~~E. However, in more complicated models we will be specifying various constraints on parameters or sets of parameters, and then it is more convenient to assign our own names to these parameters. There are two cases in which naming parameters makes life easier even in the simpler path models considered in this chapter: (a) placing simple logical or hypothesised constraints on parameters and (b) partitioning total effects into direct and indirect components. To assign your chosen name to a parameter, simply choose a name that follows the naming conventions of R and 'multiply' this to the variable. For example, if I wanted to call the path coefficient measuring the effect of A on B (A→B) as 'A.on.B' then I would change the first line in 'true.model3' from B~start(0.5)*A to B~start(0.5)*A.on.B*A.

Fixing a parameter at a specific value is straightforward. For instance, if you wanted to fix the causal effect of A on B in Figure 3.1 to, say, 0.5, then you would replace the first line of 'true.model3' from B~start(0.5)*A to B~0.5*A. Now we are not telling lavaan to *start* the estimation of this path coefficient at 0.5 using 'start()' but, rather, telling lavaan to *fix* this path coefficient at 0.5 and not change it during the maximum-likelihood estimation. In other words, the causal effect of A on B is no longer being estimated from the data but is instead being determined by your causal hypothesis.

Placing a constraint on a free parameter requires a bit more work. An example of placing a logical constraint on a free parameter is requiring a variance to be non-negative. An estimated residual variance that is negative is usually a sign of some fundamental problem in your model, but not always. If an endogenous variable is very strongly determined by its causal parents then its unexplained (residual) variance will be very close to zero and it is possible that sampling variation alone could result in the maximum-likelihood estimate of this variance 'wandering' into the forbidden part of parameter space containing negative variances. In order to prevent this we would specify an inequality constraint on a free parameter. For instance, if I wanted to ensure that the residual variance of variable E in the model specified in 'true.model3' could not become negative then I would change the last line from E~~start(0.3)* to:

```
E~~varE*E
varE>0
```

The first line above names the residual variance of E as varE. The second line above then tells lavaan to ensure that the value of varE remain positive during the maximum-likelihood iterations. The usual relational operators in R can be used (= = , >, <, < = , > = , ! =). Another reason why you might want to place constraints (in addition to the more complicated situations that will be dealt with in Chapter 7) is when your causal hypothesis does not allow you to specify an exact value for a parameter but does require that the unknown parameter value be within specified bounds.

You have already learned in this chapter how to partition the total effect between any two variables into its different components. Lavaan doesn't have any function that does this for you, and this is probably a good thing, because even moderately complicated models will have many different combinations of direct, indirect and total effects. However, there is an operator in lavaan (: =) that allows you to define new parameters that are functions of previously defined parameters, and so we can create new parameters representing specified effects and then obtain their maximum-likelihood estimates, standard errors, and so on.

Variable A in Figure 3.1 has no direct effect on variable E but does have indirect effects along two different paths (A→B→C→E and A→B→D→E). Furthermore, using the parameter values that we have simulated in this section (contained in the 'sim.data' data frame), the indirect effect of A on E along each of these paths is equal in absolute magnitude and opposite in sign in the statistical population (−0.125 and 0.125, respectively), and so the total effect, which is the sum of these two indirect effects, will be zero. We have to modify 'true.model3' in two ways in order to get lavaan to estimate these two indirect effects and the total effect. First, we will assign names to each of the path coefficients along these two paths. Second, we will use the : = operator to estimate the two indirect effects and the total effect.

```
true.model4←"
#causal relationships
B~start(0.5)*c1*A
C~start(-0.5)*c2*B
D~start(0.5)*c4*B
E~start(0.5)*c3*C+start(0.5)*c5*D+start(-0.5)*F
# fixed covariance
A~~0*F
# exogenous variances
A~~start(1)*A
F~~start(1)*F
# endogenous residual variances
B~~start(0.7)*B
C~~start(0.7)*C
D~~start(0.7)*D
E~~start(0.3)*E
# define indirect, total effects
indirect1: = c1*c2*c3
indirect2: = c1*c4*c5
total: = indirect1+indirect2
"
fit.true4←sem(true.model4,data = sim.data,fixed.x = FALSE)
summary(fit.true4)
```

```
lavaan (0.5-15) converged normally after 18 iterations
  Number of observations                              100
  Estimator                                            ML
  Minimum Function Test Statistic                  11.414
  Degrees of freedom                                    9
  P-value (Chi-square)                              0.248
Parameter estimates:
  Information                                    Expected
  Standard Errors                                Standard
                    Estimate Std.err Z-value P(>|z|)
Regressions:
  B ~
    A            (c1)    0.574   0.105   5.466   0.000
  C ~
    B            (c2)   -0.561   0.096  -5.844   0.000
  D ~
    B            (c4)    0.492   0.078   6.292   0.000
  E ~
    C            (c3)    0.493   0.046  10.753   0.000
    D            (c5)    0.391   0.055   7.079   0.000
    F                   -0.566   0.059  -9.659   0.000
Covariances:
  A ~~
    F                    0.000
Variances:
    A                    0.719   0.102
    F                    0.725   0.102
    B                    0.794   0.112
    C                    0.951   0.135
    D                    0.632   0.089
    E                    0.249   0.035
Defined parameters:
    indirect1           -0.159   0.042  -3.742   0.000
    indirect2            0.110   0.031   3.565   0.000
    total               -0.049   0.038  -1.267   0.205
```

The last three lines of the summary object give us the estimates, the standard errors of the estimates, the z-statistics and null probabilities (assuming the population values are all zero) of our two indirect paths and the total effect. In particular, we see that the two estimated indirect effects (−0.159 and 0.110) are almost equal in magnitude but opposite in sign and are different from zero. Finally, the total effect (−0.049) is not significantly different from zero (its null probability is 0.205).

5   Measurement error and latent variables

Ambient temperature affects the metabolic rate of animals. When it is cold a homeothermic animal has to burn stored energy reserves – first glycogen and fat and then, when these are exhausted, protein – in order to generate heat and maintain its body temperature. The scaling of surface area (the site of heat loss to the atmosphere) to body volume (where the heat is generated) means that small homeothermic animals, such as songbirds, can lose up to 15 per cent of their body fat in one cold night. To burn this fat the bird must increase its metabolic rate, which increases its oxygen consumption. Imagine that we conduct an experiment in which we place small birds inside metabolic chambers overnight and vary the air temperature. The hypothesised causal process is shown in Figure 5.1.

Unfortunately, we can't directly measure any of these three variables; they are unmeasured, or *latent*, and so I have enclosed them in circles following the conventions of path diagrams. If we measure the air temperature using a thermometer then we aren't directly measuring *temperature* – the average kinetic energy of the molecules in the air. Instead, we are measuring the height of a column of mercury in a vacuum and enclosed in a hollow glass tube. In fact, we can't even measure the actual height of the mercury exactly, since our observed height will include some measurement error. Nor can we directly measure metabolic rate. Typically, one measures the rate of gas exchange (oxygen decrease or carbon dioxide increase) between the air entering and leaving the metabolic chamber. If we measure oxygen consumption using an infrared gas analyser then we aren't even directly measuring *oxygen* consumption. Instead, we are measuring differences in the amount of light of particular wavelengths that is absorbed as the light passes through the air. Again, even this variable is not perfectly measured, since the observed values will also contain measurement error. When we measure the fat reserves that are burned by the birds we might actually be measuring the difference in body weight over the course of the experiment, and this too will include measurement error. One simplified representation of the actual causal process[1] is depicted in Figure 5.2.

In this causal scenario the variables that we can observe and measure are not the variables that we hypothesise to form the causal chain of interest. The causal process involves variables that we cannot directly observe. We can obtain information about these latent variables by observing other variables that are causally linked to them but

---

[1] Some readers will recognise this as a particular example of the claim in Chapter 2 that even the simplest scientific hypotheses have, hidden behind them, a whole constellation of auxiliary hypotheses.

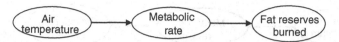

Figure 5.1 The causal structure relating air temperature, the metabolic rate of the bird and the amount of fat reserves that are used up

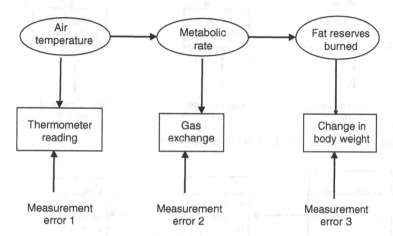

Figure 5.2 The causal structure relating air temperature, the metabolic rate of the bird and the amount of fat reserves that are used up when we conduct an experiment in the metabolic chamber and take measurements

the observed variables are also affected by other independent causes whose variation generate measurement errors. What are the consequences of this for path analysis? What are the consequences of this for causal analysis in general?

We know, based on d-separation, that if we could hold constant the actual metabolic rate of our birds then changes in the temperature of the ambient air would be independent of the changes in their fat reserves if the causal hypothesis were correct. Therefore, the partial correlation of the unmeasured variables 'Air temperature' and 'Fat reserves burned' would be zero when conditioned on the unmeasured 'Metabolic rate'. This would be true whether we used experimental controls or statistical controls. However, we cannot in fact observe the actual metabolic rate; we can only infer its constancy based on the observed rate of gas exchange. If gas exchange is perfectly correlated with metabolic rate then holding constant the rate of gas exchange would ensure that the metabolic rate was also constant. What happens if the correlations between our measured variables and the variables of theoretical interest are not perfect?

5.1   Measurement error and the inferential tests

Figure 5.3(a) shows the causal scenario as we have conceived it, and Figure 5.3(b) shows it as it would look if we were willing to ignore the fact that our measured variables are imperfect measures of the causally important variables. Looking at Figure 5.3(a),

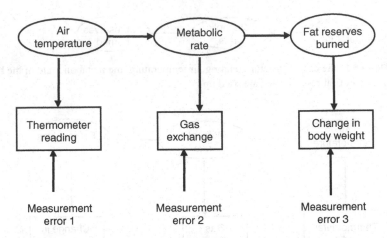

(a) The actual causal structure of the experiment.

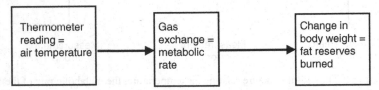

(b) The causal structure that is assumed when ignoring measurement error.

Figure 5.3 Two directed graphs showing the actual and assumed causal structure of the experiment

we see that there is only one conditional independence relationship in the basis set involving the three latent variables of interest: 'Air temperature' is independent of 'Fat reserves burned' after conditioning on 'Metabolic rate'. There are other conditional independence relationships that involve the observed variables, but these observed variables don't interest us except in how they can test the underlying causal hypothesis involving the latents. Notice that 'Thermometer reading' (one of our imperfect measures) is also independent of 'Change in body weight' (another imperfect measure) upon conditioning on 'Metabolic rate' as well. In other words, these two measured variables have the same causal implications as those involving their underlying latent variables, and so the fact that these two measured variables are not perfect indicators of their latents makes no difference in our ability to test the causal hypothesis. However, these two observed variables are not d-separated upon conditioning on 'Gas exchange' in Figure 5.3(a) (the observed variable indicating metabolic rate) even though they are conditionally independent in Figure 5.3(b), which ignores measurement error. Therefore, measurement error in the conditioning variable will introduce errors in our test of the underlying causal hypothesis. The error is not in the logic of the inferential test but in our incorrect assumption that our measure of gas exchange is a perfect indicator of metabolic rate.

We can conduct a numerical simulation based on the structural equations derived from the causal graph in order to see what happens when we include measurement error.

Table 5.1 Effects of measurement error on rates of model rejection

Measurement error			Ratio of variances			D-sep rejection rate	ML rejection rate
σ_1^2	σ_2^2	σ_3^2	X/X'	Y/Y'	Z/Z'		
0.01	0.01	0.01	0.990	0.999	0.998	0.052	0.064
10.0	0.01	0.01	0.009	0.999	0.998	0.060	0.052
0.01	0.10	0.01	0.990	0.988	0.998	0.060	0.050
0.01	0.30	0.01	0.990	0.964	0.998	0.088	0.102
0.01	0.40	0.01	0.990	0.952	0.998	0.152	0.144
0.01	0.01	10.0	0.009	0.998	0.322	0.052	0.060

Notes: The empirical rejection rates of 500 independent data sets, consisting of 1,000 independent observations (X, X', Y, Y', Z, Z') each, are shown. The true model is shown in Figure 5.3(a) with different amounts of measurement error, and the inferential tests are based on the incorrect model in Figure 5.3(b). The population variances of the latent X, Y and Z are 1, 8, and 4.75. The population variances of the observed X', Y', and Z' are $1 + \sigma_1^2$, $8 + \sigma_2^2$ and $4.75 + \sigma_3^2$.

I will represent the three latent variables (air temperature, metabolic rate and change in fat reserves) by X, Y and Z and the three corresponding observed variables by X', Y' and Z'. Here is one set of structural equations corresponding to the causal process:

$$X = N(0, 1)$$
$$X' = 1X + N(0, \sigma_1)$$
$$Y = 2X + N(0, 2)$$
$$Y' = 1Y + N(0, \sigma_2)$$
$$Z = 0.5Y + N(0, \sqrt{0.75})$$
$$Z' = 1Z + N(0, \sigma_3)$$

Since we have generated our data in accordance with the causal graph in Figure 5.3(a), we know that about 5 per cent of our simulated data sets would produce a probability of less that 0.05 when using the d-sep test. If we test 500 independent data sets, each with 1,000 independent observations, then the 95 per cent confidence interval for our empirical rejection rate would be between approximately 2 per cent and 8 per cent (Manly 1997). Table 5.1 summarises the results of these simulations as we increase the measurement errors. We see that, even when the measurement error variance of Y' is 0.3 (that is, slightly less than 4 per cent of the true variance of Y), the rejection rate is outside the 95 per cent confidence limits. As the measurement error variance increases further, the rejection rate increases rapidly. In other words, even if the hypothesised causal process involving the theoretical variables were correct, we would tend to reject it too often if we were to incorrectly assume that our conditioning variable is measured without error. Here, ignoring measurement error increases the likelihood that one will incorrectly reject a model that is correct. The effect of measurement error on the accuracy of the probabilities associated with the maximum-likelihood chi-square statistic is the same in this example (Table 5.1).

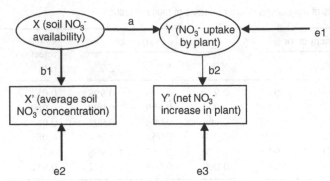

b1   b2

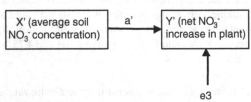

(a) The causal structure assuming measurement error.

(b) The causal structure ignoring measurement error.

Figure 5.4 Two directed graphs showing the causal structure assuming measurement error and ignoring measurement error

5.2   Measurement error and the estimation of path coefficients

The effect of measurement error on the accuracy of the estimation of the path coefficients, using either the least-squares regression methods of Chapter 3 or the maximum-likelihood regression methods of Chapter 4, are perhaps somewhat better known to biologists. Let's first look at a simple example involving only two variables (X and Y) that are measured with error. When I say 'measured with error' I don't only mean the obvious case in which the measuring device (say an analytical balance) has a certain degree of error.

Imagine that you wish to measure the nitrate availability of the soil in the rooting zone of a plant, but measure the total nitrogen content of samples of this soil only at a single time. In such a case the error of measurement will include not only the error involved in the analytic method for nitrate concentration but also the error involved in using the sample measures of total nitrogen at one point in time as a proxy variable for the total nitrogen availability in the rooting zone. Let us imagine a causal process in which the nitrate absorption rate of a plant (Y) is caused by the amount of nitrate available in the rhizosphere of its roots (X). X is estimated as the average nitrate availability of a sample of soil cores taken directly beneath the plant at one point in time. Y is estimated as the change in the net total nitrogen concentration of the plant from the time the soil is sampled until the next day. Figure 5.4 shows the causal graph assuming measurement error (a) and without measurement error (b).

The path coefficient shown as a in Figure 5.4(a) is the regression slope of Y on X. By definition, this is

$$a = \frac{Cov(X, Y)}{Var(X)} = \frac{aVar(X)}{Var(X)}$$

The true value of a' in Figure 5.4(b) can be derived from the rules of path analysis:

$$a' = \frac{Cov(X', Y')}{Var(X')} = \frac{a\,(b1)\,(b2)\,Var(X)}{b1^2 Var(X) + Var(e2)}$$

Often the measured variables (X' and Y') will scale 1:1 with the underlying latent variable – that is, a unit increase in the underlying variable will result in a unit increase or decrease of the measured variable. In such a case (or if b1 = b2) the formula can be simplified to

$$a' = \frac{aVar(X)}{Var(X) + Var(e2)}$$

From this we see that the effect of measurement error (e2) is to decrease a' relative to a. If we ignore the measurement error then a' will be a biased estimate[2] of a. The formula also shows why it is important to sample in such a way as to allow the widest variation possible in the causal variable X. Presumably the measurement error will not change with the range of X, and so as the variance of X increases the difference between a' and a will decrease. Furthermore, measurement error in the effect variable (Y) has no effect on the bias of the path coefficient.

These measurement errors have different effects on the estimation of the path coefficients and on the probabilities of the overall inferential test of the causal model. For instance, in the causal model shown in Figure 5.3, measurement error in X (the air temperature) had no effect on the probability levels estimated in the inferential test when ignoring measurement error but would bias the path coefficient from air temperature to metabolic rate. Measurement error in Y (metabolic rate) did have an effect on the estimated probabilities but would not bias the path coefficient[3] from X to Y.

5.3   A measurement model

When measurement error cannot be safely ignored, it must be explicitly included in the model and estimated. Most biologists deal with measurement error by ignoring it. Sometimes this is reasonable. After all, we are able to measure temperature, mass or carbon dioxide concentration with great accuracy. However, sometimes ignoring measurement error is not at all reasonable. Trying to estimate the fat reserves of a large free-ranging

---

[2] This is true in this simple bivariate case. When there is more than one cause of Y, each with measurement error, then the relationship between a and a' will also depend on the covariances between the measurement errors. See Bollen (1989) for the exact formulae.

[3] Measurement error in Y would bias the estimation of the path coefficient from Y to Z, if we ignore it.

mammal by palpitating its ribs and giving it a score of 1 to 4 can hardly inspire confidence in its accuracy, yet this is a common measure of 'body condition'. Similarly, we do not possess the equivalent of a thermometer that we can put into the mouth of our animals to measure their evolutionary fitness. If we try to measure fitness using indirect measures then these measures will probably possess important measurement errors.

Although they are not generally known to most biologists, methods for dealing with measurement error have been developed in the social sciences. Almost all important variables are latent in these sciences, and they can be measured only with substantial error. For instance, one might reasonably hypothesise that the degree to which a person can empathise with suffering might determine his or her career choice. It seems reasonable to suppose that a person with more empathy might choose to become a nurse rather than a mercenary soldier. Yet how can one measure 'empathy'? A common approach would be to devise a series of survey questions and develop an index of empathy based on the answers to these questions. It is obvious that choosing the answer 'A' in a multiple-choice test does not cause one to become a nurse rather than a mercenary, nor does it cause one to become more empathic. Rather, a psychologist might say that ones' empathic tendency is a common latent cause both of the answers on the survey and of one's choice of career. The survey answers are imperfect measures, or *indicators*, of the underlying latent variable, and the measurement model must separate those parts of the covariance between the answers to the personality test that are due to the underlying latent cause ('empathic tendency') from those parts of the covariance due to other causes. One simple type of measurement model is a factor model.

There is a huge literature devoted to measurement theory and to its many pitfalls. Many of these pitfalls are conceptual rather than statistical, so let's begin with an example (Dunn, Everitt and Pickles 1993) that doesn't pose any conceptual problems. You cut a number of pieces of string into different lengths and lay them on a table. Each string now has an attribute – length – that you ask four different people to measure. One person uses a ruler graduated in centimetres. One person uses his hand and measures in hand lengths. The third person uses a ruler graduated in inches and the fourth person simply eyeballs each string and tries to estimate it to the nearest centimetre. The length measurements have different units and each estimate has two different causes, one of which is the same cause in all four cases while the other is unique to each person. The common cause is the true length of the string, since each person is trying to accurately measure this same latent attribute. The unique causes consist of all those other causes that give rise to the measurement error (incorrectly calibrated rulers, tiredness, myopia, etc.).[4] Figure 5.5 shows the causal graph.

In this example, everything is in plain sight. The true length of the pieces of string, although latent, is not hypothetical. We can see the strings on the table and know that each has a fixed value of the attribute 'length'. The only uncertainty is in knowing the

---

[4] If this assumption is wrong then we have to explicitly model how these other causes relate to each other.

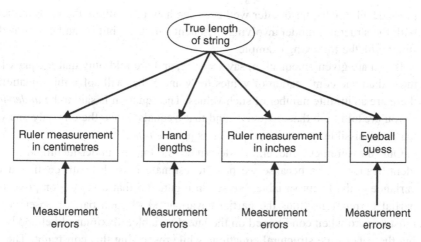

Figure 5.5 The causal structure generating the different measured lengths of the pieces of string

actual length of each string. Let the j strings (j = 1,n) measured by the four people be X_{j1} to X_{j4} and let the true length of each string be L_j. The structural equations representing this causal process are:

$$X_{j1} = \alpha_1 L_j + N(0, \sigma_1^2)$$
$$X_{j2} = \alpha_2 L_j + N(0, \sigma_2^2)$$
$$X_{j3} = \alpha_3 L_j + N(0, \sigma_3^2)$$
$$X_{j4} = \alpha_4 L_j + N(0, \sigma_4^2)$$
$$\sigma_{12} = \sigma_{13} = \sigma_{14} = \sigma_{23} = \sigma_{24} = \sigma_{34} = 0$$

These structural equations, coupled with the path diagram in Figure 5.5, states that each person's measurement (X_{jk}) is a linear function of (a) the true length of each string (L_j) and (b) other unknown causes modelled as a random variable following a normal distribution whose mean is zero and whose variance is σ_k^2. The total variance in the kth person's measure of a given string is now separated into two parts: a part that is common to everyone (the variance of the *common* cause, L) that is due to the true lengths of the strings, and a part that is unique to each person (the *unique variance*, σ_k^2) that is due to those other causes of a particular measurement. Since these unique variances are d-separated in Figure 5.5, we know that they must be uncorrelated given our hypothesis. Since there are four observed variables there are 4(5)/2 = 10 unique covariances in the covariance matrix.[5] There are only nine free parameters that we have to estimate: the four path coefficients (the four α_i), the variances of the four measurement errors (the four σ_k^2) and the variance of the latent variable. We can therefore fit this model by maximising its likelihood and then test it using the maximum-likelihood chi-square test because we have one degree of freedom left. However, before we do this we have to overcome a problem of identification. Identification is a topic that will be discussed in

---

[5] See Chapter 4.

more detail in Chapter 6, after we have seen how to combine the measurement model with the structural model involving the latent variables, but it can be intuitively understood with the following example.

If you are given an equation, say $y = x^*z$, and are told only that x equals 1, there is more than one combination of values for y and z that will solve this equation; in fact, there are an infinite number of such values. The equation is said to be *under-identified*. If you are told both that x equals 1 and that z equals 3 then there is only one value of y that is admissible: $y = 3$. The equation is just *identified*.[6]

Our measurement model, as described so far, is under-identified. The under-identification arises because we have to estimate both the path coefficients and the variance of the latent variable. We see in Figure 5.5 that d-separation predicts that the partial correlations (thus the partial covariances) of each pair of observed variables must be zero when conditioned on the latent variable. Maximum-likelihood estimation fits the data to the structural equations while respecting this constraint. The predicted covariance between the latent L and each observed X_j is given by $Cov(L, X_j) = \alpha_i Var(L)$. Since there is an infinite combination of αs and Var(L)s that can solve the equations, we must choose one by imposing an additional constraint. In reality, the imposition of this constraint consists of choosing the units that you want for your latent variable.

At this point you have a choice. If you want the latent variable to have the same units as one of your indicator variables then you can fix the path coefficient from the latent to this measure to 1. By doing this you are stating that a one-unit change in the latent variable changes the measured variable by one unit of the chosen scale. For instance, if we wish to scale our latent string lengths in centimetres then we could fix α_1 to 1. Think carefully before doing this. Your measure might systematically underestimate small values of the latent variable and overestimate large values; in this case the slope (i.e. the path coefficient) would be greater than 1. If this is the case, or if the scales of none of the measured variables are inherently more reasonable or useful than any of the others, then you can express the scale of the latent variable in units of standard deviations – i.e. you are expressing your latent variable as a standardised normal variable. This is done by fixing the variance of the latent variable to unity and allowing all the path coefficients to be freely estimated. Remember that standardisation (dividing a variable by its standard deviation so that its variance is unity) removes the original unit of the variable and replaces it with a scale of standard deviations from the mean. This has the effect of defining the scale of the latent variable by the measured variable and the path coefficient. Whether you choose a measurement scale for the latent by fixing a path coefficient or by fixing the variance of the latent makes no difference with respect to the fit of the model to the data but it does affect the numerical values of the fitted parameters.

---

[6] If we are told that x equals 1 and that two different estimates of z are 2.5 and 3.5 then the equation is *over-identified*. There is no unique solution to over-identified equations, and in any empirical problem the objective is to find the combination of estimates that gives the 'best' solution; least-squares regression or maximum-likelihood estimation are both examples of this.

Here is the full set of structural equations that are used in the likelihood maximisation using a standard deviation scale for the latent:

$$X_{j1} = \alpha_1 L + N(0, \sigma_1)$$
$$X_{j2} = \alpha_2 L + N(0, \sigma_2)$$
$$X_{j3} = \alpha_3 L + N(0, \sigma_3)$$
$$X_{j4} = \alpha_4 L + N(0, \sigma_4)$$
$$Var(X_j) = \alpha_j^2 Var(L) + \sigma_j^2 \quad j = 1, 4$$
$$Var(L) = 1$$
$$Cov(X_i, X_j) = \alpha_i \alpha_j Var(L) \quad i \neq j$$

These structural equations decompose the observed variances of the four measured variables into one part that is the same for all of them, due to the common cause of L, and one part that is different for each of them, due to the uncorrelated measurement errors. Let's simulate this causal process using the following generating equations:

$$X_{j1} = 1L + N(0, 0.5)$$
$$X_{j2} = 0.07L + N(0, 7)$$
$$X_{j3} = 0.39L + N(0, 3.3)$$
$$X_{j4} = 1L + N(0, 10)$$
$$L = N(\mu = 50, \sigma = 10)$$

Notice that the real scale for the true latent length is in centimetres in this simulation, since the path coefficients leading from the latent to X_1 and to X_3 are unity. So, I generate 100 independent 'strings', whose true length (L) is measured in centimetres. Since the first person used a ruler graduated in centimetres, the path coefficient is 1. Since she rounded her estimates to the nearest half-centimetre, I have given her measurement error a standard deviation of 0.5 cm. The second person used his hand, which was 14 centimetres long. His measurement scale was hand lengths, rounded to the nearest 'half-hand', and so the path coefficient is 0.07, with the standard deviation of the measurement error being 7 centimetres. The third person used a ruler calibrated in inches, resulting in a path coefficient of 0.39, which is the conversion from inches to centimetres. He took little care in his readings, and so the standard deviation of the measurement error was 3.3 centimetres. The last person simply visually estimated the true length in centimetres without bias, and so the path coefficient is 1. However, he was only accurate to within 10 centimetres, resulting in the standard deviation of his measurement error being 10 centimetres. Figure 5.6 shows the scatterplot matrix[7] of the 100 strings.

The measured lengths taken by the four people are all correlated since they are all trying to measure the same thing. The residual scatter in the graphs between each measured variable is due to the measurement errors of both variables in the pair, and the magnitudes of these measurement errors differ from one variable to the other. Here are

---

[7] A scatterplot matrix is like a correlation matrix except that the actual scatterplot of each pair of variables in the matrix is shown and a histogram of each variable is shown on the diagonal panels.

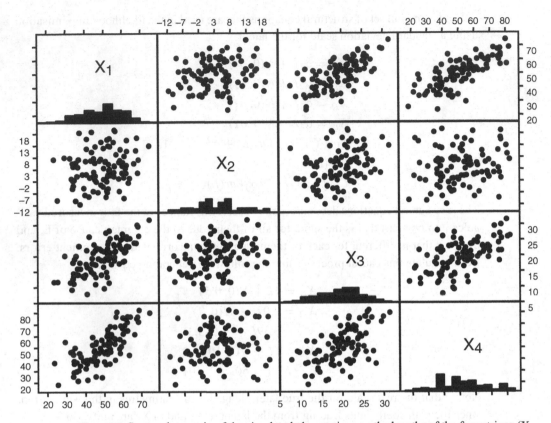

Figure 5.6 Scatterplot matrix of the simulated observations on the lengths of the four strings (X_1 to X_4)

Note: The histograms in the diagonal cells show the empirical distributions of the observations.

the maximum-likelihood estimates of the free parameters of the structural equations after fixing the variance of the latent variable to unity:

$$X_1 = 10.076L + N(0, 3.416)$$
$$X_2 = 1.829L + N(0, 7.167)$$
$$X_3 = 3.631L + N(0, 3.480)$$
$$X_4 = 12.120L + N(0, 9.075)$$

The chi-square statistic is 3.283 with two degrees of freedom,[8] producing a probability of 0.19 under the null hypothesis, telling us that the hypothesised causal structure is consistent with the data. The estimated variances of the measurement error of each variable agree well with the true values; the confidence intervals of these estimates[9]

---

[8] I said above that there was only one degree of freedom. However, fixing the variance of the latent variable to 1 means that we did not have to estimate this parameter, adding one extra degree of freedom.

[9] You can get these by multiplying the standard estimates of the path coefficients, generated as output in the summary function, by the appropriate t-value. More directly, you can input the fitted model object to the parameterEstimates() function in lavaan, which will generate both the fitted values and a chosen confidence interval.

include the true values. The estimated path coefficients are quite different from the true values. This is due to the fact that I fixed the variance of the latent variable to 1 even though we know that it is 100 (thus a standard deviation of 10) in the simulations. This means that the path coefficients are proportional to the true values with the constant of proportionality being the inverse of the true standard deviation of the latent variable. Since we know the true variance of the latent variable in this simulation, we can convert the structural equations, obtaining

$$X_1 = 1.0076L + N(0, 3.416)$$
$$X_2 = 0.1829L + N(0, 7.167)$$
$$X_3 = 0.3631L + N(0, 3.480)$$
$$X_4 = 1.2120L + N(0, 9.075)$$

These equations and the ones before are identical up to a constant $(1/10)$ and so the conversion simply changed the mean of the latent variable. Since structural equation modelling is generally concerned with the relationships between the variables, not their means, this conversion will have no consequence on the model. However, I said above that we could have fixed the scale of our latent variable to centimetres by fixing the path from the latent to X_1 to unity and allowing the variance of the latent to be freely estimated. In an empirical study we would not know the true variance of the underlying latent variable, and so this second strategy would be the one to use if the scale of the latent is important to you. One can calculate the correlation between the measured variables and the underlying latent variable in order to judge how well the measurement model has done. Box 5.1 summarises the calculations.

The accuracy with which one can estimate the underlying latent variable (up to a scaling constant) will depend both on the reliability of the measured variables and on the number of such variables used to measure the latent. However, this predictive ability is really quite secondary in the context of testing causal models with measurement error. The most important point is that a measurement model allows one to explicitly account for measurement error, therefore providing unbiased estimates of the path coefficients. Because the predicted covariances between the observed variables are functions of the path coefficients linking them, we can obtain unbiased estimates of the predicted covariance matrix and, therefore, of the asymptotic probability of the model under the null hypothesis. For those of you who like to see the algebraic details, Box 5.2 gives the generic factor model.

The minimum number of observed variables needed to fit, and test, a measurement model will depend on the number of hypothesised latent variables and the ways in which these latents are related to one another. For instance, if you have only one measured variable, the structural equation is $X_j = \alpha L + \varepsilon$. You have only one element in the covariance matrix (i.e. the variance of X) but you have three parameters to estimate: α, Var(ε) and Var(L). You can fix α to 1 to fix the scale of the latent, but this still leaves Var(L) and Var(ε). The equation is under-identified. If you can obtain an independent estimate of the error variance then you can fix Var(ε) to this value. This can sometimes be done. For instance, one could physically extract the body lipids of a sample of animals

---

Box 5.1 Calculating the correlation between measured variables and the underlying latent variable

By definition, the correlation coefficient between the latent variable, L, and its observed indicator variable, X_i, is

$$\rho_{L,X_i} = \frac{Cov(L, X_i)}{\sqrt{Var(L)Var(X_i)}} = \frac{\alpha_i Var(L)}{\sqrt{Var(L)Var(X_i)}}$$

where α_i is the path coefficient from the latent (L) to its indicator (X_i). All but $Var(X_i)$ are generated as output in the summary function, and $Var(X_i)$ is obtained directly from the data.

The coefficient of determination, ρ^2_{L,X_i}, between the latent and its indicator is

$$\rho^2_{L,X_i} = \frac{\alpha^2 Var(L)}{Var(X_i)} = \alpha_i^2 \theta_i$$

where θ_i is called the *reliability* of X_i.

If you want to obtain estimates of the latent variable ('factor scores'), up to a scaling constant, then you can form a weighting function of the observed variables. Different weighting functions have been proposed (Bollen 1989). However, no weighting function can estimate the latent without error, and so the best that we can do is to obtain an estimate of these factor scores.

The most common method, regression scores, is an ordinary least-squares estimator of the regression coefficients from the 'hypothetical' regression of the latent on the observed variables, and this is the method used in the predict() function of lavaan.

---

to obtain a precise estimate of body fat and then regress an indirect measure of this body fat to obtain the residual error variance. This will allow you to separate measurement error in subsequent data (assuming that the measurement error doesn't change), but you still cannot test this measurement model since there would be no degrees of freedom.

What about two measures of a latent? With two observed variables we have three non-redundant elements of the covariance matrix (two variances and one covariance), but we also now have five free parameters (α_1, α_2, $Var(\varepsilon_1)$, $Var(\varepsilon_2)$ and $Var(L)$). We have not solved our problem of under-identification. With three measures of a latent we have six non-redundant elements of the covariance matrix and seven free parameters. Since we have to set the scale of the latent, for instance by fixing $Var(L)$ at 1, we can now fit the structural equations, but we will have no degrees of freedom with which to test the measurement model. This is fine if we don't need to independently justify the measurement model (for instance, the relationship between a thermometer and the air temperature) but if there is any question about the causal relationships between the measured variables then we have defeated the whole purpose of modelling measurement error. Four measured variables per latent is the minimum number needed to both fit and test such a measurement model. However, this is not the case with more

Box 5.2 A standard factor model

Consider the following model (Figure 5.7) with six observed (manifest) variables and two latent variables with an unresolved covariance between them.

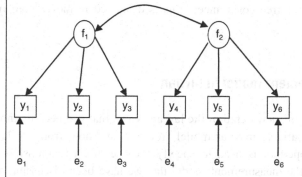

Figure 5.7 A measurement model involving two latent variables (f_1 and f_2), each measured by three observed variables (y_1 to y_6)

Here are the structural equations:

$$y_1 = a_{11}f_1 + 0f_2 + e_1$$
$$y_2 = a_{12}f_1 + 0f_2 + e_2$$
$$y_3 = a_{13}f_1 + 0f_2 + e_3$$
$$y_4 = 0f_1 + a_{24}f_2 + e_4$$
$$y_5 = 0f_1 + a_{25}f_2 + e_5$$
$$y_6 = 0f_1 + a_{26}f_2 + e_6$$

In matrix form, the equation is $\mathbf{Y} = \mathbf{AF} + \mathbf{E}$. Multiplying both sides by the transpose of $\mathbf{Y}$ gives $\mathbf{YY'} = (\mathbf{AF} + \mathbf{E})\mathbf{Y'} = (\mathbf{AF} + \mathbf{E})(\mathbf{AF} + \mathbf{E})'$. Expanding this gives $(\mathbf{AF})(\mathbf{AF})' + (\mathbf{AF})\mathbf{E'} + \mathbf{E}(\mathbf{AF})' + \mathbf{EE'}$.

Since the errors are independent of the latent factors (f_1 and f_2) and of each other, when we take expectations the following two terms are zero: $(\mathbf{AF})\mathbf{E'}$ and $\mathbf{E}(\mathbf{AF})'$. This gives $E[\mathbf{YY'}] = E[\mathbf{AFF'A'}] + E[\mathbf{EE'}]$. From this comes the standard factor model equation, $\mathbf{\Sigma_{YY}} = \mathbf{A\Phi A'} + \mathbf{\Psi}$, where

$\mathbf{\Sigma_{YY}}$ is the model covariance matrix between the observed ($\mathbf{Y}$) variables;
$\mathbf{\Phi}$ is the model covariance matrix between the latent ($\mathbf{F}$) variables; and
$\mathbf{\Psi}$ is the model covariance matrix between the errors (which don't have to be mutually independent, as in the current example). In order to avoid indeterminacies, we can set the factor variances to unity. Putting this all together for our model, we get the following matrix equation:

$$
\begin{bmatrix} \sigma_{11} & \sigma_{12} & \cdots & \sigma_{16} \\ \sigma_{21} & \sigma_{22} & \cdots & \sigma_{26} \\ \cdots & \cdots & \cdots & \cdots \\ \sigma_{61} & \sigma_{62} & \cdots & \sigma_{66} \end{bmatrix} =
\begin{bmatrix} a_{11} & 0 \\ a_{21} & 0 \\ a_{31} & 0 \\ 0 & a_{42} \\ 0 & a_{52} \\ 0 & a_{62} \end{bmatrix}
\begin{bmatrix} 1 & \phi_{12} \\ \phi_{21} & 1 \end{bmatrix}
\begin{bmatrix} a_{11} & a_{21} & a_{31} & 0 & 0 & 0 \\ 0 & 0 & 0 & a_{42} & a_{52} & a_{62} \end{bmatrix}
$$

complicated models involving more than one latent if these latents are connected by causal relationships or free covariances; this is discussed in more detail in the next chapter.

5.4   Fitting a measurement model in lavaan

There is really only one new detail of the lavaan syntax that you need to know in order to include a latent variable in your model. You already know, from the last chapter, that the tilde ($\sim$) operator is used to specify a cause–effect relationship. Since the causal structure of the measurement model that we have been discussing reflects the latent (L) as the cause and the observed indicators (x) as the effects, you might think that lavaan would represent this as L$\sim$x. You would be wrong! Lavaan uses a special operator ($=\sim$) to specify the relationship between a latent cause and its observed effects (the observed indicator variables): L $=\sim$x. Because we usually, but not always, have multiple observed indicators for each latent cause, lavaan allows us to specify this as L $=\sim$x $+$ y $+ \cdots +$ z, meaning that the observed variables on the right-hand side are all direct effects of the common latent cause on the left-hand side. What name should we give to our latent variable? By definition, a latent variable has not been directly measured, and so it will not be in our data frame. You can use any name that is valid in R so long as this name does not appear in the data frame that you specify in the sem() function. Of course, as always, the names you specify for your observed variables must match exactly with those in the data frame. Here is an example using the string data.

I have generated 100 mutually independent lines using the generating equations given below and assembled them in a data frame called 'strings', containing the following column names: 'ruler.cent', 'hand.lengths', 'ruler.inches', 'eyeball.guess'. First, we specify the model. I start by fixing the scale of the latent variable to unity so that it is measured in units of standard deviations. This can be done in two equivalent ways. The easiest and most direct way is as follows:

```
strings.model<-"
string.length=~ruler.cent+hand.lengths+ruler.inches+
eyeball.guess"
fit←sem(model = strings.model,data = strings,std.lv = TRUE)
```

The first line specifies the causal model. Each of the variable names on the right-hand side is in my data frame ('strings'). and so lavaan treats these as observed variables. The variable 'string.length' is not in my data frame, and so lavaan treats it as a latent variable. The latent indicator operator ($=\sim$) tells lavaan that the four observed variables on the right-hand side are each measures of the latent. The second line fits the model using the sem() function. There is one new argument in this function: std.lv, which means 'standard deviation of the latent'. By assigning a TRUE value to it I am telling lavaan that the variance of the latent variable in strings.model is fixed to unity.

There may be times when you want to fix the latent variances to something other than unity, or perhaps you want to fix some latent variances in the model to unity but fix the scales of other latents using the indicators. You can do this by explicitly telling lavaan (a) not to fix the path coefficient of the first indicator variable and (b) fixing the variance of the latent:

```
strings.model←"
string.length =~NA*ruler.cent+hand.lengths+ruler.inches+
eyeball.guess
string.length~~1*string.length"
fit←sem(model = strings.model,data = strings)
```

Note that the path coefficient associated with the first indicator variable ('ruler.cent') was multiplied by NA. This is necessary because if the std.lv argument in the sem() function is not specified as TRUE then lavaan by default fixes the scale of the latent by fixing the path coefficient of the first indicator variable to unity. By specifying 'NA*ruler.cent' I am telling lavaan not to fix this path coefficient to unity. Next, I must explicitly tell lavaan to fix the variance of the latent 'string.length' to unity:

```
string.length~~1*string.length.
```

If, instead, I wanted to fix the scale of the latent 'string.length' to centimetres then I would allow its variance to be freely estimated but would fix the path coefficient of 'ruler.cent' to 1. The easiest way to do this is simply

```
strings.model←"
string.length =~ruler.cent+hand.lengths+ruler.inches+
eyeball.guess"
fit←sem(model = strings.model,data = strings)
```

Unless specifically stated, lavaan assumes that the latent scale should be fixed by a 1:1 relationship between it and its first indicator variable. If you want to fix the scale using a different indicator from the first one, or if you want to fix the scale using a different fixed path coefficient, then you have to explicitly tell lavaan to do this. For instance, if I wanted to fix the scale of 'string.length' to inches, I would (a) freely estimate the path coefficient of 'ruler.cent', (b) fix the path coefficient of 'ruler.inches' to unity and (c) not assign TRUE to 'std.lv':

```
strings.model←"
string.length =~NA*ruler.cent+hand.lengths+1*ruler.inches+
eyeball.guess"
fit←sem(model = strings.model,data = strings)
```

Here is the output from the summary() function when I fix the scale of the latent to centimetres (i.e. fixing the path coefficient of 'ruler.cent' to unity). You will see that the model converged normally, the data do not show significant deviations from the model

$(X^2_{ML} = 2.056, 2\ df, p = 0.358)$ and the parameter estimates agree with those simulated to within sampling variation.

```
lavaan (0.5-16) converged normally after 55 iterations
    Number of observations                              100
    Estimator                                            ML
    Minimum Function Test Statistic                   2.056
    Degrees of freedom                                    2
    P-value (Chi-square)                              0.358
Parameter estimates:
    Information                                    Expected
    Standard Errors                                Standard
                    Estimate  Std.err  Z-value  P(>|z|)
Latent variables:
    string.length =~
        ruler.cent      1.000
        hand.lengths    0.036    0.080    0.452    0.651
        ruler.inches    0.427    0.048    8.815    0.000
        eyeball.guess   1.166    0.132    8.816    0.000
Variances:
        ruler.cent      3.621    6.586
        hand.lengths   54.520    7.711
        ruler.inches   11.356    2.004
        eyeball.guess  84.598   14.929
        string.length  87.486   14.452
```

We can get the proportion of the total variance of each indicator that is explained by the latent variable (the coefficients of determination) using the inspect() function:

```
inspect(fit,"r2")
    ruler.cent hand.lengths ruler.inches eyeball.guess
    0.960251220  0.002116242  0.584299879   0.584373183
```

We see, as expected, that the first person, who measured the strings to within 0.5 centimetres, has the highest correlation with the true string lengths while the estimates of the second person, who simply estimated the lengths to within a half a hand, were very poor. We could obtain the factor score estimates – i.e. our best estimates of the true string lengths – using the predict(fit) function.

5.5   The nature of latent variables

So far I have described latent variables simply as variables that we have not directly measured but that we can directly observe. In the previous examples there was no question that the animals really did have lipid reserves or that the strings really did have a length. Our only concern was in accurately measuring these variables. In such situations the development of the measurement model involves choosing measurable indicator variables that are all linearly related to the same latent variable. Ideally, the *only* causal relationships between these indicators will be through the common effect of the latent variable. If there exist other causal relationships between the measured variables, through other latents or not, then these must also be included in the model.

Often nature is not that accommodating. What happens if we want to model latent variables that we *cannot* directly observe? In such cases, even the existence of the latent variable is hypothetical. The invocation of such theoretical entities presents much more difficult choices, since we cannot rely on direct observation to know if such things even exist, although the actual modelling is no different. Nonetheless, the history of science is littered (or enriched, depending on your philosophical view) with such things. When Gregor Mendel invoked recessive and dominant alleles of genes to explain patterns of inheritance in pea seeds, he did not measure or observe such things. Rather, he inferred them because the ratios of the resulting phenotypes agreed with the binomial proportions that would result if such things existed.[10] Genes were latent variables – and still are; no one has ever directly observed a gene. Atoms, too, are latent; the periodic table was developed by inferring atomic structures from the numerical regularities that resulted from experiments. Ernst Mach, who was mentioned in Chapter 3 as one of the phenomenalists who influenced Pearson's views, initially refused to accept the reality of shock waves caused by bullets going faster than the speed of sound. He accepted such waves only when he was able to devise an experiment in which a camera was rigged to take a picture just as a bullet cut a fine wire covered in soot, revealing a V-shaped pattern.[11] As these 'successful' latent variables attest, scientists have regularly invoked things that can only be indirectly observed through the use of proxy measures. The problem is that scientists have also invoked 'unsuccessful' latent variables. A classic example is the 'ether', through which light waves were supposed to cross outer space. The use of latent variables in measurement models or SEM is not so much a statistical controversy as a scientific and philosophical one. Think carefully before including latent variables in your models and be prepared to justify their existence.

Much of my personal discomfort with latent variable models comes from the causal claims that many (by no means all) latent variables make. It is one thing to invoke a theoretical unmeasured variable and quite another to demonstrate that such an entity has both a reality in nature and has causal efficacy. Choosing, developing and justifying such latent variables is, perhaps, the most difficult aspect of structural equation modelling. I don't know of any set of rules that can unfailingly guide us in this task either. The exploratory methods, described in Chapter 8, can help to alert us to the existence of latent variables. The statistical tests based on maximum likelihood allow us to compare our data with such hypothesised latent variable models and therefore to potentially reject them. However, scientists generally demand stronger evidence that an

---

[10] There has been a long-running debate as to whether Mendel 'cooked' his data by ignoring outliers, since the observed and predicted ratios are so remarkably close as to be highly unlikely to occur by chance.

[11] Needless to say, he did not really observe the shock waves, only the indirect effect on the soot particles.

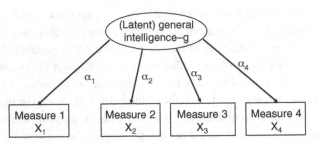

Figure 5.8 A measurement model for 'general intelligence' – a latent variable
Note: The errors for each of the observed measures (X_1 to X_4) are not shown but will exist unless the measures are all perfectly correlated with this latent variable.

acceptable statistical fit before accepting the physical reality of unmeasured variables. Before continuing further, it is again useful to briefly look at the history of latent variable modelling in statistics. The hornet's nest of confusion involving latent variable models is due in part, I believe, to the historical link between latent variable models and factor models in the social sciences.

In 1904 the English psychologist Charles Spearman combined the new psychometric work of Alfred Binet on human intelligence with correlation coefficients. In his provocatively titled paper 'General intelligence objectively determined and measured' (Spearman 1904), he hypothesised that the observed measures of intelligence between people, obtained from test questions, were all correlated because they were all due to a general latent intellectual capacity ('g') that varied from person to person. If we have four different measures of intelligence (from an IQ test, say) then the causal graph would look like Figure 5.8.

Now, given this structure, the population correlation between any two observed variables, say X_1 and X_2, would be $\rho_{12} = \alpha_1 \alpha_2$. It follows that the following three equations must be true if the model in Figure 5.8 is true:

$$\rho_{12}\rho_{34} - \rho_{13}\rho_{24} = 0$$
$$\rho_{13}\rho_{24} - \rho_{14}\rho_{23} = 0$$
$$\rho_{14}\rho_{23} - \rho_{13}\rho_{24} = 0$$

Spearman called these 'vanishing tetrads', because each involves four correlation coefficients, and they are, in modern terms, constraints on the correlations due to the causal structure of the model. He argued (incorrectly, as explained in Chapter 8) that data obeying such vanishing tetrads constituted evidence for this unmeasured latent cause ('generalised intelligence'). Spearman apparently viewed this latent variable as a real, causally efficacious attribute of people. As more measured variables were added, and more complicated latent structures were hypothesised, one could derive the vanishing tetrads that were implied by the model, but this quickly became difficult to do, both conceptually and computationally.

In the 1930s Harold Hotelling invented principal components analysis (PCA) and L. L. Thurstone invented factor analysis. These methods were not based on any explicit causal model; quite the contrary. Thurstone viewed science in much the same way as Pearson did: that it consisted of erecting 'constructs' that could describe the data as simply as possible. Whether or not such 'constructs' actually existed in nature was irrelevant; they could economically summarise the patterns of correlation, and a single 'construct' could replace a large number of observed variables. These constructs, or factors, had the property that they could partition the variances of each measured variable into one part (the construct) that was common to all and one part that was unique to each measured variable. Described in this way, we appear to be right back to our measurement model, as described above. However, such factor models (like principal component models) could perform this trick quite independently of whether the 'common variance' was really due to some unmeasured common cause. Moreover, the method, if drawn as a graph, always has the arrows going from the construct (factor, common variance) to the measured variables. Such a structure was a requirement of the method. Interpreted as Thurstone intended, this was not a problem, since the constructs were simply mathematical functions designed to summarise data. Interpreted as causal models (as Thurstone most emphatically did not intend), factor models had the bizarre property of requiring that the direction of causality always went from the latent construct to the observed variables. The obvious advantage of factor analysis or principal components analysis[12] over Spearman's method of vanishing tetrads was that these former methods were easier to use and based on standard formulae. One had only to plug the data into the equations and out popped the construct or the principal component axis.

Thus, vanishing tetrads became an historical footnote, and factor analysis (with its requirement that the arrows go from the construct to the measured variables) took their place in psychometrics. Jöreskog (1967; 1969) applied maximum-likelihood methods to factor analysis to develop an inferential test for such models, and then extended this to allow for cause-and-effect relationships between the latent variables based on econometric simultaneous equations models, giving rise to structural equation models (Jöreskog 1970; 1973). Although there is no longer any mathematical requirement that the arrows always go from the latents to the measured variables – as was the case with factor analysis – the formalism of factor analysis persists in SEM, along with some of its philosophical origins.

As an example, consider the description given in Bollen's (1989) influential book on SEM. He states that the measurement process begins with a concept and defines a concept as an idea that unites phenomena under a single term. He gives the example of 'anger', which provides the common element tying together attributes such as screaming, throwing objects, having a flushed face, and so on. The concept of anger, he says, 'acts as a summarising device to replace a list of specific traits that an individual may exhibit'. This is a close paraphrase of Thurstone's original description of a 'factor', but remember that Thurstone's factor analysis was explicitly acausal. To Bollen's rhetorical question 'Do concepts really exist?' he answers that '[c]oncepts have the same reality or lack of reality as other ideas... The concept identifies that thing or things held in common. Latent variables are the representations of concepts in measurement models.' If we are dealing with purely statistical models, devoid of causal implications, then

---

[12] PCA, another multivariate data summary method, requires that the path coefficients always go in the opposite way from factor analysis.

such a view might be fine. If our models are statistical translations of causal processes then the latent variables in our models must be something more than a mathematical summary; latent variables must represent variables with physical reality having causal relationships to the measured variables.

One of the early controversies amongst geneticists at the turn of the twentieth century concerned the inheritance of size differences in different body parts. Herman Chase, an influential geneticist at the time and Sewall Wright's student, argued that there was a single 'size factor' that was inherited and that determined the allometric scaling of different body parts. Part of this argument was based on correlation coefficients, calculated by Wright while still a graduate student, relating five different bone measurements of rabbits. Charles Davenport (1917), studying human stature, argued that the patterns of correlation between different lengths of different body parts suggested that these attributes of size were inherited independently. In 1918 Wright published 'On the nature of size factors' (Wright 1918), based on rabbit measures, in which he calculated a series of partial correlations. On the basis of these calculations he concluded that his own supervisor was wrong and that '[t]hese three correlations[13] suggest the existence of growth factors which affect the size of the skull independently of the body, others which affect similarly the length of homologous long bones apart from all else, and others which affect similarly bones of the same limb'. Since no one knew what these 'size factors' were, the entire argument concerned the number of latent variables controlling the inheritance of size in different body parts. The following example shows how one can test such claims.

5.6   Horn dimensions in bighorn sheep

Marco Festa-Bianchet and his students have been following a population of bighorn sheep from the Rocky Mountains of Alberta for many years. Horn size is very important in this species, because, among other things, it affects the ability of males in combat during the rut and therefore the evolutionary fitness of these males. Every year from 1981 until 1998[14] the researchers measured the total length and the circumference at the base of the two horns of the captured sheep. As one might expect, these four variables are highly correlated and display the sort of allometric scaling patterns that are so ubiquitous in biology. Are these four variables simply responding to a single latent 'size factor'? In other words, are the patterns of correlation between the four variables simply due to a single common unmeasured cause that determines increases in linear dimensions, as Chase might have supposed?

Figure 5.9 shows this hypothesis, translated into a causal graph. To fix the scale of the latent, I fixed the variance of the latent variable to 1. The data do not follow a multivariate normal distribution, even after a Ln-transformation, as shown by Mardia's

---

[13] Actually, they were partial correlations.

[14] These measurements are still ongoing as of the time of writing, 2014, but the analysis that follows is based only on values taken during those years.

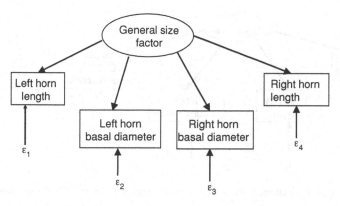

Figure 5.9 The hypothesised measurement model relating four observed attributes of the horns of male bighorn sheep

normalised coefficient of kurtosis. Because of this, I use a robust estimation method for the chi-square statistic (the Satorra–Bentler chi-square); these statistics are explained in detail in Chapter 6. The data are clearly not consistent with the single common latent model in Figure 5.9, since the Satorra–Bentler chi-square statistic is 759.106 with two degrees of freedom. The probability of observing this by chance is far lower than one in a million.

A look at the residuals shows why the fit is so bad. The residuals of the two length measures are highly correlated, indicating that there is something else that is affecting length independently of the basal circumference. Perhaps the 'General size factor' causes two 'Specific size factors' — one for length and one for circumference? In Chapter 6 I explain how one can statistically test these ideas. However, I can't point to any specific biological mechanism to justify this proliferation of hypothetical unmeasured variables.

At this point I apply the 'squirm' test. When the hypothesised latent variable begins to resemble 'a summarising device to replace a list of specific traits that an individual may exhibit' rather than a physical 'thing' with causal efficacy then I begin to squirm. The statistical model has wandered too far away from any causal model to which it was supposed to be a translation. I cannot justify one of the auxiliary assumptions (that the latent variable is not simply a statistical construct) beyond reasonable doubt. Each person will have his or her own tolerance for this, but in my experience most biologists (and reviewers!) have a very low 'squirm' tolerance indeed. My own (highly personal) opinion is that this is a good thing.

5.7 Body size in bighorn sheep

Body size is another important attribute of bighorn sheep. Large animals are less likely to fall prey to predators. Animals that have been able to amass sufficient fat reserves in the autumn are more likely to survive the severe winter conditions at the top of a mountain in Canada. Larger males are more successful in the rut and therefore get to

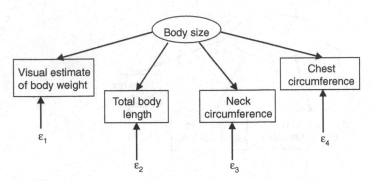

Figure 5.10 The hypothesised measurement model relating four observed size attributes of bighorn sheep

copulate with more of the females. The reproductive success of a female is affected by her fat reserves. Now imagine that you are a field biologist, perched at the top of a steep rocky slope with a temporarily subdued animal, and you need to estimate its body size. You do not have a balance (and have to keep your own balance!) but you can quickly take measurements of attributes associated with body size. Can you construct a measurement model that will be able to estimate the unmeasured 'body size'?

The following analysis, based on data provided by Festa-Bianchet, are from four indirect measures of body size based on 248 observations of bighorn sheep. The observed variables are the total length of the animal (nose to tail), the circumference of the neck, the circumference of the chest just behind the front legs and a visual estimate (which sounds better than 'guess') of the body weight in kilograms. These data, transformed to their natural logarithms, are consistent with multivariate normality, based on Mardia's coefficient. The measurement model consists of a single latent variable, which I have labelled 'Body size', which is the single common cause of the four indirect measures listed above. As always, one can set the scale of the exogenous latent variable either by fixing its variance to unity or by fixing the path coefficient from it to one of the observed variables. Since 'Body size' is usually interpreted as meaning a mass, I have therefore chosen to fix the path coefficient from the latent to the estimated body weight at unity. This means that the latent 'Body size' is measured in Ln(kilograms) (the units of the estimated body weight). Figure 5.10 shows the directed graph for this measurement model.

The chi-square statistic for this model is 0.971, with two degrees of freedom, giving a probability under the null hypothesis of 0.615. The data are perfectly consistent with the model. Here are the structural equations and the proportion of the variation of each measured variable that is accounted for by the latent 'Body size':

$$Ln(Estimated\ weight) = 1Ln(Body\ size) + N(0, 0.023) \quad R^2 = 0.893$$
$$Ln(Total\ length) = 0.370Ln(Body\ size) + N(0, 0.003) \quad R^2 = 0.911$$
$$Ln(Neck\ circumference) = 0.424Ln(Body\ size) + N(0, 0.005) \quad R^2 = 0.883$$
$$Ln(Chest\ circumference) = 0.387Ln(Body\ size) + N(0, 0.001) \quad R^2 = 0.982$$
$$Ln(Body\ size) = N(0, 0.191)$$

We see that the estimated error variance of the 'Ln(Estimated weight)' is 0.023 and the latent 'Body size' accounts for 89.3 per cent of the variance of this estimated weight. The guesses were not so bad after all. In fact, these guesses of the true body weight appear to be just as tightly correlated with the latent body size as the measures of neck circumference. The observed variable that was most highly correlated ($R^2 = 0.982$) with the latent body size was the chest circumference. If only one measurement is to be taken, the chest circumference should be the one to use.

If we now wish to test a causal model involving body size in a new data set we can either use one of the measured variables and explicitly include our estimates of the error variance or else use all four measured variables. The advantage of using all four variables is that we do not have to assume that the error variances will remain the same from one study to the next; we can instead estimate them. This might be a wise decision if, for instance, different people take these measurements from one study to the next and some people make more measurement errors than others.

The problem of different people making systematically biased estimates is the likely explanation for the lack of fit that occurs when another measured variable is included in the measurement model described above. This variable is the length of the hind foot. The chi-square statistic for a new model that includes this new variable is 25.808, with five degrees of freedom, for a probability of about 1×10^{-4}. This five-variable measurement model can be made to fit the data only by letting the length of the hind foot co-vary freely with the weight estimate and the neck diameter.[15] In other words, there are other causes, independent of the latent 'Body size', that are generating associations between these three measured variables. Although it is possible that these other causes are related to the biology of these animals, it seems more likely that these other causes are due to the way the data were collected.

The measurements were taken over many years by many different people – mostly graduate students with different levels of ability in fieldwork. Unlike the other length measurements, the length of the hind foot requires that the foot and hoof be consistently extended to the same degree. These measurements must be taken quickly while the animal is still subdued. Imagine that you are sitting at the edge of a steep precipice on the top of a mountain, with an adult bighorn sheep about to wake up. It is safe to assume that the care with which the foot is extended will vary from person to person. So, if there are any systematic biases between people in how they measure the three variables in question (the length of the hind foot, the weight estimate and the neck diameter), this would be a cause of correlations between them independent of differences in 'Body size'.

5.8   The worldwide leaf economic spectrum

Wright et al. (2004) compiled published data from over 2,500 plant species worldwide involving a series of leaf traits involved in the capture, use and cycling of resources (the

[15] Actually, one permits covariation between the error variables of these measured variables.

'economy' of the leaf). The traits of interest to us are the specific leaf mass (the ratio of leaf dry mass to projected surface area), the maximum net photosynthetic rate expressed per unit dry leaf mass (A_m), the leaf lifespan (LL) and the leaf nitrogen content per unit dry mass (N_m). Using principal components analysis they found that a large proportion of the total variation of these variables was captured by a single principal component (hence a 'spectrum') with very little difference between major habitat types or growth forms. Using the exploratory methods described in Chapter 8 I found that (a) there is no path model involving only these variables than can fit the data, (b) there is evidence for a latent common cause of these four variables but (c) it is not a simple measurement model as we have considered so far (Shipley et al. 2006). What might this latent variable be? Whatever it is, it cannot be something that is unique to only certain species or habitats. This next example involves a latent variable model whose development is more theoretical.

All leaves are composed of cells. In these plant cells, most physiological activity, including photon capture in the chloroplasts and the nitrogen-containing photosynthetic enzymes, occurs in the cytoplasm while most of the dry mass is found in the cell wall. Imagine that you could increase the total volume of the leaf that is occupied by cell walls relative to the volume that is occupied by cytoplasm without changing the total volume of the leaf. For instance, this could be done by having more, but smaller, cells or by increasing the thickness of the cell wall at the expense of the cytoplasmic volume. Different tissue types in the leaf (mesophyll, collenchyma, sclerenchyma, xylem) vary greatly in the ratio of cytoplasm to cell wall volumes. Such an increase in the volume of cell walls versus cytoplasm while maintaining the total leaf volume constant should decrease the total amount of carbon fixed and decrease the total amount of nitrogen in the leaf, because there is less cytoplasm overall, while increasing the total amount of dry mass; that is, it would decrease the average amount of carbon fixed per gram dry mass and the average amount of nitrogen per gram dry mass. Changing this ratio would also change the specific leaf mass of the leaf, because SLM is the product of two more fundamental properties of a leaf: its thickness and its tissue density (i.e. its dry mass per volume). Thus, we can expect correlated changes in three of our variables (A_m, N_m and SLM) by changing the ratio of the leaf volume occupied by cytoplasm versus cell walls.

The last variable, leaf lifespan, requires a bit more theory. Species vary enormously in the typical lifespans of their leaves, and Kikuzawa (1995) asked: 'What is the lifespan of a leaf that maximises the net return (in terms of carbon fixed) to the plant?' He produced a model predicting this optimal lifespan as a function of two leaf properties: the maximum net photosynthetic rate (A_{max}) and the 'construction cost' (C). In this model, the optimal leaf lifespan decreases with A_{max} and increases with increasing C. The construction cost is the amount of energy and resources that the plant must invest in order to construct the leaf. It would be very difficult to add up the energy cost of building each component of a leaf. Whatever these costs, the total cost must be related to the total number of atoms of carbon, nitrogen and other limiting nutrients that the plant must assemble together to construct different components. Because the number of such atoms per unit volume (i.e. density) is much higher in the cell wall than in the cytoplasm (because much of the cytoplasm is water), it is reasonable to assume that it

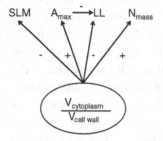

Figure 5.11 The causal model for the leaf economic spectrum
Notes: A single latent variable, hypothesised to be the ratio of the leaf volume occupied by cytoplasm to cell walls, is a common cause of the four observed variables: specific leaf mass (SLM), maximum net photosynthetic rate (A_{max}), leaf lifespan (LL) and leaf nitrogen content (N_{mass}). There is also a direct causal effect of A_{max} on LL.

costs more *per unit volume* to construct the cell wall than to construct the same volume of cytoplasm. If so, then the construction cost will increase as the volume of cell walls increases relative to the volume of the cytoplasm.

Putting together the hypothesised relationships between our four observed variables and the unmeasured (i.e. latent) ratio of the volumes occupied by cytoplasm and cell walls, and the theoretical model of Kikuzawa, we end up with the model shown in Figure 5.11. In fact, this model fits the data well (Shipley et al. 2006). Does this mean that this latent variable model is consistent with the data? Yes. Does this mean that the latent variable in the model is the ratio of the volumes of cytoplasm to cell walls? No; this is only an interpretation. The next step would be to actually estimate these volumes. Such estimates, obtained from image analysis of photographs of thin slices of leaves and extrapolated to three dimensions, would necessarily contain measurement error, and so we would include this new variable as another indicator of the latent. The more such indicator variables that we can include in our model that are logically connected to the ratio of the volumes of cytoplasm to cell walls but not to alternative interpretations of the latent, the more confident we can be about this interpretation.

5.9   Name calling

A certain Henry P. Crowell of Ravenna, Ohio, bought a bankrupt mill in 1881 and 'went into the business of convincing people to consume what previously only poverty-stricken Scotsmen, Germans and horses had eaten' (Burke 1996). How did he accomplish this feat of marketing?[16] Since people associated the word 'Quaker' with honesty and a healthy lifestyle, he simply called his new product 'Quaker Oats'. Thus began one of the longest-running breakfast cereals in the United States. A certain mouthwash

---

[16] To show what Crowell was up against, consider that the first edition (1755) of Samuel Johnson's classic *Dictionary of the English Language* defined 'oats' as a grain that sustained horses in England and people in Scotland.

company proudly proclaimed that its product, besides making one's breath taste fresh, also cured 'halitosis'. The dictionary definition of halitosis is 'bad-smelling breath'. So, what is the latent variable, shown in Figure 5.10, that is the single independent 'cause' of the estimated body weight, the total length, the neck circumference and the chest circumference of the sheep? I have labelled it 'Body size', but is this misleading advertising? Just as saying that something 'cures halitosis' evokes connotations beyond simply 'curing bad breath', does calling my latent variable 'body size' evoke connotations beyond simply saying 'that theoretical variable that has the property of making the partial covariances between each unique set of measured variables equal to zero, when conditioned on it'?

Remember Bollen's (1989) claim that '[l]atent variables are the representations of concepts in measurement models', and that '[t]he concept identifies that thing or things held in common'. It is certainly reasonable to state that 'body size' is that which is common to weight, length and circumference of the body (the measured variables). However, if Figure 5.10 is to be interpreted as a description of a causal process then the latent variable also represents a single common *cause* of body weights, lengths and circumferences. The causal claim must be that there is a single biological process that determines all these bodily dimensions. It would obviously be better if we knew enough about the genetic and developmental processes determining body size that we could label our latent variable as 'hormone X' or 'gene Y'. If we can't then we could at least label it as the 'unknown cause of body size'. So long as both you and I understand the name 'body size' as being a short form of saying this, then we can properly translate between the causal claim and the statistical model.

It is particularly important that we choose our words carefully when dealing with latent variables, and the burden of clarity is on the person proposing the model, not on the reader. In the model for the leaf economic spectrum the naming of the latent variable as 'the ratio of the volumes of cytoplasm to cell walls' derives from an underlying theoretical causal argument and points to a property of the leaf that we can independently verify. On the one hand, we know, independently of the statistical model, that leaves are composed of both cytoplasm and cell walls and that the volumes occupied by these vary between species. We don't have any metaphysical problems accepting that such a ratio exists in real leaves. On the other hand, we don't have any good evidence, beyond the latent variable model itself, that the latent actually is 'the ratio of the volumes of cytoplasm to cell walls', and it might be something completely different. Until (and if) such evidence is found, it is better to simply call it a 'latent variable' and leave the name-calling to a part of the text that can elaborate on the tentative nature of the name. If you see a latent variable in a structural equation and its meaning and causal justification are not clearly explained, think of bad breath.

6   The structural equation model

The structural equation model is commonly described as the combination of a measurement model and a structural model. These terms derive from the history of SEM as being a union of the factor analytic, or measurement, models of psychology and sociology and the simultaneous structural equations of the econometricians. In its pure form it therefore explicitly assumes that every variable that we can observe is an imperfect measure of some underlying latent causal variable and that the causal relationships of interest are always between these latent variables. As in many other things, purity is more a goal than a requirement. Using the example in Chapter 5 of the effect of air temperature on metabolic rate (Figure 6.1), the things that we can measure (the height of the mercury in the thermometer or the change in CO_2 in the metabolic chamber) always contain measurement error (ε_i). The measurement model, shown by the dashed squares in Figure 6.1, describes the relationship between the observed measures and the underlying latent variables (the average kinetic energy of the molecules in the air and the metabolic rate of the animal). The structural model, shown by the dashed circle in Figure 6.1, describes the relationship between the 'true' underlying causal variables. If we have only one measured variable per latent variable and we assume that the measured variable contains no measurement error (i.e. the correlation between the measured variable and the underlying latent variable is perfect) then we end up with a path model. If we have a set of measured variables for each latent variable and we do not assume any causal relationships between the latent variables then we have a series of measurement models. If we have more complicated combinations, in which we assume causal relationships between the latent variables, then we have a full structural equation model. Therefore, if you have understood Chapters 1 to 5, you already know how to construct and test a structural equation model; you simply have to put the pieces together.

The goal of this chapter is therefore to deal with some technical details that I have ignored up to now. The first detail is the problem of identification. In models involving more complicated combinations of latent and observed variables, how can we make sure that no model parameters are under-identified? The second involves composite latent variables (Grace and Bollen 2008) – that is, latent variables that are the hypothesised combined outcome of several observed variables. The third detail involves the robustness of SEM to violations of two important assumptions: large sample sizes and multivariate normal distributions. What happens when our data do not agree with these assumptions, and what can be done about it?

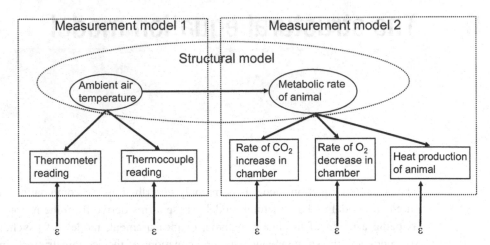

Figure 6.1 The relationship between the measurement model and the structural model in the causal structure relating the ambient air temperature and the metabolic rate of an animal

6.1   Parameter identification

We have already met the problem of under-identification in Chapter 5. Intuitively, a model is under-identified when more than one combination of parameter values can account for the same pattern of covariance. If a model is under-identified then you can't trust the parameter estimates, their standard errors, the chi-square value or its probability level. Because under-identification can arise in so many different ways, lavaan does not have a specific error message to warn us of this problem. If you are told that a parameter estimate 'is a linear combination' of some other set of parameters, or that an estimated variance estimate is negative or 'set at zero', then this can mean that the model is under-identified. However, depending on the model details, lavaan can even report normal convergence and print out the model fit and a null probability with an under-identified model. One tell-tale indication of under-identification is that the standard errors of the parameter estimates cannot be calculated. However, a lack of standard errors of parameter estimates can also arise for reasons other than under-identification. For instance, consider the following script to fit the measurement model in Figure 5.5 (the string lengths), which is under-identified. The model is under-identified because it doesn't fix either the variance of the latent variable (the 'std.lv' argument is set to FALSE) or any of the path coefficients.

```
mod←"string.length = ~NA*ruler.cent+hand.lengths+
ruler.inches+eyeball.guess"
sem(model = mod,data = strings,std.lv = FALSE)
```

The lavaan output tells us that it 'converged normally after 39 iterations' and prints out the ML chi-square statistic and its null probability. It also prints out the parameter estimates of all nine free parameters, but it doesn't print out the standard errors of these estimates and, therefore, doesn't report z-values or null probabilities of the estimates. The only indication that this model is under-identified is the warning message:

```
Warning message:
In lav_model_vcov(lavmodel = lavmodel, lavsamplestats =
lavsamplestats, : lavaan
WARNING: could not compute standard errors!
```

A model can be *structurally* under-identified or *empirically* under-identified. Structural under-identification, which exists in the above example, means that the model will be under-identified for any combination of parameter estimates – i.e. the problem is in the way the model itself is constructed. You will want to ensure that your model is not structurally under-identified before collecting data in order to avoid wasting your time. Empirical under-identification means that the model is under-identified only for some particular sets of parameter estimates; the problem is not in the general construction of the model but, rather, with the particular values found in the data. These points will be illustrated with examples later. Let's start with some useful rules for avoiding this problem.

6.2 Structural under-identification with measurement models

Following the notions introduced in Chapter 5, let's call a measurement model any factor-analytic model consisting of a set of latent variables and a set of observed indicator (measurement) variables that are each caused by at least one of the latent variables. The latent variables can be allowed to freely covary (i.e. there can be curved double-headed arrows between them) but there are no cause–effect relationships between the latent variables (i.e. there can be no arrows from one latent to another). Bollen (1989) has summarised three rules to help in judging if a measurement model is structurally identified. These rules cover many types of measurement models, but not all. All these rules assume that the scale of each latent variable has been fixed, as described in Chapter 5, either by fixing one of the path coefficients to 1 or by fixing the variance of the latent variable to 1.

Rule 1: there must be positive degrees of freedom in your model. In other words, the number of free parameters (t) in the model[1] must be smaller than, or equal to, $n(n+1)/2$, where n is the number of observed variables in the model.

This rule is necessary for identification. If the rule doesn't hold in your model then you can be sure that the model is not identified. Unfortunately, this rule doesn't ensure that your model will be identified; even if the rule holds the model might still be under-identified. In fact, the under-identified model that we fitted just previously satisfied this first rule, since it had $t = 9$ free parameters (four path coefficients, the error variances of the four observed variables and the error variance of the latent) and four observed variables, thus $9 < (4)(5)/2$. The following two rules are sufficient (i.e. if they hold then

---

[1] That is, free path coefficients, free error variables and free covariances either between the latents or between the error variables.

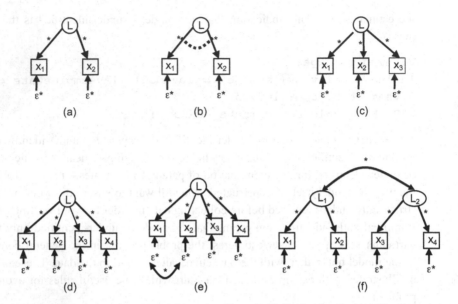

Figure 6.2 Six measurement models used to illustrate Bollen's rules for identification

the model is identified) but not necessary (i.e. there are still under-identified models that violate these rules).

Rule 2: a measurement model is identified if, along with rule 1,

(1) there are at least three indicator variables per latent variable;
(2) each indicator variable is caused by only one latent variable; and
(3) there are no correlations between the error variables.

Rule 3: a measurement model is identified if, along with rule 1,

(1) there is more than one latent variable;
(2) there are at least two indicator variables per latent variable;
(3) each indicator variable is caused by only one latent variable;
(4) each latent variable is correlated with at least one other latent variable; and
(5) there are no correlations between the error variables.

To understand how these rules work, let's look at Figure 6.2, which shows six different measurement models. The scale of the latent variables (L) have all been set by fixing the variances of the latent variables to unity. Free parameters (the path coefficients, the variances of the error variables and the covariances indicated by curved double-headed arrows) are shown by an asterisk (*). The model in Figure 6.2(a) is under-identified; there are four free parameters (t = 4) and two observed indicator variables (n = 2) but rule 1 states that t ≤ n(n+1)/2. If we constrain the values of the two path coefficients to be the same during the iterative procedure that minimised the maximum-likelihood chi-square statistic (shown in Figure 6.2(b) as the dashed line between the two free path coefficients) then we have only three free parameters, and the model is just identified.[2] This trick, while allowing us to identify the model, doesn't really allow us to get unbiased estimates of the measurement error or of the two path coefficients.

The model in Figure 6.2(c) is just identified. There are $t = 6$ free parameters and $n = 3$ observed variables, therefore $6 = 3(4)/2$ and rule 1 is fulfilled. Since there are no degrees of freedom left, we cannot test such a model using the maximum-likelihood chi-square. Such a model can always be fitted even if the causal assumption of a single common latent cause is wrong, and we can't know if the causal assumption is reasonable or not based on statistical criteria.

The model in Figure 6.2(d) is over-identified. There are $t = 8$ free parameters and $n = 4$ observed variables, therefore $8 < 4(5)/2$ and rule 1 is fulfilled. Since there are $4(5)/2-8 = 2$ degrees of freedom left, we can also test such a model using the maximum-likelihood chi-square. Such a model can always be fitted but if the causal assumption of a single common latent cause is wrong then we would obtain a significant probability estimate of the measured maximum-likelihood chi-square statistic, and could therefore reject the model. Both the measurement models for the bighorn sheep horns and for the body dimensions, studied in Chapter 5, were of this form, and we saw that the model for the horn dimensions was clearly rejected ($p < 10^{-6}$) while the model for the body dimensions was not rejected ($p = 0.615$).

The model in Figure 6.2(e) is like the model in Figure 6.2(d) except that it has a free covariance between the error variables of X_1 and X_2. Rule 1 is still satisfied, since $t = 9$, $n = 4$ and $9 < 4(5)/2$. Rule 2 is not satisfied; although there are at least three indicator variables per latent (there are four) and each indicator variable is caused by only one latent, there is also a correlation between two of the error variables. Rule 3 cannot be applied either, since there is only one latent variable. However, rules 2 and 3 are sufficient conditions, not necessary conditions. We can't state that the model is definitely not identified, only that we can't tell one way or the other. In fact, the model in Figure 6.2(e) is structurally identified.

The model in Figure 6.2(f) looks a bit like two models from Figure 6.2(a) combined. Remember that model A wasn't identified because rule 1 was violated. What about model F? There are $t = 9$ free parameters and $n = 4$ observed variables. Since $9 < 4(5)/2$, rule 1 is satisfied. Rule 2 is not satisfied (there are only two indicator variables per latent), but rule 3 is satisfied. Therefore, the model in Figure (f) is structurally identified.

This model also provides a good example of how a model can be structurally identified but empirically under-identified. If, in reality, the covariance between the two latent variables (the curved, double-headed arrow) is close to zero then the estimated covariance in the data might be zero due to sampling fluctuations. If this occurs then the maximum-likelihood procedure will be trying to fit two independent measurement models, each with only two indicators per latent. Since each separate measurement model has four free parameters (two error variances and two path coefficients) but only two indicator variables, rule 1 would be violated in this particular case.

---

[2] In fact, this model is equivalent to the so-called *major axis* (errors-in-variables) regression of Sokal and Rohlf (1981). *Reduced major axis* regression is obtained by simply standardising x_1 and x_2 to unit variance and a zero mean before fitting the model.

Recall the measurement model for the length and basal diameter of the left and right horns of the bighorn sheep. We were quite confident that the correlations between these four measures were not due to a single common unknown cause, because the measurement model with a single latent variable was strongly rejected. Perhaps the observed correlations are due to two correlated latent causes, as shown in Figure 6.2(f)? In bilaterally symmetrical organisms the left and right halves of the body should be mirror images in terms of size and shape. You have only to look into a mirror (when no one else is watching) to see that no one is really perfectly bilaterally symmetrical. Various environmental perturbations during embryonic development can cause random deviations from perfect bilateral symmetry, and the degree of this 'fluctuating asymmetry' is sometimes used as an index of pollution load or other forms of environmental stresses. Perhaps the model with a single latent cause of horn dimensions was rejected because there were additional causes of the left and right horns besides a single 'size' factor that generated deviations from bilateral symmetry? This hypothesis produces the model in Figure 6.2(f), and we know that this model is both structurally identified and has one degree of freedom left to test the model. The single 'size' factor is reflected in the free covariance between the two latent variables. The two latent variables, according to our present hypothesis, should represent the different causes of the left and right horns generating deviations from bilateral symmetry. The two latent variables in Figure 6.2(f) would then represent the causes specific to the left and right horns. When I fit this model to the data (using the Satorra–Bentler chi-square, since there is significant multivariate kurtosis in the data), I get a chi-square value of 146.1205 with one degree of freedom. Clearly, this model, too, is wrong.[3]

If we think of how the measurements were taken we are led to another measurement model with two latent variables. The horns are strongly curved, and their length is measured with a measuring tape that has to properly follow the curve of the horn. It is possible that longer horns, with a more pronounced curve, would be systematically underestimated as the exasperated researcher tries to make the measuring tape follow the curve of the horn before the sheep regains control. The longer the horn, and the more it is curved, the greater the degree of underestimation, since the measuring tape will have more chance to slip down a bit along the horn. A similar systematic bias might occur for the two measures of basal diameter, since this measurement too requires a subjective decision as to where the base of the horn begins. If these speculations are correct then each of the length measures and each of the diameter measures will have a separate cause (i.e. the way that they are measured) besides a common 'size' factor due to bilateral symmetry during development. When I fit this model to the data (again using the Satorra–Bentler chi-square, since there is significant multivariate kurtosis in the data), I get a chi-square value of 3.948 with one degree of freedom, giving a probability level of 0.05. This model has an ambiguous probability level and its true value is probably

---

[3] I use this example simply for pedagogical reasons. In reality, the notion of fluctuating asymmetry is that the deviations are random from individual to individual. In some individuals the asymmetry will be on the left side and on others the deviation will be on the right side. In this case there would be no systematic difference in the population, and this would not generate a latent cause that is systematic to either left or right horns. The causes of the asymmetry would simply be subsumed into the unique error variances.

higher given the large positive multivariate kurtosis of the data (Mardia's coefficient of multivariate kurtosis is 27.1); this point will be further discussed in the section on non-normality of the data. Therefore, I conclude that there is not sufficient evidence to reject it.

The path coefficients from the latent 'diameter' to the two diameter measures were 0.522 and 0.519 for the left and right horns, respectively. Since the square of the diameter of a circle is proportional to its area (a dimension of 2), one would expect these path coefficients to be 0.5. The path coefficients from the latent 'length' to the two length measures were 0.875 and 0.869, respectively, with standard errors of about 0.03 for the left and right horns. Since length has a dimension of 1, one would expect these path coefficients to be 1. An approximate 95 per cent confidence interval of the length path coefficients is therefore about $0.87 \pm 2(0.03)$ and 1.0 is clearly outside this interval. Therefore, the measurement model suggests that the lengths of the longer horns were systematically underestimated. The covariance (and correlation, since their variances were fixed at unity) between the two latents for 'length' and 'diameter' was 0.9995. If we accept this two-factor measurement model then all these points suggest that the horn dimensions are caused by a single latent 'size' factor, but that there was a systematic bias in measuring the longer horns that introduced a second latent cause of the lengths independent of the diameters.

Now apply your personal 'squirm' test. Does my interpretation of these latent variables seem reasonable to you? The acceptable fit of the measurement model with two latents says nothing about *what* these two latent variables represent. The explanation that I have outlined, that the two latents represent the systematic errors made in measuring lengths and diameters and the covariance between the latents is due to the common 'size' factor, is an *interpretation* of the latent variables. This interpretation of the latent variables in the model is not supported by any statistical evidence; rather, my evidence comes from how the variables were measured and the sorts of errors of measurement that might occur. The next step would be to search for an independent confirmation of this explanation. For instance, if the explanation is correct then the two latent variables should disappear and be replaced by a single latent variable once we measure horn length in a way that does not systematically underestimate longer horns. One way would be to photograph the horns and then measure the lengths using image analysis.

Davis (1993) describes a way of testing for identification in much more complicated measurement models, applicable to any measurement model in which each observed indicator variable is caused by only one latent variable. This method (the FC1 rule) requires that the scale of each latent be fixed by fixing one path coefficient to 1 rather than fixing the scale by fixing the variance of the latent variable to 1. Box 6.1 summarises the FC1 rule.

6.3   Structural under-identification with structural models

Obtaining identification of the measurement model is necessary to fit a structural equation model. However, SEM also includes the causal relationships between the latent

Box 6.1 The FC1 rule

(1) For each latent variable, L_i, in the measurement model, construct a square binary matrix $\mathbf{P}_i$ with q_i rows and columns; here q_i is the number of observed indicator variables of L_i. Each element (p_{ij}) of $\mathbf{P}_i$ has a 1 if the error variables of indicators i and j are d-separated or if the covariance between them has been fixed.

(2) Form the matrix $\mathbf{D}_i^1 = \mathbf{P}_i \cdot \mathbf{P}_i'$. Iteratively multiply $\mathbf{D}_i^{j+1} = \mathbf{P}_i \cdot \mathbf{D}_i^j$ until you get the matrix $\mathbf{D}_i^{q_i-1} = \mathbf{P}_i \cdot \mathbf{D}_i^{q_i-2}$.

(3) The first requirement for identification is that every element of $\mathbf{D}_i^{q_i-1}$ be non-zero in the row corresponding to the indicator that defines its scale. This must be true for all latent variables.

(4) The second requirement for identification of the full measurement model is that, for every pair of latent variables whose covariance is to be estimated (i.e. that are not d-separated or whose covariance is not fixed), there must be at least one pair of indicator variables (one for each latent) whose error variables are zero (i.e. d-separated) or fixed.

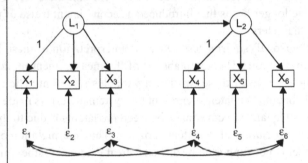

Figure 6.3 A structural equation model used to illustrate the FC1 rule for identification

I will show how this rule works with reference to a structural equation model. In this model (Figure 6.3) there are two latent variables, and so we need two $\mathbf{P}$ matrices:

$$\mathbf{P}_i = \begin{bmatrix} 0 & 1 & 0 & 0 & 1 & 1 \\ 1 & 0 & 1 & 1 & 1 & 1 \\ 0 & 1 & 0 & 1 & 1 & 0 \end{bmatrix} \quad \mathbf{P}_2 = \begin{bmatrix} 0 & 1 & 1 & 0 & 1 & 1 \\ 1 & 1 & 1 & 1 & 0 & 1 \\ 1 & 1 & 0 & 0 & 1 & 1 \end{bmatrix}$$

Note that $p_{12} = p_{21} = 1$ because ε_1 has a fixed (zero) covariance with ε_2, and similarly for ε_4 and ε_5 in $\mathbf{P}_2$. There are three indicator variables for each latent, and so $q_1 = q_2 = 3$. We must form $\mathbf{D}_1^{3-1}$ and $\mathbf{D}_2^{3-1}$ – i.e. $(\mathbf{P}_1 \cdot \mathbf{P}_1') \cdot \mathbf{P}$ and $(\mathbf{P}_2 \cdot \mathbf{P}_2') \cdot \mathbf{P}_2$. These are

$$\mathbf{D}_1^2 = \begin{bmatrix} 2 & 5 & 2 & 4 & 7 & 5 \\ 5 & 4 & 5 & 7 & 9 & 7 \\ 2 & 5 & 2 & 5 & 7 & 4 \end{bmatrix} \quad \mathbf{D}_2^2 = \begin{bmatrix} 4 & 7 & 5 & 2 & 5 & 2 \\ 7 & 9 & 7 & 5 & 4 & 5 \\ 5 & 7 & 4 & 2 & 5 & 2 \end{bmatrix}$$

Now, the first requirement is that every element in the row of each matrix representing the scaling variable must be non-zero. The scaling variable of the first latent variable is X_1, and so every element in row 1 of the first matrix must be non-zero.

> The first part of this first requirement is fulfilled. The scaling variable of the second latent variable is X_4, and so every element in row 1 of the second matrix must be non-zero. The second part of this first requirement is also fulfilled.
>
> The second requirement is that there be at least one pair of error variables (one from each of the two latents) whose covariance is fixed (in this case to zero). Error variables ε_2 and ε_5 fulfil this second requirement. Therefore, the model is structurally identified.

variables. In fact, you can think of the structural model as the 'path' model that is imbedded in the full model. A path model is therefore also a structural model. The rules for ensuring structural identification that I will describe come from Rigdon (1995). Rigdon's rules do not apply to models in which there are cyclic relationships involving more than two variables (for example, if X causes Y causes Z causes X). On the other hand, these rules are both necessary and sufficient for acyclic or block-acyclic structural models (a 'block-acyclic' model is defined below). This means that any acyclic or block-acyclic structural model that satisfies these rules is guaranteed to be structurally identified and any such structural model that does not satisfy these rules is guaranteed to be non-identified.

The first step is to conceptually divide the structural model into segmented blocks. The model is fully segmented when (a) there are no cyclic relationships between the blocks and (b) each block contains the minimum number of variables needed to satisfy (a). In other words, if a set of variables in the model do not have a cyclic relationship then each variable defines a separate block. If, on the other hand, a set of variables does define a cyclic relationship (for example, A causes B causes C causes A) then they must be included in the same block. If, once this has been done, there are more than two variables in any block then the identification status of the model can't be determined. If this is not the case then the identification status of the whole structural model can be determined by verifying the identification status of each block. If each block is identified then the whole structural model is also identified.[4] In evaluating these blocks, we don't need to consider the exogenous variables. Figure 6.4 illustrates these points.

In Figure 6.4(a) there are seven variables. Since variables x_3 and x_4 have a reciprocal (cyclic) relationship, they must be included in the same block. Since variables x_5 and x_6 have correlated errors, they too must be included in the same block. Variables x_1 and x_2 also have correlated errors and form a block, but they are exogenous in this model and so we don't have to worry about them. Finally, x_7 is in a block all by itself. Therefore, Figure 6.4(a) can be decomposed into four blocks, the causal relationships between the blocks have no cyclic patterns, and the model fulfils the requirements for Rigdon's test.

In Figure 6.4(b) there are three variables (x_1, x_2 and x_3) that possess a feedback relationship. Therefore, all three variables must be included in a single block. The last variable, x_4, forms a second block. Because there are more than two variables in one of the blocks, we cannot determine the identification status of this model using Rigdon's rules.

---

[4] Such a model is called a *block-recursive* model.

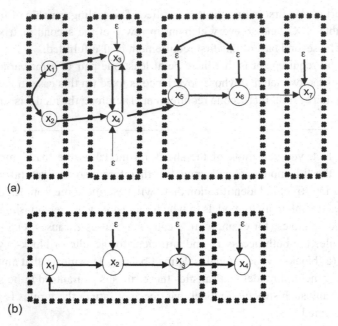

(a)

(b)

Figure 6.4 An illustration of Rigdon's rules for structural identification with cyclic models

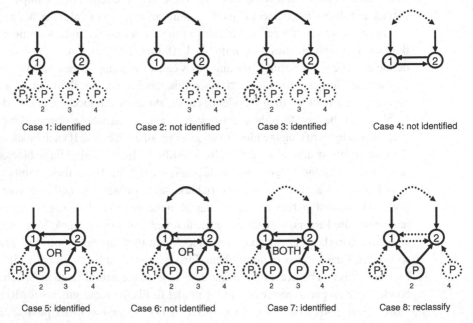

Case 1: identified   Case 2: not identified   Case 3: identified   Case 4: not identified

Case 5: identified   Case 6: not identified   Case 7: identified   Case 8: reclassify

Figure 6.5 Eight different cases used to evaluate Rigdon's rules for structural identification

Once the structural model has been reduced to these blocks, you simply have to determine the identification status of each block. To do this, refer to Figure 6.5, which shows eight different patterns. To interpret these diagrams, you will need some notational conventions. The two variables indicated as '1' and '2' are the two variables in the block (if there is only one variable in the block then it is automatically identified). The variables indicated as 'P' are causal parents. If an arrow and a circle are shown with solid lines then the two 'P' variables must be present. If an arrow and circle are shown with broken lines then the two 'P variables can be present but their existence is irrelevant to determining the identification status of the block. Finally, 'OR' means that at least one of the two 'P' variables must be present, and 'BOTH' means that both 'P' variables must be present.

Now, look at each block of the structural equation that contains two variables in a cyclic relationship and classify it as belonging to one of the eight cases in Figure 6.5. If any of these blocks are not identified then the model is also not identified. The only complication is with Case 8 in Figure 6.5. To determine the identification status of a block belonging to this case, ignore the common causal parent of 1 and 2 and then see which of the other seven cases corresponds to the block while ignoring the common causal parent.

6.4   Representing composite variables using latents

So far I have presented latent variables only in the context of measurement models. In other words, the latent variable is viewed as the unobserved cause and the observed variables are viewed as observed effects of the latent. Such latent variables are sometimes called 'reflective' latents. This is the way that latent variables are overwhelmingly represented in structural equation modelling, and readers of this literature would be forgiven for thinking that this is the only way to model the relationship between latents and observed variables. Not so.

The association between latent variables and measurement models in SEM is probably due to the fact that this statistical methodology was primarily used and developed by researchers in the social sciences. The observed data in these sciences are often responses to survey questions. In other words, one hypothesises that people have certain unobservable psychological or sociological propensities and that these *latent* propensities cause them to express certain observable choices when presented options on survey questionnaires. These observable choices are consequences of the unobservable propensities. However, the causal structure (latent→observable) is imposed by nature, not by the statistical method. In some situations that are not uncommon in biology the latent variable of interest can be a consequence of observed variables ('formative' latent variables). Grace and Bollen (2008) call these 'composite latents' and describe how to fit and interpret them.

For example, imagine that an ecologist wants to link a series of soil properties related to soil fertility and the point along a gradient of soil fertility where different plant species reach their maximum abundance. She conceptualises 'soil fertility' as the combined effect of those soil properties that determine the level of net primary productivity (grams of biomass produced per square metre per day) that a soil can support at a given volumetric water content, temperature and irradiance level. Given this conception of soil fertility it would be very difficult (perhaps impossible in practice) to directly measure

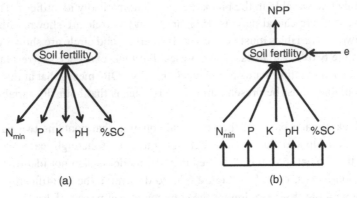

Figure 6.6 Two alternative conceptions of the notion of 'soil fertility'

it in the field. Instead, she measures a series of soil properties from the field samples (nitrogen mineralisation rate, available phosphorus and potassium concentration, pH, percentage of silt and clay), as well as the above-ground net biomass production over a single growing season. How should she structure her causal model? If she constructs a measurement model with these observed variables as indicators of the latent 'soil fertility' (Figure 6.6(a)) then she is implicitly claiming that there exists some causally efficacious property in soils ('soil fertility') that varies between samples and whose variation *causes* coordinated changes in all her observed variables. But this is not how she conceived her latent variable. Rather, the observed soil properties are supposed to jointly cause the 'fertility' level of the soil while the measured above-ground net primary productivity over a single growing season is an imperfect measure of this latent ability (Figure 6.6(b)). The measured above-ground net primary productivity is an imperfect measure because it does not include below-ground biomass production, as it is based on only a single season of growth, and because it reflects variation in properties of the environment (water availability, temperature and irradiance) that are not part of her concept of 'soil fertility'.

Grace and Bollen (2008) call the structure in Figure 6.6(b) an 'M→L' block, meaning a block of variables in which manifest (observed) variables jointly cause the latent variable. They call the classical measurement model shown in Figure 6.6(a) an 'L→M' block, in which the latent variable jointly causes the set of observed variables. The causal structure in Figure 6.6(b) is an M→L block because it has two properties: (a) a series of observed (M) variables jointly cause variation in the latent (L) and (b) there are other unobserved causes of the latent besides these observed variables that are represented by the error variance of the latent. The second property arises because the ecologist does not think that the chosen observed causes of 'soil fertility' are exhaustive. Finally, Grace and Bollen define an 'M→C' block as a causal structure in which a set of observed variables jointly causes a latent variable, but this set of observed variables is the only cause of the latent; because of this, the latent variable has no error variance. In other words, the latent is *defined* as the sum of the observed variables and is called a 'composite' latent by Grace and Bollen (2008). An example of a composite variable is the 'importance value' in plant community ecology, which is defined as the sum of (a) the proportion of the individuals of a species to the total number of individuals in the community, (b) the proportion of sampling points in which a species is observed and (c) the proportion of the total area or basal area occupied by the species. Because the 'importance value' is defined as the sum of these three variables, we can predict its value without error once we have the values of these three variables. Combinations of different blocks give rise to more complicated models. The message here – and it is an important message – is that the underlying causal hypothesis must structure the model. Do not think that you must always combine latents and observed variables in the form of a measurement (L→M) block just because most SEM models have done this in the past. A structural equation model is a flexible tool that you should structure to fit you causal hypothesis, not a preformed mould into which you must structure your causal hypothesis. However, every latent variable must still have at least one observed measurement variable. In other words, you can't have a latent that is caused by other variables but that doesn't cause anything in the model.

6.5   Behaviour of the maximum-likelihood chi-square statistic with small sample sizes

Some of my in-laws like to make home-made wine. A superficial glance at bottles of these 'wines' might convince you that they are the real thing. When you taste them you realise that they vary along a gradient from 'gut-rot' to 'drinkable' to 'divine'. As with latent variables, giving something a name doesn't make it so. The 'maximum-likelihood chi-square statistic' (MLX^2) is the statistical equivalent of home-made wine. It is not really distributed as a chi-squared variate at all and, unfortunately, its true sampling distribution is unknown. However, as the size of the sample of independent observations increases, the sampling distribution of this statistic becomes closer and closer to the theoretical chi-square distribution. At very small sample sizes the MLX^2 statistic is like gut-rot wine; it bears an approximate resemblance to the true χ^2 distribution but there is no confusing the two. At moderate sample sizes the MLX^2 is like drinkable home-made wine; it is a reasonable approximation of the real thing unless it is to be used for a special occasion. It is only when sample sizes are very large that one cannot distinguish between the two. So, how big is 'big enough', and what can be done if one's sample is not big enough? In this section I will discuss the effects of sample size on the MLX^2 assuming that the data follow a multivariate normal distribution. To explore these questions I will use simulations drawn from the path model shown in Figure 6.7, with all exogenous variables being drawn from a standard normal distribution (i.e. zero mean and unit standard deviation).

Figure 6.8 shows the empirical sampling distribution of the MLX^2 statistic, based on 1,000 independent data sets. I fixed all path coefficients to their theoretical values (0.5) and all the error variances to their theoretical values (1). This way, the only free parameters were the variances of X_1 and X_2, and the model covariance matrix could be determined without iteratively minimising the MLX^2. There were therefore 13 degrees of freedom, and the curve shown in Figure 6.8 is the theoretical χ^2 distribution with 13 degrees of freedom. Figure 6.8(a) shows the distribution of the MLX^2 statistic in the 1,000 data sets with 10 observations each. It is clear that this empirical distribution

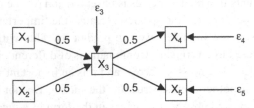

Figure 6.7 The path model used to simulate the data that are summarised in Figure 6.8.

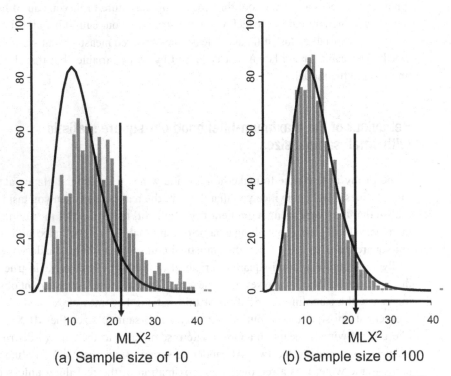

(a) Sample size of 10   (b) Sample size of 100

Figure 6.8 Empirical distributions of simulated data based on sample sizes of 10 or 100 observations per data set
Notes: The solid curve shows the theoretical chi-squared distribution and the arrow shows the 95 per cent quantile corresponding to the 5 per cent probability level.

is not well approximated by the theoretical χ^2 distribution; the 95 per cent quantile, corresponding to a 5 per cent significance level, is shown by the arrow. Figure 6.8(b) shows the distribution of the MLX^2 statistic in the 1,000 data sets with 100 observations each. The empirical 90 per cent, 95 per cent, 97.5 per cent and 99 per cent quantiles in the second simulation were 19.68, 21.94, 24.33 and 27.13, respectively, corresponding to theoretical probabilities of 0.103, 0.056, 0.028 and 0.012. Now, the empirical and theoretical distributions are quite close. In this case, if we assume that the MLX^2 is truly distributed as a χ^2 distribution, we will make only insignificant errors.

In general, small sample sizes result in conservative probability estimates. In other words, if you have a small data set then the true probability level will probably be larger

Table 6.1 Empirical quantiles and theoretical probability levels using the model shown in Figure 6.7

Sample size	50% quantile	90% quantile	95% quantile	97.5% quantile	99% quantile
10	16.18 (p = 0.24)	26.80 (p = 0.01)	30.58 (p = 0.004)	33.70 (p = 0.001)	38.65 (p = 0.0002)
30	13.55 (p = 0.41)	21.38 (p = 0.07)	24.72 (p = 0.025)	27.00 (p = 0.012)	29.81 (p = 0.005)
50	12.79 (p = 0.46)	20.48 (p = 0.08)	22.47 (p = 0.05)	24.43 (p = 0.027)	27.52 (p = 0.011)

Notes: Empirical quantiles from 1,000 independent data sets, with different numbers of observations ('Sample size'), are shown, along with the theoretical probability levels assuming a χ^2 distribution with 13 degrees of freedom.

than the value obtained when assuming a χ^2 distribution. In this case, if your model produces a MLX^2 value that is judged to be significant using the χ^2 distribution then you have an ambiguous result, and you will have to use a different method of estimating the true probability level. For instance, Table 6.1 shows the empirical quantiles and theoretical probability levels using the model shown in Figure 6.7. For this particular model a sample size of only 30 provides a passable estimate of the tail probabilities, but with somewhat conservative probability estimates and a sample size of 50 it is quite acceptable. In general, the more free parameters in the model that need to be estimated, the larger the sample size required. More complicated models may require sample sizes of 200 or more. One rule of thumb is that there should be at least five times more observations than free parameters (Bentler 1995).

What can be done if your sample size is too small to confidently assume that the sampling distribution of the MLX^2 statistic is close to the theoretical χ^2 distribution? If there are no latent variables in your model then you can use the d-sep method described in Chapter 3. If there are latent variables then you will need another way. One way is to use bootstrap methods; since this method is also useful in cases in which the variables have other distributional problems, the bootstrap will be described later. Another way around the problem of small sample sizes (but not non-normal distributions in general) is to use Monte Carlo methods, as used in the simulations reported in Table 6.1.

The first step is to fit your model using the sem() function, then obtain the MLX^2 statistic (call it X), the degrees of freedom (df) and the parameter estimates of the free parameters. You can get all this information from the summary() function or from the inspect() function. The inspect() function of lavaan extracts all the important information in the sem object and `inspect(myFit,"fit")` extracts all the information related to model fit, where 'myFit' is the name of the object that contains the output from the sem() function. By typing `inspect(myFit,"fit")[2]` you will extract only the chi-square statistic.

Next, write a script in R that reproduces your structural equations using the parameter estimates from your fitted data and use the `rnorm(n,mean = 0,sd)` function to generate random normal values for each error variance, where 'n' is the size of your actual data set, 'mean' is the mean of the random variable (which will always be zero for error variables) and 'sd' is the standard deviation of the random variable as output from your model; note that this function demands the standard deviation (i.e. the square root of the residual variance), not the variance, which is what is returned from the summary() function. Simulate a large number N (say 1,000) data sets, each with the sample size (n) of your original data. Next fit each simulated data set to your original model with the same pattern of free and fixed parameters and save the calculated MLX^2 values of each run. Finally, count the number (x) of these simulated MLX^2 values that are greater than the value (X) obtained in your original data. The proportion x/N will estimate the null probability (p) of observing the data, given your model. Since these simulated data sets are mutually independent and large in number you can also obtain a 95 per cent confidence interval (Manly 1997) around p by referring to a normal distribution whose mean is $(x - Np)$ and whose standard variation is $Np(1 - p)$. Thus, the 95 per cent confidence interval is $p \pm 1.96Np(1 - p)$.

An advantage of the above simulation is that, by replacing the normal random number function (rnorm) by a random number generator following a different probability density,[5] you can simulate non-normal data. A disadvantage is that this can take rather a long time. There is a much faster way to estimate the sampling distribution of the maximum-likelihood chi-square statistic for small sample sizes, although this next method is limited to multivariate normal data. Since the MLX^2 statistic requires that we calculate the determinant, and the inverse, of the model covariance matrix it is useful to choose a matrix for which this can be easily done. The determinant of a square matrix whose non-zero values are all on the diagonal is simply the product of these diagonal values. Similarly, the inverse of such a diagonal matrix is simply a diagonal matrix whose diagonal values are the inverse of the original matrix. We therefore simulate data from a model consisting of v mutually independent variables, each of which is drawn from a standard normal distribution. The predicted covariance matrix, Σ, of such a model has non-zero values only on its diagonal. There are $v(v + 1)/2$ non-redundant elements. If we estimate the variance of q of the v variables, which will be on the diagonal of Σ, then there will be $v(v + 1)/2-q$ degrees of freedom. So, here are the steps needed to estimate an empirical probability level for a MLX^2 statistic of X.

(1) Given a desired degrees of freedom (df), find the smallest integer value of v such that

$$v \geq \frac{-1 + \sqrt{1 + 8df}}{2}$$

For example, if df = 9 then we need the smallest integer value of v such that

$$v \geq \frac{-1 + \sqrt{1 + 8(9)}}{2} = 3.8$$

Thus, v = 4. Find the integer value of c such that

$$c = \frac{v(v + 1)}{2} - df$$

So, if df = 9 and v = 4 then c = 1.

[5] Many different probability densities and distributions are in the stats package. Search for 'Distributions {stats}'.

(2) Construct a model covariance matrix Σ with v rows and columns. Leave the first c diagonal elements blank; these will be filled by the sample estimates of the variances of the first c of the mutually independent normally distributed random variables in step 4. Define all other diagonal elements (the remaining variances) to be 1 and all non-diagonal elements to be 0. This is the population covariance matrix of v mutually independent standard normal variates, of which the variances of the first c of these variables have been estimated from the data. This model covariance matrix will have df degrees of freedom. For instance, if your actual data set has n = 30 observations and you have calculated v = 4 in step 1, you would generate four mutually independent vectors of 30 standard normally distributed random numbers using 'rnorm(n = 30, mean = 0, sd = 1)', estimate the variance of the first vector and place this estimate in the first element of the diagonal matrix. The remaining three diagonal elements of this matrix would be equal to 1.

(3) Now, repeat step 2 a large number of times (say N = 1000). Each time, besides performing step 3, which results in a matrix Σ, create a second matrix (the sample covariance matrix, S), which is also a diagonal matrix whose diagonal elements contain the sample variances of the four vectors of normally distributed random numbers. Now calculate the maximum-likelihood chi-square statistic $MLX_i^2 = (n-1)(Ln|\Sigma| + tr(S_i\Sigma^{-1}) - Ln|S_i| - c)$.

(4) Count the number (x) of the N MLX^2 values that are greater than the value of the MLX^2 value obtained in your real data (X).

(5) The estimated empirical probability of your data will be x/N and the 95 per cent confidence interval of this estimate can be calculated as described before.

6.6 Behaviour of the maximum-likelihood chi-square statistic with data that do not follow a multivariate normal distribution

A biologist, a physicist and a statistician are shipwrecked on a deserted island. Besides themselves, only a crate of canned food has been washed ashore. After staring hungrily at the cans for a number of hours, the biologist suggests that they break open some cans with a large rock. The physicist considers this solution to be inelegant and suggests instead that they climb to just the right height in a palm tree. She explains that the kinetic energy, as the can hits the ground, should be just enough to crack it open without losing any food. Glancing over to the statistician, who has just finished writing some equations in the sand, they see him shaking his head in disapproval at their inexact methods. He announces that he has just found an exact solution and points proudly at his equations. 'Now,' he begins, pointing to the first equation, 'assume that we have a can opener . . . '

Sometimes we don't have the statistical equivalent of a can opener. Knowing the assumptions of a statistical test is important but knowing what might happen if the assumptions are wrong can be just as important. Another assumption of the maximum-likelihood chi-square statistic is that the data follow a multivariate normal distribution.[6]

---

[6] Actually, the assumption is that the endogenous variables follow a multivariate normal distribution. Exogenous variables (i.e. ones that are not caused by any others in the model) don't have this restriction (Bollen 1989).

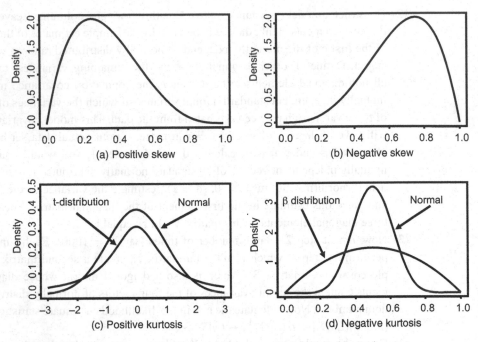

Figure 6.9 Examples of curves showing positive or negative skew and kurtosis

We require methods of both testing, and relaxing, this assumption. First, let's look at how to test for departures from multivariate normality.

The normal distribution is fully characterised by its mean and variance. Departures from normality can be characterised by non-zero skew and kurtosis. The skew measures the degree of asymmetry of the distribution. A negative skew occurs when a univariate distribution has a longer tail to the left and whose mode is to the right of centre. A positive skew occurs when a univariate distribution has a longer tail to the right and whose mode is to the left of centre (Figure 6.9). An index of skew for a series of n observations of a random vector x is

$$g_{1,1} = \frac{\sum_{i=1}^{n}(x_i - \bar{x})^3}{ns^3}$$

where s is the standard deviation of x. Box 6.2 summarises the calculations for using this statistic to test for skew in large (>150) sample sizes. D'Agostino, Belanger and D'Agostino (1990) provide more exact formulae that can be used for sample sizes as low as eight.

Kurtosis measures the concentration of values near the mean and at the extremes, relative to intermediate values. For symmetrical unimodal distributions, positive kurtosis indicates heavy tails and peakedness relative to the normal distribution and negative kurtosis indicates light tails and flatness (DeCarlo 1997). A familiar distribution with positive kurtosis is Student's t-distribution. With this distribution the kurtosis increases as the degrees of freedom decrease; Figure 6.9(c) shows a t-distribution with three degrees

---

Box 6.2 Measures of skew and kurtosis

Univariate skew: the expected value of $g_{1,1}$ for a normal distribution is 0. The following statistic is approximately distributed as a standard normal variate for large $(N > 149)$ sample sizes, and values greater than 1.96 in absolute value would indicate skew:

$$z = g_{1,1} \sqrt{\frac{(N+1)(N+3)}{6(N-2)}}$$

For tests applicable to small samples, the reader is directed to D'Agostino, Belanger and D'Agostino (1990) or Bollen (1989).

Univariate kurtosis: the expected value of $g_{2,1}$ is 3 for normally distributed variables. For this reason, many computer programs often report the centred version of the $g_{2,1}$ statistic $(g_{2,1} - 3)$ even though this is not always well documented. The following statistic follows a standard normal distribution only in very large samples $(N > 1500)$, but at least it provides a rough guide (a more complicated statistic, applicable to small sample sizes $(N < 19)$, is given by D'Agostino, Belanger and D'Agostino 1990 and Bollen 1989):

$$E(g_{2,1}) = \frac{3(N-1)}{(N+1)}$$

$$Var(g_{2,1}) = \frac{24N(N-2)(N-3)}{(N+1)^2(N+3)(N+5)}$$

$$z = \frac{(g_{2,1} - E(g_{2,1}))}{\sqrt{Var(g_{2,1})}}$$

The 'moments' package in R does these calculations. In using these tests on all variables in your model, you should use a Bonferonni correction to the significance levels. If you want to test at an overall level of α then test each of the v variables at a level of α/v. For instance, if you want to test at a 95 per cent level $(\alpha = 0.05)$ then test each variable at a level of 0.01 (i.e. 0.05/v).

Multivariate measures of skew and kurtosis: the above measures of skew and kurtosis are applied separately to each variable. Since it is possible for the joint distribution to have skew or kurtosis even though each individual variable shows no evidence of this, Mardia (1970; 1974) has developed multivariate analogues of these statistics. These statistics are based on a matrix of squared Mahalanobis distances. For a single variable (x_i) with n observations, the squared Mahalanobis distance is simply

$$\frac{1}{\sigma_i^2} \sum_{j=1}^{n} (x_{ij} - \bar{x})^2$$

If each of the j observations consists of a series of v variables then the resulting data set, **X**, has n rows and v columns. The squared Mahalanobis distance matrix for the entire data set is **X'SX**, where **S** is the covariance matrix of **X**. Looking at the Mahalanobis distance for each observation helps to identify outliers in the multivariate space.

Based on the squared Mahalanobis distance, Mardia's multivariate measure of skew with v variables is

$$g_{1,v} = \left(\frac{1}{n^2}\right) \sum_{i=1}^{n} \sum_{j=1}^{n} (X'SX)^3$$

If the data follow a multivariate normal distribution then the expected value of this statistic is 0. The statistic

$$\frac{n \cdot g_{1,v}}{6}$$

asymptotically follows a chi-squared distribution with v(v+1)(v+2)/6 degrees of freedom if the data are multivariate normal. Mardia's multivariate measure of kurtosis is

$$g_{2,v} = \left(\frac{1}{n}\right) \sum trace((X'SX)^2)$$

where 'trace' means the diagonal elements. If the data follow a multivariate normal distribution then the expected value of $g_{2,v}$ is v(v+2) and the variance is 8v(v+2)/n. The statistic

$$\frac{(g_{2,v} - v(v+2))}{\sqrt{8v(v+2)/n}}$$

asymptotically follows a standard normal distribution. Bollen (1989) provides more complicated test statistics that are applicable to small data sets.

Note: The 'MVN' package in R 'MultiVariate Normal' calculates Mardia's statistic.

of freedom as well as a standard normal distribution. An index of kurtosis is

$$g_{2,1} = \frac{\sum_{i=1}^{n}(x_i - \bar{x})^4}{ns^4}$$

Box 6.2 summarises the steps for identifying significant deviations from normality with respect to kurtosis.

If any of the variables in your model have significant skew or kurtosis then the joint multivariate distribution also has significant skew or kurtosis. However, it is possible for the multivariate distribution to have either skew or kurtosis even though each variable, taken singly, is normally distributed. This means that we also require a multivariate version of our measures of skew and kurtosis. Such measures, along with their tests, are given by Mardia (1970; 1974). The calculations are explained in Box 6.2.

Most biologically oriented statistics texts describe the Box–Cox method of choosing a transformation to make data more closely follow a normal distribution. This is because statistical tests involving means (t-tests, ANOVA, etc.) are more sensitive to skew, and the Box–Cox method helps to reduce skewness in data. However, tests involving variances and covariances, such as those used in SEM, are more sensitive to kurtosis

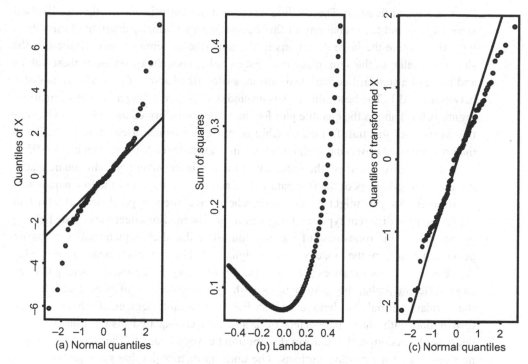

Figure 6.10 Transformations for kurtosis

Notes: Panel (a) shows a normal quantile plot for 100 values drawn from a t-distribution with there degrees of freedom. Panel (b) shows the value of the sum of squared distances between the quantiles of these values and those of a standard normal distribution for varoius values of λ using Jobson's transformation. Panel (c) shows the normal quantile plot after transforming the data using the best value of λ.

than to skew (Mardia, Kent and Bibby 1979; Jobson 1992). Jobson (1992) describes a modified power transformation that is designed to reduce kurtosis:

$$y = SIGN \frac{(|x - x_M| + 1)^\lambda - 1}{\lambda} \quad \lambda \neq 0$$
$$y = SIGN \ln(|x - x_M| + 1) \quad \lambda = 0$$

SIGN is the sign of the original value of $(x - x_M)$ and X_M is the median value of X.

To find the value of λ that reduces the kurtosis best, I calculate the sum of squared differences between a series of quantiles of y (say, 5 per cent, 10 per cent, 50 per cent, 90 per cent and 95 per cent) for different values of λ and the quantiles of a normal distribution with the mean and standard deviation[7] of y. The value of λ that minimises this sum of squared differences will best reduce the kurtosis of the original values. Most statistics packages allow one to plot the empirical quantiles (cumulative per cent) against these theoretical quantiles. Using these graphs, you can try different values of λ and choose the one in which the resulting graph looks most like a straight line. Figure 6.10(a) shows a quantile plot for 100 values drawn from a t-distribution with

---

[7] Alternatively, you can standardise your variable first and then refer to the quantiles of a standard normal distribution. These quantiles can be found in most tables of the standard normal distribution.

three degrees of freedom. These values have a centred kurtosis of 4.12, the standardised value is 9.06 and the probability of this occurring in a normally distributed variable is 6.6×10^{-5}. Notice the large deviations in the tails (the extreme values). Figure 6.10(b) shows the value of the sum of squared distances between the quantiles of these values and those of a standard normal distribution, as described above, for various values of λ between 0 and 1. The best value of λ is around 0, thus demanding a Ln-transformation. Figure 6.10(c) shows the quantile plot for the values transformed using $\lambda = 0$. The centred kurtosis of this transformed variable is −0.10, the standardised value is −0.23 and the probability of observing such a kurtosis in a normally distributed variable is 0.95.

Non-normality can affect the accuracy of the maximum-likelihood chi-square statistic, the standard errors of the free parameters and the estimation of the free parameters themselves. As you might expect, researchers have spent a good deal of effort in exploring how different types and degrees of non-normality affect these statistics. To get an idea of the robustness of the maximum-likelihood chi-square statistic, I again generated data from the model shown in Figure 6.7. The path coefficients were fixed at 0.5. The exogenous variances (i.e. the variances of x_1, x_2, e_3, e_4 and e_5) were generated from different probability distributions with different degrees of skew and kurtosis: the standard normal distribution, a t-distribution with three degrees of freedom, a beta distribution with shape parameters of 2 and 5, a chi-squared distribution with two degrees of freedom and a uniform distribution between 0 and 1. The t-distribution has no skew but strong positive kurtosis. The uniform distribution has no skew but strong negative kurtosis. The chi-squared distribution with two degrees of freedom has both strong positive skew and kurtosis. The beta distribution with shape parameters of 2,5 has positive skew and negative kurtosis. For each distributional type I simulated 1,000 independent data sets of either 50, 100 or 200 observations. Average values of Mardia's multivariate estimates of skew and centred kurtosis for these data were also estimated. The results are shown in Table 6.2.

The first thing to notice is that distributions with strong kurtosis produce conservative probability levels; there is a tendency for models to be rejected more often than they should be when assuming a theoretical χ^2 distribution. This is the same result as we saw for small sample sizes. The second thing to notice is that even models generated from very non-normal distributions produce quite acceptable probability levels as sample sizes increase. This shows the asymptotic robustness of the maximum-likelihood chi-square test statistic. If the errors are *independently* distributed of their non-descendants in the model, the test statistic should asymptotically follow a chi-square distribution. The robustness conditions hold under independence and not necessarily under 'uncorrelatedness';[8] for example, if the variance of the error variable changes systematically with respect to any of its causal parents then this would undermine independence.[9] The robustness of the maximum-likelihood chi-square statistic depends on many different attributes of the model and the data: the number of free parameters, the distributional properties of each variable and (especially) the non-independence of the error variables with respect to their non-descendants in the model.[10]

---

[8] I am sorry for using such an ugly word.
[9] This is similar to heteroscedastic error variances in ordinary regression.
[10] Except, of course, when the non-independence is explicitly modelled.

Table 6.2 Consequences of non-normality on the ML chi-square statistic

N	Type	Quantiles (theoretical probability)				Kurtosis
		50	90	95	97.5	
50	Normal	10.00	16.58	18.73	21.17	−1.26
		(0.440)	(0.084)	(0.044)	(0.020)	
50	t(3df)	9.72	17.87	20.89	24.81	17.83
		(0.465)	(0.057)	(0.022)	(0.006)	
50	Beta(2,5)	9.92	17.00	19.20	21.23	−1.51
		(0.448)	(0.074)	(0.038)	(0.020)	
50	χ^2(2 df)	9.91	17.24	20.11	23.41	11.98
		(0.448)	(0.069)	(0.028)	(0.009)	
50	Uniform	9.48	16.59	19.92	22.04	−5.38
		(0.487)	(0.084)	(0.030)	(0.015)	
100	Normal	9.45	15.87	18.34	20.76	−0.70
		(0.490)	(0.103)	(0.050)	(0.023)	
100	t(3df)	9.25	16.98	19.48	22.16	34.37
		(0.509)	(0.075)	(0.035)	(0.014)	
100	Beta(2,5)	9.68	16.43	19.10	20.79	−1.13
		(0.469)	(0.089)	(0.039)	(0.023)	
100	χ^2(2 df)	9.30	17.09	19.48	22.09	17.86
		(0.504)	(0.072)	(0.035)	(0.015)	
100	Uniform	9.24	16.35	18.53	20.09	−5.68
		(0.509)	(0.090)	(0.047)	(0.028)	
200	Normal	9.38	16.19	18.50	20.67	−0.17
		(0.497)	(0.094)	(0.047)	(0.024)	
200	t(3df)	9.25	16.45	19.61	22.74	64.24
		(0.508)	(0.087)	(0.033)	(0.012)	
200	Beta(2,5)	9.46	16.42	18.44	20.77	−0.76
		(0.489)	(0.088)	(0.048)	(0.023)	
200	χ^2(2 df)	9.64	16.67	18.37	22.04	23.51
		(0.472)	(0.082)	(0.049)	(0.015)	
95% confidence intervals for probabilities		0.531 to 0.479	0.119 to 0.081	0.064 to 0.036	0.035 to 0.015	

Notes: These are simulation results (1,000 data sets each) based on sample sizes of N with exogenous variables drawn from a standard normal, a t-distribution with three degrees of freedom, a beta distribution with shape parameters of 2 and 5, a uniform distribution between 0 and 1, and a χ^2 distribution with two degrees of freedom. Shown are the 50 per cent, 90 per cent, 95 per cent and 97.5 per cent quantiles of the 1,000 maximum-likelihood statistics for each simulation, as well as the theoretical probabilities assuming a χ^2 distribution with 10 degrees of freedom. The average of Mardia's multivariate centred estimate of kurtosis is also shown.

6.7 Solutions for modelling non-normally distributed variables

Since non-normality can cause problems with the maximum-likelihood chi-square statistic, a number of alternative ways of fitting the model have been devised. Different ways of estimation methods in lavaan are specified via the 'estimator' argument in the sem() function. Besides the maximum-likelihood estimator (estimator = "ML"), which is the default method, the lavaan package has five different estimators. Most commercial SEM programs will include statistics based on generalised least squares, elliptical estimators and distribution-free estimators, as well as a method of correcting for non-normality that produces 'robust' chi-square statistics and confidence intervals.[11] The most popular, and best studied, method that produces estimates that are robust to deviations from non-normality comes from Satorra and Bentler (1988) and is obtained in lavaan by setting the estimator argument of sem() to 'MLM' (i.e. estimator = "MLM").

There now exists an extensive literature that uses Monte Carlo methods to explore the relative merits of these different solutions for non-normality. Different studies have explored the effects of sample size, the number of free parameters, the model type (measurement models, path models, full structural models) and distributional violations (kurtosis, skew and non-independence of errors and their causal non-descendants). Hoogland and Boomstra (1998) have done a meta-analysis of these studies. Their main findings are the following.

(1) With respect to sample size, they recommend that there be at least five times as many observations as there are degrees of freedom in the model.

(2) When the observed variables have an average positive kurtosis of 5 or more, the sample size may have to be increased up to 10 times the degrees of freedom.

(3) The generalised least-squares chi-square statistic has an acceptable performance for a sample size that is two times smaller than the sample size needed for an acceptable performance of the maximum-likelihood chi-square statistic. This estimator is obtained in lavaan via estimator = "GLS".

(4) With small samples the standard errors of the estimates of the free parameters are biased. Positive kurtosis results in estimates of the standard errors that are smaller than they should be. Negative kurtosis results in estimates of the standard errors that are larger than they should be.

(5) The degree of skew has little effect on the bias of the estimators.

(6) The asymptotic distribution-free estimator should not be used except for very large sample sizes (> 1,000).

(7) The Satorra–Bentler robust estimator, upon which is based their robust (S-B) chi-square statistic and standard errors, largely corrects for excessive kurtosis and for problems in which the errors are not independent of their causal non-descendants. This is particularly important for models that include latent variables and measurement models, since the S-B chi-square statistic can correct for cases in which the latent variables and the measurement errors are not independent.

Basically, unless your data are very strongly kurtotic and your sample sizes are very low, you can still perform a reasonable test of your causal model. As a last resort, you can use bootstrap methods (Bollen and Stine 1993). The bootstrap is related to Monte Carlo methods except that, rather than sampling from some theoretical

---

[11] Another correction method is given by Browne (1984). This consists of dividing the ML chi-square statistic by the ratio of Mardia's multivariate measure of kurtosis to its expected value given normality. Little simulation work seems to have been done on this correction.

> **Box 6.3** Steps required to generate a bootstrap distribution of the maximum-likelihood chi-square statistic
>
> Here are the steps to take in order to generate a bootstrap sampling distribution and perform an inferential test.
>
> (1) Given your original data set (Y), with N rows and p variables centred about their means, calculate the sample covariance matrix (S) and obtain the predicted model covariance matrix (Σ) and the maximum-likelihood chi-square statistic, χ^2.
>
> (2) Calculate the Cholesky factorisation of S and Σ to give $S^{-1/2}$ and $\Sigma^{1/2}$.
>
> (3) Form a new data set: $Z = YS^{-1/2}\Sigma^{1/2}$.
>
> (4) Randomly choose N observations from Z with replacement to form a bootstrap sample Z^*. Form the covariance matrix from this bootstrap sample, fit the model to these data and save the bootstrap value of the maximum-likelihood chi-square statistic (χ^{2*}).
>
> (5) Repeat step 4 a large number of times (at least 1,000).
>
> (6) Count the proportion of times that χ^{2*} is greater than χ^2. This proportion is the empirical estimate of the probability of observing the data given the model. Note that this probability does not assume any particular sampling distribution.

distribution (multivariate normal or otherwise), you sample from your own data to build up an empirical sampling distribution (see Manly 1997 for a discussion of bootstrap methods in biology). Box 6.3 summarises the steps required to generate a bootstrap distribution of the maximum-likelihood chi-square statistic. Bootstrapped estimates are obtained in lavaan in one of two ways. While fitting the model in the sem() function you can get bootstrapped standard errors of your free parameters by specifying the argument se = "boot" and you can get a bootstrapped null probability for the chi-square statistic by specifying the argument test = "boot". For instance, using our strings example again, to get these bootstrapped values we would specify

```
sem(model=string.length,data=strings,std.lv=TRUE,se="boot",
test="boot")
```

6.8   Alternative measures of 'approximate' fit

This next section deals with various methods of assessing the degree of 'approximate' fit between data and a theoretical model. I don't like these methods, and don't advise you to use them either, for reasons that I will explain below. However, they are popular with many users of SEM and are produced in lavaan. These measures of approximate fit are generally used once the model has already been rejected, and their purpose is to determine the degree to which the rejected model is 'approximately' correct.

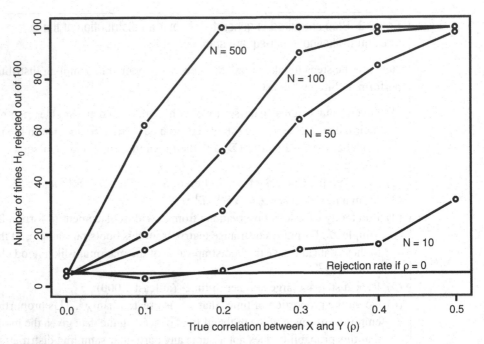

Figure 6.11 The proportion of the 100 data sets of various sample sizes (N) for which the null hypothesis ($\rho = 0$) was rejected at $\alpha = 0.05$ when the true population value of the correlation coefficient took various values between $\rho = 0$ and $\rho = 0.5$

The origin and rationale behind the use of these approximate fit indices come from a consideration of statistical power. The power of a statistical test can be defined as the probability that the test will reject the null hypothesis when it is indeed false. To illustrate this notion, imagine that we wish to test the null hypothesis that two random variables, X and Y, are uncorrelated ($H_0:\rho = 0$). I generated 100 independent data sets each with 10, 50, 100 or 500 observations in which the true population correlation coefficient was either 0, 0.1, 0.2, 0.3, 0.4 or 0.5. We know that, if H_0 is true, we should reject about one out of 20 tests at the $\alpha = 0.05$ level. If we had perfect statistical power then we should also reject all data sets in which ρ is different from 0. In other words, we should reject a proportion α when the null hypothesis is correct and proportion 1 whenever ρ deviates, however slightly, from 0. Figure 6.11 shows the actual proportion of the 100 data sets for which the null hypothesis ($\rho = 0$) was rejected at $\alpha = 0.05$.

In Figure 6.11 we see that when the sample size is very small (N = 10) then, even when the null hypothesis is false (i.e. when the true correlation between X and Y is not zero), the null hypothesis won't be rejected a large proportion of the time; even when $\rho = 0.5$ only 33 out of 100 tests[12] rejected the null hypothesis that $\rho = 0$. As the number of observations per data set increases then the number of times that the test correctly rejects the hypothesis that $\rho = 0$ also increases. The curves in Figure 6.11 are called power functions, and the proportion of times that a test will reject $H_0:\rho = 0$

[12] This is a simple example of why failing to reject a null hypothesis is not the same as showing that it is true.

when, in fact, $\rho = 0$ is called the *power* of the test. From Figure 6.11 we can see that if we have 50 observations then the test has at least a 90 per cent chance of rejecting our null hypothesis (thus, a power of 0.9) when ρ is greater than about 0.5. If we have 100 observations then we have a power of 0.9 as soon as ρ is greater than about 0.3, and if we have 500 observations then we have a power of 0.9 as soon as ρ is greater than about 0.16. In other words, as the sample size increases we have a greater and greater chance of detecting a smaller and smaller difference between the hypothesised value and the true value. Even minuscule differences (say, $\rho = 0.01$) will be almost surely detected at very large sample sizes – i.e. the null hypothesis would almost always be rejected as soon as the true correlation is even slightly different from zero.

Usually, more power is a good thing. Tests of structural equation models, based on the chi-square distribution, also have power properties. The justification for using alternative tests of fit is based on the premise that statistical power is not always such a good thing. If you remember the section in Chapter 2 dealing with the logic of inference in science then you will recall that no hypothesis is ever really tested in isolation. Every hypothesis contains within it many other auxiliary hypotheses. In the context of testing structural equation models with reference to a chi-square distribution we are really interested in knowing if the causal structure of the model is wrong. Unfortunately, when we conduct our statistical test we are testing all aspects of the model: the causal implications, the distributional properties of the variables, the linearity of the relationships, and so on. Now, when we add the notion of statistical power to our argument we realise that, as sample size increases, we run a greater and greater risk of rejecting our models because of very minor deviations that might not even interest us. This point was raised early in the history of modern SEM by Joreskog (1969).

What might these uninteresting and minor deviations be? These can't be minor deviations from multivariate normality since the maximum-likelihood chi-square statistic is asymptotically robust against non-normality. In any case, we have already seen ways of dealing with this. Small amounts of non-linearity could be one such minor deviation that would not interest us. If some parameter values (for instance, path coefficients or error variances) are fixed to non-zero values in our model then small deviations from these fixed values might be another minor difference that would not interest us. For instance, we might have only a single indicator of some latent variable whose error variance we fix at 1.1, perhaps based on previous experience. If the true error variance of this indicator was 1.15 and we use a large enough sample size then our model would be rejected. However, the principal 'minor deviation' that is evoked in the justification for measures of approximate fit is a minor deviation in the causal structure of the model. The theoretical objective of the various indices of approximate fit is therefore to somehow quantify the degree of these deviations. The various alternative fit indices attempt to quantify the degree of such deviations by measuring the difference between the observed covariance matrix and the predicted (model) covariance matrix. The most popular fit indices do this in a way that standardises for differences in sample size.

At first blush, then, these indices of approximate fit have a seductive quality. Would it not be nice, after having found that one's preferred causal explanation (as translated by the structural equation model) has been rejected, to be able to say: 'But it is almost right!

The remaining lack-of-fit is only due to minor errors that are not really very important anyway.' This, I suspect, is the real (psychological) objective of these fit indices. Even this weakness of the flesh could be tolerated if there was any justification for the implicit assumption that minor errors in specifying the causal structure will translate into only minor differences between the observed and predicted covariance matrices. Unfortunately, no such one-to-one relationship has ever been demonstrated for these indices of approximate fit. To me, evoking such an argument of approximate fit to justify accepting a causal model is like the old joke about the drunk in the car park.[13] The alternative fit indices measure different aspects of the ability of the *observational* model (the structural equations) to *predict* the data, not the *explanatory* ability of the *causal* model. As such, the indices of approximate fit commit the sort of subtle error of causal translation that I discussed in Chapter 3: small (but real) differences between the observed and predicted covariances of the observational model do not necessarily mean only small (but real) differences between the actual causal structure and the predicted causal structure.

Now that I have given my reasons why you should not use these alternative fit indices, you can read the justifications of those who promote them and decide for yourself (Bentler and Bonnett 1980; Browne and Cudeck 1993; Tanaka 1993). Below, I describe two of the more popular alternative fit indices. The book by Bollen and Long (1993) contains a number of chapters that deal with these alternative indices of approximate fit.

6.9   Bentler's comparative fit index (CFI)

Let us go back to the maximum-likelihood chi-square statistic for a moment. This statistic, and its inferential test, measure exact fit between the observed and predicted covariance matrices. The logic is that if the data are generated by the process specified by the structural equations (and therefore the causal structure of which these equations are a translation) then the observed and predicted covariance matrices will be identical except for random sampling variation. If this assumption is true then the maximum-likelihood chi-square statistic will asymptotically follow a chi-square distribution with the appropriate degrees of freedom (υ). Actually, it is more precise to say that this statistic will asymptotically follow a *central* chi-square distribution (χ^2_υ) with the appropriate degrees of freedom. The central chi-square distribution is a special case of a more general chi-square distribution called the *non-central* chi-square distribution. The non-central chi-square distribution ($\chi^2_{\upsilon,\lambda}$) has two parameters: the degrees of freedom (υ) and the non-centrality parameter (λ). A central chi-square distribution is simply a non-central chi-square distribution whose non-centrality parameter (λ) is zero.

Now, if the degree of mis-specification of the model covariance matrix is not zero (as assumed in the test for exact fit) but is small relative to the sampling variation

---

[13] You enter a car park late at night and see a drunk on his knees underneath the only street light. He explains that he is looking for his car keys. 'Are you sure that you lost your keys here?' you ask. 'No,' he answers. 'In fact, I have no idea where they are, but at least here I have enough light to see.'

in the observed covariance matrix, the maximum-likelihood chi-square statistic actually asymptotically follows a non-central chi-square ($\chi^2_{\upsilon,\lambda}$) distribution and the non-centrality parameter (λ) measures the degree of mis-specification. The expected value of the non-central chi-square distribution is simply the expected value of the central chi-square distribution plus the non-centrality parameter: $E[\chi^2_{\upsilon,\lambda}] = E[\chi^2_{\upsilon}] + \lambda = \upsilon + \lambda$. In practice, the non-centrality parameter is estimated as the value of the maximum-likelihood chi-square statistic minus the degrees of freedom of the model (i.e. the expected value that the maximum-likelihood chi-square statistic would have if there were no errors of mis-specification). Because the non-centrality parameter cannot be less than zero, negative values are replaced with zero. Therefore, $\lambda = \max\{(\text{MLX}^2 - \upsilon), 0\}$.

The Bentler comparative fit index uses this fact to measure by how much the proposed model has reduced the non-centrality parameter (thus, the degree of mis-specification) relative to a baseline model. The most common baseline model is one that assumes that the variables are mutually independent. If λ_i is the estimate of the non-centrality parameter for the model of interest and λ_0 is the estimate of the non-centrality parameter for the baseline model, the comparative fit index is defined as

$$CFI = \frac{\lambda_0 - \lambda_i}{\lambda_0}$$

If the model of interest fits exactly then the expected value of its non-centrality parameter (λ_i) would be zero and the CFI value would be 1.0. Therefore, the CFI index varies from 0 (the proposed model fits no better than the baseline model) to 1. The sampling distribution of this index is unknown, and users of this index consider a value of at least 0.95 as being an acceptable 'approximate' fit. There is no theoretical justification for this value; it is simply a rule of thumb. The extractor function (fitMeasures) in lavaan prints out a large number of different indices of approximate fit. If you specify the argument 'cfi' within this function then only the comparative fit index is generated as output. To get this index for our strings model we would specify fitMeasures(fit, "cfi"), where 'fit' is the name of the object created by the sem() function.

Actually, the description above is for the sample-based CFI. Although this is the index usually reported in most commercial SEM programs, it is known that the sample-based CFI is a biased estimator of the population-based CFI. The result of this bias is to exaggerate the degree of misfit. Steiger (1989) explains how to calculate the unbiased estimator of the population CFI from the information provided by most commercial programs. If p is the number of observed variables in the model, df is the degrees of freedom and n is the sample size then first get the model fit index, or calculate it as

$$\hat{F} = \frac{(X^2 - df)}{n - 1}$$

The unbiased estimator of the population CFI is

$$\frac{p}{p + 2\hat{F}}$$

6.10 Approximate fit measured by the root mean square error of approximation (RMSEA)

Another popular measure of approximate fit was developed by Steiger (1990) and expanded by Browne and Cudeck (1993). This measure also relies on the non-centrality parameter. The root mean square error of approximation (RMSEA, ε) is defined as

$$\varepsilon = \sqrt{\frac{\lambda}{n\upsilon}} = \sqrt{\frac{\max\{MLX^2 - \upsilon), 0\}}{n\upsilon}}$$

where λ is the non-centrality parameter and υ is the degrees of freedom of the model. If we propose a null hypothesis for the RMSEA (H_0: $\varepsilon_a \leq a$) then we can test this hypothesis using the non-central chi-square distribution and produce confidence intervals around it. If your favourite statistics program doesn't have this probability distribution then you can use the algorithm given by Farebrother (1987). This is obtained in R via the `pchisq(q=,df=,ncp=, lower.tail=F)` function, with the non-centrality parameter being specified by the 'ncp' argument.

Of course, if your null hypothesis is that $\varepsilon_a = 0$ then you are doing a test of exact fit with reference to the central chi-square distribution. Here are the steps.

(1) Specify the null hypothesis H_0: $\varepsilon_a \leq a$.
(2) Calculate the maximum-likelihood chi-square statistic and the non-centrality parameter $\lambda^* = n \cdot \upsilon \cdot \varepsilon_a^2$.
(3) Find the probability of having observed MLX^2 given a non-central chi-square distribution with parameters υ, λ^*.
(4) If the probability is less than your chosen significance level, reject the null hypothesis and conclude that ε_a is greater than that specified in the null hypothesis.

An obvious problem with this test is in choosing the null hypothesis. Remember that these indices of approximate fit are used when one has already rejected the null hypothesis of exact fit (i.e. $\varepsilon_a = 0$). We already know that there is something wrong with the model. Browne and Cudeck (1993) recommend the null hypothesis of $\varepsilon_a \leq 0.05$ but this is only their rule of thumb. Again, there is no compelling reason for choosing this value as a reasonable level of 'approximate' fit. This index of approximate fit is obtained by specifying 'rmsea' in the fitMeasures() extractor function. Thus, to get this index in our strings example, we would specify `fitMeasures(fit, "rmsea")`.

Quite apart from using the RMSEA to measure 'approximate' fit there is a very useful property of the inferential test for this fit statistic. If we have not been able to reject our model then it is still important to be able to estimate a confidence interval for the RMSEA. In such a case the confidence interval will have a lower bound of 0. The upper bound will reflect the statistical power of our test. A large upper bound indicates that the test had little statistical power to reject alternative models. A 90 per cent confidence interval for RMSEA would not reject the null hypothesis of exact fit at the 5 per cent level. This interval can be calculated as the values of λ for a non-central chi-square distribution whose 5 per cent and 95 per cent quantiles equals the calculated MLX^2 statistic (Browne and Cudeck 1993).

6.11   Missing data

We sometimes have 'holes' in our data sets because of missing values in some variables; in fact, we sometimes have rather complicated patterns of 'missingness' in which we lack values on different variables for different observations. What do we do when this happens? This depends on the type of missing data. Little and Rubin (2002) define three different types of missing data, with acronyms of MCAR ('missing completely at random'), MAR ('missing at random') and NMAR ('not missing at random'). In what follows, the pattern of missingness refers to the pattern of holes for a given variable or set of variables. For instance, imagine that you have a variable (x) whose values in your data set are x = {2.1, 3.2, NA, 5.0, NA, 1.1}, where 'NA' is the special code in R for a missing value. If we code a missing value by 0 and an observed value by 1 then the pattern of missingness for variable x is {1, 1, 0, 1, 0, 1}. The mice package in R (van Buuren and Groothuis-Oudshoorn 2011) contains the md.pattern() function, which takes your data frame or matrix as its argument and generates this pattern of missingness as output.

Data for a given variable are missing completely at random (MCAR) when the fact that a value is missing or not is unrelated to the actual values of that variable or to the values of any other variable. Of course, it is not possible to statistically determine if those values of a variable that are missing are systematically different from those that are observed for this same variable because . . . well . . . we don't know the values of the missing observations! On the other hand, we often know why we missed some observations. If we missed them because we couldn't measure values in certain ranges then the pattern of missingness is related to its value (NMAR). If we missed them because we were sick, or the measuring device broke down or it was raining then we might be willing to assume MCAR unless the variable we are measuring is related to our health or precipitation patterns.

A variable is missing at random (MAR), but not completely at random, if its pattern of missingness is unrelated to its own values but is related to other variables in the data set – in other words, if we can predict the pattern of missingness of this variable given the values of other variables. It is possible to determine this by conducting a logistic regression in which the pattern of missingness (the '1's and '0's) of the variable is regressed on the values of the other variables.

If you have variables that are not missing at random then I don't know of any statistical solution. However, if your variables are MAR then there are solutions. In Chapter 4 I introduced the notion of maximum-likelihood estimation. This used information about the covariances (and means, if intercepts are included) to obtain parameter estimates for the model covariance matrix. A variant, called 'full-information maximum likelihood' (FIML), finds the parameter estimates for each observation that maximise the likelihood for each observation, after which these are combined to produce the maximum-likelihood estimates for the model covariance matrix. Because FIML applies to each observation then, if the patterns of missingness in the data set are MCAR or MAR, we can proceed in the presence of missing values. To do this in lavaan, you simply have to include the missing argument in the `sem()` function: `sem(..., missing="fiml")`.

6.12　Reporting results in publications

There are a number of 'best practices' that you should follow when publishing the results of an SEM analysis. First, if you can't publish the actual data set used in the analysis (perhaps as an electronic appendix) then at least publish the observed covariance matrix and its sample size. It is always possible to calculate the observed correlations from the covariance matrix but you can't get covariances from a correlation matrix unless you also provide the variances. This will allow others to compare their independently obtained results to your model. Without such independent comparisons, any published SEM model can be viewed only as provisional. Equally importantly, since you are presumably going through all this effort in order to learn something, the best way of doing this is to obtain independent replications so that you can know when the model works and when it doesn't. If others have access to your data then they can quantitatively compare your model to their data via multigroup models (Chapter 7). In the absence of your actual data (or, at least, the covariance matrix of these data) then the only thing that others can independently evaluate is the causal structure of your model.

Since the purpose of an SEM model is to test or describe a causal process, it is important to provide the causal reasoning for the model. In other words, readers must be able to understand the hypothesised reason behind each arrow in the path diagram and whether this reason is justified by some theoretical notion or by a purely empirical one. If you have latent variables in the path diagram then such explanations are even more important, since such reasoning will be critical in interpreting the meaning of the latents. The reasoning behind an a priori SEM model would probably be given in the introduction, since it is an hypothesis that will be tested by the data. The reasoning behind a more exploratory SEM analysis would probably be given in the discussion section, since it is an hypothesis that is derived from your data. You can even have both types of SEM model in the same paper if you start with an a priori SEM model that is then rejected and modified. You must always clearly label your models as confirmatory or exploratory.

You should always provide sufficient summary information about the model fit so that others can properly evaluate it. This includes, at a minimum, (a) the sample size (to judge statistical power), (b) the type of fit statistic used to judge the fit between the data and the model and its value, (c) the degrees of freedom of the model, (d) the null probability associated with the fit statistic and (e) the estimates of the free parameters and their standard errors. It is usually a good idea to also provide both the actual values of these estimates and their standardised values so that one can directly compare these values even if they have different measurement units.

Path diagrams should follow certain conventions. Observed variables are either enclosed in boxes or else not enclosed at all. Latent variables should be enclosed within circles. Constants (for instance, if you include intercepts) are generally enclosed in triangles. Directed effects are shown as unidirectional arrows while free covariances are shown as curved double-headed arrows. Often you will also want to include the values of fixed and free path coefficients, free covariances and error variances on the path diagram. Estimated values of free parameters should include the standard errors within parentheses. Remember to tell readers if your parameter values are standardised or not.

6.13   An SEM analysis of the Bumpus house sparrow data

Natural selection was in the air during the last decade of the nineteenth century. According to H. C. Bumpus (1899), natural selection was literally *in the air* during a New England snow and ice storm one cold night. Many house sparrows (*Passer domesticus*) were immobilised during that storm, and 136 of the unfortunate birds were collected and transported to the Brown University Anatomical Laboratory. Of these, 72 birds (51 males and 21 females) subsequently recovered but 64 birds (36 males and 28 females) died. Bumpus determined the sex of all 136 birds and also measured nine phenotypic attributes of each bird, alive or dead. He used these data to show the selective elimination of individuals in a population based on their characteristics.

These data have been subsequently analysed by many different people.[14] In particular, Lande and Arnold's influential paper on the statistical estimation of selection gradients (Lande and Arnold 1983) used this particular data set as an example. The method of Lande and Arnold was essentially an application of multiple regression of a suite of correlated characters on a measure of evolutionary fitness. The regression coefficients were interpreted as causal measures of the selection gradient. In Chapter 2 we saw the problems that can occur when we use multiple regression in such a context.

Pugesek and Tomer (1996) reanalysed the Bumpus data using SEM. Besides two binary variables representing sex (male/female) and survival (alive/dead) they used seven other observed variables of various body measurements, transformed to natural logarithms: femur, tarsus, humerus, sternum, wing and head length and skull width. They began with the measurement model involving all birds, living or dead. The first model that they considered (Figure 6.12(a)) was that all seven length measures were due to a single latent variable. Since there were two sexes they actually used a two-group model with across group constraints (see Chapter 7). This first measurement model did not fit well (MLX^2 = 43.59, 28 df, p = 0.03). Their second measurement model was a three-factor model (Figure 6.12(b)). Pugesek and Tomer interpret these latents as a latent general 'size' factor that is a common cause of all seven body measurements, a latent 'leg size' factor that is an additional common cause only of the femur and tarsus lengths, and a latent 'head size' factor[15] that is an additional common cause only of the

---

[14] Pugesek and Tomer (1996) provide a brief history of these analyses.

[15] Since the two specific latent variables had only two observed variables each, identification of the model was obtained by constraining the two path coefficients of each latent to be equal. This is simply a mathematical trick, and means that the actual values of the path coefficients cannot be interpreted.

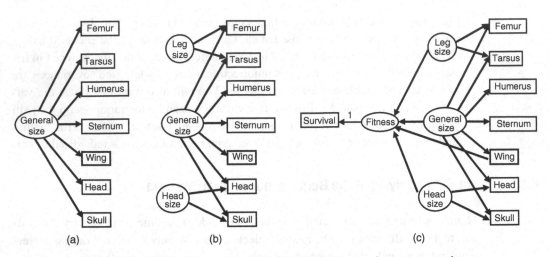

Figure 6.12 Pugesek and Tomer's model of the classic Bumpus house sparrow data

head length and skull width. As we have seen before, giving the latents these names doesn't necessarily mean that the names are accurate. Whatever the source of the latent variables, the model provided a good fit (MLX^2 = 28.82, 24 df, p = 0.227). A series of nested models (see Chapter 7) showed that there were no significant differences between the males and females in any of the free parameters. Fixing all these free parameters to be equal in the two sexes provided a final measurement model with an acceptable fit (MLX^2 = 49.29, 40 df, p = 0.149).

The next step was to relate the measured and latent variables to survival. Pugesek and Tomer allowed the three latent 'size' variables to be direct causes of a fourth latent variable, which they call 'Fitness', that then determines the death or survival of the individual bird. Since they fixed the path from the latent 'Fitness' to the observed 'Survival' at 1, and the residual error of 'Survival' at zero, the latent 'Fitness' variable is redundant, since it will be perfectly correlated with 'Survival'.[16] This two-group model provided a marginal fit to the data (MLX^2 = 62.72, 48 df, p = 0.075) but the authors added an edge from 'Wing length' directly to the latent 'Fitness' that significantly improved the fit of the model (MLX^2 = 52.37, 46 df, p = 0.241). Finally, a nested sequence of models showed no significant differences in the path coefficients or error variances leading into, and out of, the latent 'Fitness' variable, and so these were constrained to be equal in the males and females.

The final model is shown in Figure 6.12(c) (the parameter values of this final model can be found in Figure 8 of Pugesek and Tomer 1996). The path coefficients (based on standardised variables) allow one to determine by how much a change in one morphological variable will change the probability of survival of the bird. For instance, from their model one can calculate that an individual whose general size was one standard

---

[16] Unless there was some detail that was omitted from the paper, it was not necessary to fix the error variance of 'Survival' to 0.

deviation larger than average increased its changes of survival by 0.564 standard deviations more than the average. Based on their model, it seems that larger birds were less likely to die during the storm than the smaller birds, birds whose legs and head were even larger than average given their general size were even less likely to die (although the path coefficients from these latent variables were not significant at the 5 per cent level) but birds whose wings were shorter that average given their general size were less likely to die.

7 Multigroup models, multilevel models and corrections for the non-independence of observations

Like successful politicians, good statistical models must be able to lie without getting caught. For instance, no series of observations from nature are *really* normally distributed. The normal distribution is just a useful abstraction – a myth – that makes life bearable. In constructing statistical models we pretend that the normal distribution is real and then check to ensure that our data do not deviate from it so much that the myth becomes a fairy tale. In the last chapter we saw how far we could stretch the truth about the distributional properties of our data before our data called us a liar. The goal of this chapter is to describe how SEM can deal with two other statistical myths that people often tell with respect to their data. These two myths are (a) that the observations in our data sets are generated by the same causal process (causal homogeneity) and (b) that these observations are independent draws from this single causal process.

Consider first the myth of causal homogeneity. It is easy to imagine cases in which different groups of observations might be generated by partially different causal processes. For instance, a behavioural ecologist studying a series of variables related to aggression and social dominance in primates would not necessarily want to combine together the observations from males and females, since it is possible that the behavioural responses of males and females are generated by different causal stimuli. When we sample from populations with different causal processes, either in terms of the causal structure or of the quantitative strengths between the variables, and we wish to compare the causal relationships across the different groups, we require a model that can explicitly take into account these differences between groups. Such modelling is called *multigroup* SEM, and this, in turn, requires the notion of *nested* models.

The assumption of the independence of observations can often be violated as well, because the observations are nested in space or time. The process of speciation itself suggests one way in which we can get non-independence of observations (Felsenstein 1985; Harvey and Pagel 1991). The attributes of organisms, if they have a genetic component, will often tend to be more similar to those of close relatives than to genetic strangers. The process of speciation therefore generates a hierarchical structure to data when we combine observations from different families, populations or species. If we ignore this hierarchical structure, and therefore ignore the non-independence of the observations, then we will obtain incorrect probability estimates.

Similarly, when combining observations that come from different geographical locations, observations taken close together will often share more similar environmental conditions than those taken further apart. If these environmental conditions have causal

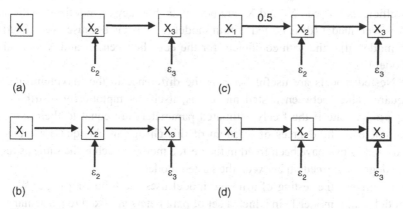

Figure 7.1 Four path models used to illustrate the concept of nesting

relationships to your modelled variables, and not all the causally relevant environmental variables are explicitly included in your model, then this too will generate
non-independence between your observations. There are two ways of dealing with
such problems. One way is to explicitly include this nested structure in the model via
multilevel SEM. The second way is to remove the phylogenetic or spatial dependencies
from your data and then model the causal structure that remains. We will look at both
approaches.

7.1   Nested models

Given two SEM models with the same set of variables, one model is *nested* within the
second one if (a) all the fixed parameters in the first model are also fixed to the same
values in the second but (b) some of the free parameters in the first are still fixed in the
second. In other words, the fixed parameters in the first model are a subset of the fixed
parameters in the second model. The notion of nesting can be grasped most easily by
comparing some path diagrams. In Figure 7.1 model (b) is nested within (a) and model
(d) is nested within (c).

Model (a) has two fixed parameters. The path coefficients for the edges between X_1
to X_2 and between X_1 to X_3 have each been fixed to zero; therefore, there is no edge
between X_1 and X_2 or between X_1 and X_3. There are two fixed parameters in model (a),
all others being freely estimated.[1] There is only one fixed parameter in model (b) – the
path coefficient for the edge between X_1 and X_3 is still fixed to zero – and all others,
including the path from X_1 to X_2, are freely estimated. So, the fixed parameters of model
(b) are a subset of those in model (a), and model (b) is nested within model (a).

Model (c) also has two fixed parameters. The path coefficient for the edge between X_1
and X_3 is still fixed to zero and the path coefficient for the edge from X_1 to X_2 has been
fixed to 0.5. Note that model (c) is not nested within model (a); it is true that the path

---

[1] The free parameters representing the variances are not shown in Figure 7.1.

coefficients between X_1 and X_2 are both fixed but they are not fixed to the same *value*. However, model (d) is nested within model (c). This is because every fixed parameter in model (d) – the path coefficient for the edge between X_1 and X_3 – is also fixed in model (c).

Nested models are useful because the difference in the maximum-likelihood chi-square values between nested models is, itself, asymptotically distributed as a chi-squared variate if the freely estimated parameters are equal to their associated fixed parameters. The degrees of freedom of this change in chi-square are the number of parameters that have been freed in the nested model, which is the same as the change in the degrees of freedom between the nested models.

Intuitively, the testing of a nested model uses the following logic. We start with a model (call it model 1) in which a set of parameters are fixed to particular values (zero or otherwise). Now we define a new nested model (call it model 2) by freeing some previously fixed parameters but without changing anything else relative to model 1. If we allow some of these previously fixed parameters to be freely estimated, but these newly freed parameters really do have the values to which they had previously been fixed, then the only difference in the estimated covariance matrices between models 1 and 2 will be due to random sampling variation. If this is true then the difference between the maximum-likelihood chi-square statistics will also follow a chi-square distribution with degrees of freedom equal to the number of previously fixed parameters that have been freed in the nested model 2. Here are the steps.

(1) Fit the model at the top of the nested sequence, then obtain its chi-square value (X_1^2) and its degrees of freedom (df_1).
(2) Fit the model at the bottom of the nested sequence, then obtain its chi-square value (X_2^2) and its degrees of freedom (df_2).
(3) Calculate the change in the chi-square value and the change in the degrees of freedom: $\Delta X^2 = X_1^2 - X_2^2$ and $\Delta df = df_1 - df_2$.
(4) Determine the probability of having observed this change in the chi-squared value $(\Delta \chi^2)$ assuming that the freed parameters in the second (nested) model are equal to those in the first model, except for random sampling variation.
(5) If this probability is less than the chosen significance level, conclude that the freed parameters were not the same as those fixed in the first model.

Tests of nested models are used in a number of different research contexts. One reason might be if you want to test for the equality of a set of parameters to some theoretical values but don't care if the model as a whole is acceptable. Two exploratory methods in SEM (the Wald and Lagrangian multiplier tests) are based on this logic. Perhaps the most useful application of nested models is in the context of multigroup models and multilevel models.

7.2   Dealing with causal heterogeneity: multigroup models

An assumption of the tests for structural equation models that have been described so far is that all the observations come from the same statistical population. In other words, we are assuming that the same causal process has generated all our observations even if we don't know what this causal process might be. Often we know (or suspect) that this is not the case. For instance, if we are studying attributes related to reproductive success then we might suspect that different causal processes are at work for males and females. Even if the causal structure is the same in males and females, it is possible that the two sexes differ in the numerical strength of the causal relationships. If we were to combine males and females into one data set then we would obtain incorrect parameter estimates and might incorrectly reject the model even though the qualitative structure of the model is correct. Perhaps our data come from three different geographical regions and we are not willing to assume that the same causal forces (with the same numerical strengths) apply to the observations in these different regions. Perhaps our data come from groups that we have subjected to different experimental treatments. All these examples require us to explicitly include the group structure into our analysis. Such analyses are called *multigroup SEM*.

The first impulse (which is not always wrong) is to analyse the data in each group separately. The real strength of multigroup SEM is the ability to statistically compare between groups and determine which parts of the models in each group (i.e. which parameters) are the same and which parts differ. In this sense, multigroup SEM is analogous to ANOVA, except that, rather than testing for differences in the means between groups, we are testing for differences in the covariance structure between the groups. To do this we construct a series of nested multigroup models.

A multigroup model can be fitted with a minor modification of the method that you already know. Since the standard structural equation model is simply a multigroup model with only one 'group', let's start there. With only one group we have only one observed covariance matrix, S_1. We then set up the model covariance matrix (Σ_1) using covariance algebra and iteratively find values of the free parameters of Σ_1 that minimise the maximum-likelihood chi-square statistic: $(N_1 - 1)(Ln|\Sigma_1(\theta_1)| + trace(S_1\Sigma_1^{-1}(\theta_1)) - Ln|S_1| - p_1)$. This is the same formula that you saw in Chapter 4, except that I have added subscripts to emphasise that we are referring to group '1'. When our data are divided into g groups with $N_1, N_2, \ldots, N_g$ observations in the different groups then we have g sample covariance matrices ($S_1, S_2, \ldots, S_g$) and also g population covariance matrices ($\Sigma_1, \Sigma_2, \ldots, \Sigma_g$). Each population covariance matrix can potentially have different sets of free and fixed parameters, or even different sets of variables. We iteratively choose values of all these free parameters simultaneously to minimise:

$$[(N_1 - 1)(Ln|\Sigma_1(\theta_1)| + trace\left(S_1\Sigma_1^{-1}(\theta_1)\right) - Ln|S_1| - p_1)]$$
$$+ [(N_2 - 1)(Ln|\Sigma_2(\theta_2)| + trace\left(S_2\Sigma_2^{-1}(\theta_2)\right) - Ln|S_2| - p_2)]$$
$$+ \cdots [(N_g - 1)(Ln|\Sigma_g(\theta_g)| + trace(S_g\Sigma_g^{-1}(\theta_g)) - Ln|S_g| - p_g)]$$

Although this equation looks intimidating, it is simply the sum of the maximum-likelihood chi-square statistics for each group.

The value of this multigroup maximum-likelihood chi-square statistic at the minimum also asymptotically follows a chi-squared distribution, with degrees of freedom equal to the sum of the degrees of freedom of the model in each group. Even if a particular parameter is free in each group, we can constrain the fitting procedure to choose the *same* value for all groups (this generates g − 1 extra degrees of freedom). In this way we are stating that, although we don't know what the numerical value of the free parameter is, it must be the same numerical value in all groups. Viewed in this way, we see that multigroup models define a continuum. If we propose the same causal structure and the same numerical values for all free parameters across the groups then we get the same result as if, for each group, we had centred each variable around its group mean and then put all our group-centred data into one big group. At the other extreme, if we allow all the free parameters to differ between groups then we get the same result as if we had tested each group separately and summed the maximum-likelihood chi-square statistics and degrees of freedom. By constraining the estimation of different sets of free parameters across groups we can define a series of nested models. In this way, we can test for the equivalence of various free parameters in the different groups. If we do this more than once then we should adjust our significance level using a Bonferonni correction.[2]

The following example comes from Meziane (1998). Although it is a path model without latent variables, the logic and approach are identical to full SEM models. The study consisted of 22 species of herbaceous plants grown under controlled conditions in four different environmental conditions: a high (N) and low (n) nutrient concentration in hydroponic culture crossed with a high (L) and low (l) light intensity. This gave four different groups of data corresponding to the four different environments: NL, Nl, nL, nl. Two leaves on each plant were harvested and a series of four morphological attributes were measured: the water content of the leaf, the thickness of the lamella, the thickness of the midvein and the specific leaf area (the ratio between the projected leaf area and its dry weight). The values of the two leaves per plant were averaged. Due to a few missing values, there were a total of 80 independent observations in the final data set. A previously published study (Shipley 1995) had described a path model relating these variables, and one objective of Meziane (1998) was to see if the previous path model could be applied under different environmental conditions. If Meziane had simply combined the data from all four environments and tested his path model then he would have implicitly assumed that the different environments had no effect on the relationships between the four variables. By 'no effect' I mean both that the causal structure of the relationships (their presence or absence) and their numerical strengths do not differ between environments. If you remember that each variable is centred around its mean in the data set, combining the data from all four environments would also implicitly require that the treatments did not affect the mean values of the variables either. By separating the data into the four groups, the variables are centred around their respective group means. In this way, the treatment effects on the means are removed and only the relationships between the variables are analysed.

In his multigroup analysis Meziane specified four models, each with the same causal structure (i.e. the same DAG) but each potentially differing in the numerical strengths of the free parameters. I have shown this in Figure 7.2, in which I have included the free parameters.

---

[2] If we do this t times with a significance level of α then we should test the change in the chi-square of each test at a significance level of α/t.

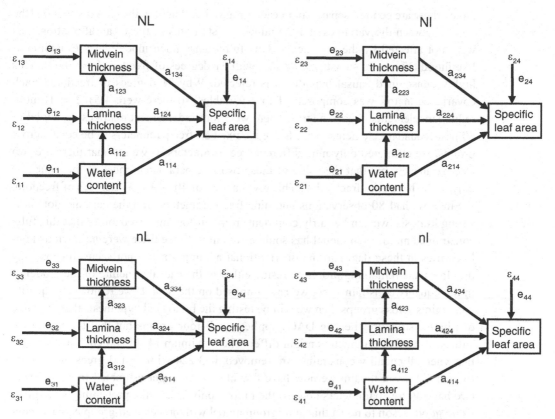

Figure 7.2 Meziane's four-group path model relating four attributes of the leaves of herbaceous plants
Note: Each group refers to plants grown in different environments of hydroponic nutrient solution and light intensity.

In this model there are five free path coefficients and four free error variances in each of the four models. There are therefore from nine (if all free parameters are constrained to be equal across groups) to potentially 36 different free parameters to estimate (if no free parameters are constrained to be equal across groups). Using the rule of thumb requiring five times more observations than free parameters, we see that any multigroup model with more than 16 free parameters will not be well approximated by a chi-square distribution and will have true probability values that are somewhat higher than those obtained using a chi-square distribution. The data, after transforming to natural logarithms, had reasonably low values of Mardia's multivariate index of multivariate kurtosis (-4.37, -2.70, -3.34 and 1.40 for the NL, Nl, nL and nl groups, respectively).

The first step was to fit the data to the most constrained model – namely that in which all nine free parameters are forced to be equal across the four groups.[3] This is equivalent to hypothesising that the treatments might have affected the mean values

---

[3] Note that, although the free parameters are constrained to be equal, the variables are still centred about their group means, not the overall mean of all data taken together. Therefore, differences in the means between the groups are still removed.

(since these are centred separately in each group), but that both the causal structure (the DAG) between the variables and the numerical strengths of these causal relationships were not affected by the treatments. This fully constrained model gave a maximum-likelihood chi-square statistic of 48.271 with 31 degrees of freedom (p = 0.02). This highly constrained causal hypothesis is rejected. Why 31 degrees of freedom? Each covariance matrix was composed of four variables, so there were 4(5)/2 = 10 non-redundant elements in each matrix. There were four independent matrices for a total of 40 non-redundant elements in all. Since we fixed all free parameters to be equal across groups, we estimated only nine different free parameters; if we say that there are two free parameters but that the value of these two free parameters must be the same, we only really have one 'free' value. This gives a total of 40 − 9 = 31 degrees of freedom.

Since we had 80 observations and nine free parameters, and the data did not show strong kurtosis, we can be fairly confident from our low null probability that this fully constrained multigroup model has some errors in it. Since there were no obvious non-linearities in these data and the distributional assumptions do not seem to cause any problems, the remaining problems reside either in the causal structure of the model or in the equality constraints that we have imposed on the data. If we remove all equality constraints across groups then we will be testing the biological hypothesis that the same qualitative structure (i.e. the DAG) applies to all four groups but that the numerical values of the free parameters might differ. The maximum-likelihood chi-square statistic, when all equality constraints are removed, is 3.224 with four degrees of freedom (p = 0.52). Even though we now have few observations per estimated free parameter (we have 36 free parameters now, so the ratio is only 2.2:1), the probability level gives us no good reason to reject this multigroup model with no between-group equality constraints for the parameter estimates.[4]

Since the lack of fit that was detected in the fully constrained multigroup model appears to be in the equality constraints between the four groups, we can use a series of nested models to detect which of the equality constraints is unreasonable. If we remove only one between-group equality constraint, for instance if we allow only one free parameter to differ between groups, then this new model will be nested within the fully constrained model. There will be three degrees of freedom fewer in this new model since we now have to independently estimate the value of this free parameter in all four groups, rather than just estimating one value for all four groups. The difference in the two maximum-likelihood chi-square statistics, compared to a chi-square distribution with three degrees of freedom (the difference in the degrees of freedom between the fully constrained model and the new model), will test for a difference in the value of this parameter between the four groups. We do this nine times, once for each separate direct effect, each time removing the equality constraint for a different free parameter. Since we have done this test nine times at a significance level of 5 per cent, we adjust the overall significance level to 0.05/9 = 0.0056.

Table 7.1 summarises the results. The first column lists the free parameter whose between-group equality constraint has been removed. The second column lists the maximum-likelihood chi-square statistic for this new model (always with 28 degrees

---

[4] Remember that the effect of small sample sizes is to produce conservative probability estimates.

Table 7.1 The results of comparisons of a series of nested models based on the four-group model shown in Figure 7.2

Free parameter whose between-group equality constraint was released	MLX^2	Change in MLX^2 (ΔMLX^2)	Probability of ΔMLX^2
None	48.271		
Variance of leaf water content	42.141	6.130	0.105
Error variance of specific leaf area	45.995	2.276	0.517
Error variance of lamina thickness	46.387	1.884	0.597
Error variance of midvein thickness	47.195	1.076	0.783
Path coefficient from leaf water content to specific leaf area	39.411	8.860	0.031
Path coefficient from lamina thickness to specific leaf area	27.710	20.561	0.0001
Path coefficient from midvein thickness to specific leaf area	35.122	13.149	0.004
Path coefficient from leaf water content to leaf lamina thickness	44.34	3.931	0.269
Path coefficient from lamina thickness to midvein thickness	47.646	0.625	0.891

Notes: The first row gives the results of the fully constrained model assuming all free parameters are the same in the four groups. The remaining rows show the result of relaxing one constraint at a time.

of freedom). The third column lists the change in the maximum-likelihood chi-square statistic relative to the model with all between-group equality constraints applied. The fourth column lists the (asymptotic) probability for the change in the maximum-likelihood chi-square statistic.

From Table 7.1 we see that only two of the path coefficients (those between the thickness of the lamina and midvein and the specific leaf area) differ between the four groups given our chosen significance level. Note that, although the path coefficient from leaf water content to specific leaf area had a probability level of 0.031, we had to adjust our individual significance levels to $0.05/9 = 0.0056$ in order to maintain an overall significance level of 0.05. Our final multigroup model fixes all free parameters except for these two path coefficients to be equal across groups. This model has a maximum-likelihood chi-square statistic of 21.463 with 25 degrees of freedom, giving a probability level of 0.667. The confidence interval of the RMSEA for this model is (0, 0.074). Since the original purpose of the analysis by Meziane (1998) was to see if the original path model that I had proposed (Shipley 1995) could be applied to plants growing in different resource environments, the conclusion is that the model appears to apply in its general structure, but that the numerical strengths of the two thickness measures on the specific leaf area change in the different environments. Of course, given the rather small number of observations and therefore low power, we must temper this conclusion, since small differences between groups might not have been detected.

In interpreting the results of a multigroup SEM it is important to remember that we are only testing for differences in the relationships between the variables within each group. Differences in the mean values of the variables between the groups are never detected, because the variables are centred about their mean values within each group. In the example from Meziane (1998), the mean values of every one of the variables differed between the groups, based on analyses of variance. In other words, the different levels of nutrients and light intensities did cause changes in the average values of the leaf attributes (from the ANOVA) and did change the numerical strength of the effects of the two thickness measures on specific leaf area, but did not appear – given the rather modest level of statistical power available – to change the numerical strengths of the other relationships, the error variances or the causal structure (i.e. the topology) between the variables.

Fitting multigroup models in lavaan, when the causal structure (i.e. DAG) is the same across groups, is quite easy. If you have the actual data then you must have a variable in your data set that indexes the group to which each multivariate observation belongs. This grouping variable can be either a character or numeric. After specifying the model using the model syntax rules that you already know and then saving this model object, you then use the sem() function with a new argument: group = . For instance, if your model is saved in an object called 'my.model' and your data set, including a variable indicating group membership called 'my.groups', is saved in an object called 'my.data' then the most basic multigroup model can be fitted[5] using

sem(model = my.model, data = my.data, group = my.groups).

If you have access only to the covariance matrices of each group then you must create a list containing each covariance matrix and a list or vector containing the sample sizes of each matrix. For instance, `combined.cov←list(my.group1 = ,` `my.group2 = )` and `combined.n←list(my.group1 = 52,my.group2 = 46)` would create a list called 'combined.cov' that contains the covariance matrices of the two groups and a list called 'combined.n' that contains the sample sizes associated with each covariance matrix. You would then use a command such as `sem(model = my.model, sample.cov = combined.cov,sample.` `nobs = combined.n)`. The sem() function will automatically fit a multigroup model, since there is more than one sample covariance matrix entered.

The above command will fit the data in my.data to the DAG specified in my.model separately for each group but will force each free parameter to be equal across the groups. Now suppose that we want to fit a multigroup model that is the same as the above model but with only some types of free parameters constrained to be equal. We can do this using the group.equal argument of the sem() function. The Appendix lists the different choices of types of free parameters. For instance, if we want only the path coefficients between observed variables to be constrained to be equal across groups, but not any other type of free parameter (for example, the error variances of the endogenous variables or free covariances), then we would specify

```
sem(model = my.model, data = my.data, group = my.groups,
group.equal = "regressions")
```

---

[5] It is possible to restrict the multigroup model to only a subset of the groups named in your data set via the group.label= argument; see the Appendix.

meaning that only the free parameters linked by the '~', or 'regression' operator of lavaan, are to be equal across groups. The group.equal argument is a character vector and so you can combine different types of free parameters. The Appendix to this book provides more details of the various arguments. If you want to constrain only some of the path coefficients to be equal then you would use an additional argument of the sem() function, called 'group.partial', which takes a vector of character strings containing the labels of the parameters that should be free in all groups; this is used to override the group.equal argument for some specific parameters.

There is a more flexible way, which is also sometimes easier, of constraining free parameters across groups. Remember that each free parameter in your model has a name. The default name of any free parameter is a concatenation of the names of the two variables defining the free parameter plus the operator; for instance, if you have a line in your model specification such as y~x then this defines a free parameter which is the path coefficient (called a 'regression' coefficient in lavaan) whose default name is y~x. However, you can provide explicit names for any free parameter by 'multiplying' the right-hand variable by a character name. For instance, if your model specification has a line such as y~a*x then a becomes the name of the path coefficient linking y and x. If you have a multigroup model and you had this line in your model specification then the path coefficient linking y and x would have the same name in all groups. If you want to give a different name to this path coefficient in different groups then you would provide a vector of names. For instance, if you had three different groups and you had a line in your model specification such as y~c(a1,a2,a3)*x then the names of the path coefficient linking y and x in groups 1, 2 and 3 would be a1, a2 and a3, respectively.

Why is this important? It is important because if a free parameter has the same name in different groups then its value is constrained to be equal across these groups. If the name of the free parameter is different in different groups then the value is not constrained to be equal. This means that the pattern of names that you give to your different free parameters within and across groups determines the pattern of equality constraints that you impose on your data. In this way you have complete freedom to specify the combinations of free parameters that you want to be constrained to equality and which combinations you want to be separately estimated. The sem() function automatically determines the correct number of degrees of freedom. If you make sure that your competing models are nested then you can test different patterns of equality constraints by comparing the change in the chi-square statistic and the change in the degrees of freedom.

As a simple example, imagine that we wish to test a DAG of the form x→y→z in a data set called 'my.data'. Measures of these three variables have been taken on many individuals in each of three different species, and so we also have a variable in my.data called 'species' and having three different values (sp1, sp2, sp3) in order to indicate to which species each individual belongs. We have no reason to believe that the variance of x, or the error variance associated with y, should be the same between the three species but we do expect the error variance associated with z to be the same; in other words, we believe that the importance of y in causing z relative to the other possible independent causes of z should be the same in all three species. We don't expect the quantitative effect of x on y (the path coefficient) to be the same between species.

Finally, we wish to distinguish between two competing causal hypotheses, one of which claims that the quantitative effect of y on z is the same in all three species and one that does not require this equality. We will therefore fit three nested models in lavaan: (1) a model in which all free parameters are equal between groups, (2) a model relaxing these equality constraints as specified above except that the path coefficient associated with y→z is constrained to be equal and (3) the same as model 2 but with the y→z path coefficient unconstrained. In fact, model (2) is the model used to generate the data. Here is model 1:

```
model1←"
# path x→y constrained to equality across groups
y~c(a,a,a)*x
# path y→z constrained to equality across groups
z~c(b,b,b)*y
# variance of x and residual variances constrained to equality
x~~c(Varx,Varx,Varx)*x
y~~c(Vary,Vary,Vary)*y
z~~c(Varz,Varz,Varz)*z
"
fit1<-sem(model=model1,data=my.data,group="species",fixed.x=
FALSE)
```

When we fit this model we obtain an overall chi-square statistic of 290.01 with 13 degrees of freedom. Why 13 degrees of freedom? We have three separate covariance matrices, each with six unique elements, for a total of 18 unique elements. Because this first model constrains all free parameters to be equal between groups (since we have given the same names to these free parameters in each group) we have only have five different free parameters to fit. These are the five different parameter names defined in the model (a, b, Varx, Vary, Varz). Thus, the degrees of freedom are $18 - 5 = 13$.

Here is model 2:

```
model2←"
# path y←x is not necessarily equal across groups, thus
# different names
y~c(a1,a2,a3)*x
# path z←y must be equal across groups, thus same name
z~c(b,b,b)*y
# variance of x, residual variance of y not equal across groups
x~~c(Varx1,Varx2,Varx3)*x
y~~c(Vary1,Vary2,Vary3)*y
# residual variance of z must be equal across groups
z~~c(Varz,Varz,Varz)*z
"
```

When we fit this model we obtain a chi-square statistic of 6.18 with seven degrees of freedom (p = 0.52). We now have 11 different free parameters to estimate (a1, a2, a3, b, Varx1, Varx2, Varx3, Vary1, Vary2, Vary3, Varz), and so we end up with 18 − 11 = 7 degrees of freedom. Since this model is nested within model 1, we can compare the change in the chi-square statistics between the two models (290.01 − 6.18 = 283.83) given the change in the degrees of freedom (13 − 7 = 6). Clearly, at least one of the free parameters that we allowed to vary between groups in model 2 did actually differ between groups.

Here is model 3:

```
"
#path y←x is not necessarily equal across groups, thus
# different names
y~c(a1,a2,a3)*x
# path z←y not necessarilyt equal across groups, thus
different names
z~c(b1,b2,b3)*y
# variance of x, residual variance of y not equal across groups
x~~c(Varx1,Varx2,Varx3)*x
y~~c(Vary1,Vary2,Vary3)*y
# residual variance of z must be equal across groups
z~~c(Varz,Varz,Varz)*z
"
```

When we fit this third model we obtain a chi-square statistic of 4.98 with five degrees of freedom. We have lost a further two degrees of freedom because, by not constraining the path coefficient associated with y→z to be equal across groups (since we have given this free parameter a different name in each group), we had to estimate each one separately. Since this model is nested within model 2, we can compare the change in the chi-square statistics between the two models (6.18 − 4.98 = 1.20) given the change in the degrees of freedom (7 − 5 = 2). The null probability of obtaining a chi-square value of 1.20 with two degrees of freedom is only 0.54, and so we have no good evidence[6] that this path coefficient did actually differ between the three species.

Finally, it is possible to specify different causal models in different groups. In order to do this, you must enter the different models for the different groups with each unique model preceded by the name of the groups and a colon (:). For instance,

```
My.mod<-"
Male group:
y~a*x
z~b1*y
Female group:
y~a*x + c*z
".
```

---

[6] $1-\text{pchisq}(1.2,2) = 0.5488116.$

The above model specifies two groups ('Male group' and 'Female group'), names the path coefficients (a, b1, c), and requires the value of the path coefficient for y←x to be the same in the two groups.

7.3   The dangers of hierarchically structured data

Let's now turn to the problem of analysing data when the observations are not independent. To illustrate the problems caused by partially dependent data, consider first the following analysis: I hypothesise that men (i) have shorter hair and (ii) bluer eyes than women. To test these hypotheses I sample from my head and from my wife's head and record our eye colours. To ensure 'replication' I randomly choose a hair from my head, measure its length and then record my eye colour and sex. I then do this for my wife. I then repeat this 20 times, resulting in a data set with 40. I then conduct two t-tests on these observations using (40 − 2) degrees of freedom for each test. Of course, I find a highly 'significant' difference between the two groups ('men' and 'women') – and, of course, the probability level associated with this test would be profoundly wrong. Most beginning students in statistics realise that such a test is nonsense, though they can't explain why. The problem is that, although I have 40 observations, I don't have 40 degrees of freedom. A large proportion of the total variation in hair length and all the total variation in eye colour resides at the level of individual people, of which there are only two, not at the level of individual observations, of which there are 40. Clearly, I do not have 40 *independent* observations of the three variables (eye colour, hair length, sex).

If two values of some variable X (say X_1 and X_2) are independent then knowing the value of one observation (X_1) tells us nothing about the likely value of the next one (X_2). Two independent values give us two 'pieces' of information and n independent values give us n 'pieces' of information. If, in some group, every individual had exactly the same values of X then, as soon as you knew X_1, you would also know the values of all other observations of X in the group. No matter how many observations of X that you took from such a group, you would have only one 'piece' of non-redundant information.

Now, imagine that we create two groups. We randomly choose 20 values of X to form the first group, and these values are independently and normally distributed with a mean of 1 and a standard deviation of 0.5. We randomly choose 20 values of X to form the second group, and these values are also independently and normally distributed but with a mean of 2 and a standard deviation of 0.5. If the values within each group were exactly the same then we would still have only two 'pieces' of information, but this is not the case. If knowing that an observation came from a particular group told us nothing about what values it might have, we would have 40 'pieces' of information, but this is not the case either. So, we have more than 2 and fewer than 40 'pieces' of information.[7] This is the nature of hierarchically structured data, and we require

---

[7] The ratio of the variation of a variable at a given level in a hierarchy to its total variation is given the unfortunate name of 'intraclass correlation' (Muthén and Satorra 1995). If we let the estimate of the variance of a mean, when ignoring the hierarchical nature of data, be Var$_{SRS}$ ('simple random sampling'), the correct

a way of determining how many 'pieces' of information (i.e. degrees of freedom) different variables possess at different levels of the hierarchy. Hierarchies and partial dependence are common in nature. Because of this, we need to either explicitly incorporate these into our models or else remove such partial dependence from our data before continuing with the modelling of nature. One source of such partial dependence is the phylogenetic structure of the natural world. Species sharing a recent common ancestor will probably have more similar trait values than species that are more distantly related, and this is why researchers who test evolutionary hypotheses using data involving different species wish to remove, or at least to account for, the degree of phylogenetic relatedness between species. Similarly, sites that are geographically closer will probably have more similar values of some environmental variables than sites that are more geographically distant, and this is why researchers testing ecological hypotheses involving several sites of varying distances might wish to remove, or at least to account for, the physical distances between sites. When such processes give rise to hierarchically organised data with only a few[8] well-defined hierarchical levels, we can explicitly include such a structure into our causal models via multilevel models. When this is not the case, we can sometimes at least remove the partial phylogenetic or spatial dependencies.

Removing phylogenetic signals

There is a very large literature dealing with phylogenetic effects on the covariance between phenotypic traits and of removing such phylogenetic signals. Such methods are complicated by the fact that one must make certain assumptions about the way in which evolution and speciation occurs. For instance, Felsenstein's (1985) method of phylogenetically independent contrasts (PICs) assumes a process of Brownian motion and linear relationships between traits, while Martins and Hansen's (1997) method of phylogenetic generalised least squares (PGLS) method can incorporate other assumptions as well. All such methods require that you already know the phylogenetic relationships between the species that compose your data. Errors in the assumed phylogeny will be carried into the phylogenetic corrections. I won't go into the details of these methods, and if you want to use them in the context of causal models then you should master this literature before proceeding. Given that such phylogenetic methods exist, the important point for our purposes is that such methods replace the measured covariance matrix of the traits with a 'corrected' one that removes the phylogenetic signal. Once we have such a corrected covariance matrix then we can use this in our structural equation model.

For example, Santos and Cannatella (2011) used the PICs method to obtain a phylogenetically corrected covariance matrix of seven traits related to poison frogs. Once

---

variance estimate be Var_C, the number of observations per group be c and the intraclass correlation be ρ, the relationship between them is $Var_C = Var_{SRS}(1 + (c - 1)\rho)$. A similar formula exists for the variance of a linear regression slope with hierarchical structure. If ρ_ε is the intraclass correlation for the residuals and ρ_X is the intraclass correlation for the predictor, Scott and Holt (1982) have shown that $Var_C = Var_{SRS}(1 + (c - 1)\rho_\varepsilon\rho_X)$.

[8] There is no limit to the number of hierarchical levels in principle but, in practice, multilevel models involving more than a few levels become difficult to fit and interpret.

such a corrected covariance matrix is obtained then you can simply input this matrix, along with the sample size, directly into the sem() function of lavaan by using the `sample.cov` = and `sample.nobs` = arguments.[9] Santos and Cannatella could have used the PGLS method to do the same thing since, under the conditions of that study, PICs and PGLS are statistically equivalent (Blomberg et al. 2012).

If your causal model involves latent variables then the above approach is the only way to proceed. If your causal model doesn't involve latent variables then you can also very easily use the PGLS method to test your model using the d-sep test from Chapter 3, as described by von Hardenberg and Gonzalez-Voyer (2012). The PGLS method has the advantages that different assumptions about the process of evolutionary change, besides simple Brownian motion, and variables having non-normal distributions can be included. The d-sep method allows one to use whatever inferential test of (conditional) independence is appropriate for the data. Combining the PGLS method of phylogenetic corrections and the d-sep method of testing path models consists simply of using the PGLS regressions to test the hypotheses of conditional independence that are specified by the basis set.

Incorporating hierarchical structure

Before going on to the mechanics of fitting such models, or even of interpreting them, it is useful to have a simple concrete example of such a structure. A good example is the relationship between seed size and seedling relative growth rate. Relative growth rate (RGR) is the amount of new biomass produced by a plant over a unit of time, relative to the amount of biomass that the plant had as an initial 'capital' at the beginning of the time period. If we plot the weight of different seeds of individuals within a single species against the relative growth rate of the seedling that emerges, we often find a positive relationship between the two. For individuals within a given species, having larger seeds translates into more rapidly growing seedlings, with all the attendant benefits. However, when we compare across species we see that both average seed size and average potential RGR vary much more between species than within species. For instance, the seeds of some orchids are almost microscopic, while it might require two hands to hold the seeds of a coconut palm. Under constant environmental conditions and resource levels, the variation of seedling RGR within a single species usually varies by less than 10 per cent of the variation in the mean RGR values between different species. Curiously, the relationship between the *average* seed size and the *average* seedling RGR of different species usually shows a negative relationship. Figure 7.3 plots some simulated data showing this pattern. There are 10 simulated species, each having a different plotting symbol. Figure 7.3(a) shows the relationship between the two variables for one species, along with its regression line. Figure 7.3(b) shows the relationship between the two variables when the data for all 10 species are combined.

---

[9] See the Appendix for more information on doing this.

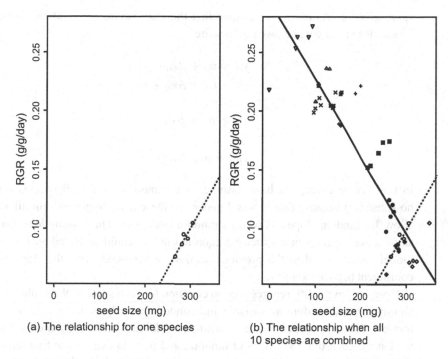

Figure 7.3 Hypothetical relationships between the relative growth rate (RGR) of 50 plants and their seed size
Note: In panel (b) the relationship is positive within every species but the overall trend shows a negative relationship because the relationship between the species' means is negative.

This example, although very simple, demonstrates many of the challenges of analysing data that are hierarchically ordered. It is clear that part of the variation in each variable, and the covariation between the two, is generated by differences between individuals within each species, as shown by Figure 7.3(a). It is also clear that part of the variation and covariation involving these variables is generated by differences between the species means. We would like to model this covariation both within and between species, taking into account, via the degrees of freedom, the fact that individuals within a given species tend to resemble each other more than individuals of different species do. Finally, we would like to model how the different levels in the hierarchy interact and constrain each other.

Since there are two levels to this data hierarchy, let's call 'level 2' the level of species and call 'level 1' the individual level. If we had only one species, then we could write the regression equation as

$$y_i = \alpha + \beta x_i + \varepsilon_i$$

In this equation, the subscript i refers to each individual, α is the value of the average individual when x = 0 (i.e. the intercept), β is the slope linking y and x, and ε_i is the amount by which the value of y for the ith individual deviates from its expected value given x_i. As usual, we assume that these deviations are random. Since we have 10 species in our data set, we could write

$$y_{i1} = \alpha_1 + \beta x_{i1} + \varepsilon_{i1}$$
$$y_{i2} = \alpha_2 + \beta x_{i2} + \varepsilon_{i2}$$
$$\dots$$
$$y_{ij} = \alpha_j + \beta x_{ij} + \varepsilon_{ij}$$
$$\dots$$
$$y_{i10} = \alpha_{10} + \beta x_{i10} + \varepsilon_{i10}$$

In this simple example I have assumed a common slope for all species (thus β has no subscript) because that is how I generated these data. In general, multilevel models allow for random slopes as well as random intercepts. The assumption of a common slope across species in a structural equation model could be tested with a multigroup model, as described in the previous section, remembering that the slope is the path coefficient between x and y.

Here, we have a different regression equation for each species. We could, at this point, simply introduce a dummy variable and conduct an analysis of covariance. In the context of SEM, this would be the equivalent of doing a multigroup model. However, if we have chosen our 10 species at random, and want to extrapolate to a larger population of species, then our regression intercepts (α_j) are, themselves, random variables, and we might want to model how these species-level random variables change as well. In this case, the 10 intercept terms (α_1 to α_{10}) are random variables from some larger statistical population, which we can model as $\alpha_j = \alpha + \delta_j$. Here, α is the overall intercept term for the entire population of species and δ_j is the random deviation of the intercept for the jth species from this overall intercept term. Putting this all together we obtain $y_{ij} = (\alpha + \delta_j) + \beta x_{ij} + \varepsilon_{ij}$. Here, the i subscript refers to the level-1 units (the individual plants) and the j subscript refers to the level-2 units (the different species). Rearranging, we obtain $y_{ij} = \alpha + \beta x_{ij} + \delta_j + \varepsilon_{ij}$. This equation expresses each y_{ij} as a function of a systematic component ($\alpha + \beta x_{ij}$), and two random components due to (a) the differences in the mean values of y_{ij} between the species (δ_j) and (b) a random component due to the differences of individuals within each different species (ε_{ij}).

Up to now the model that we have developed is perhaps familiar to some readers, since it is simply a variance components model with a between-species variance component (the variance of the δ_j) and a within-species component (the variance of ε_{ij}). However, since the intercepts (α_j) are random variables, the relationships between these intercepts may be determined by other, species-level variables. For instance, in Figure 7.3 it is clear that, as the mean value of seed size for a given species increases, the mean value of RGR for that species decreases.

Secondary succession in plant communities starts with some major disturbance event, such as a field that has been cultivated and then abandoned. Over time the relative abundance of each species changes as different species re-invade the site. Because of this one often finds that particular species tend to be most abundant in abandoned fields of a specific age. Immediately following a major disturbance event one typically finds annual species that have rapid relative growth rates and that produce a large quantity of small

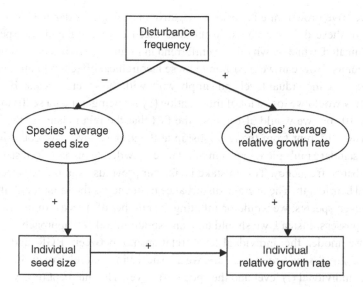

Figure 7.4 The causal process determining the relationship between seed size and relative growth rate

Notes: The causal process determining the relationship between seed size and relative growth rate operates at two levels. At the level of individuals, seed size causes relative growth rate. However, the average seed size and relative growth rate of each species are caused by the typical disturbance frequency of the habitat occupied by each species.

seeds. As secondary succession proceeds dominance shifts to species with larger seeds and slower relative growth rates (Grime 1979). This is a selection process in which different frequencies and intensities of density-independent mortality (the result of disturbance events) select for different suites of plant attributes. As such, selection pressures represented by such variables as 'average time since the last major disturbance event' affect species-level properties by determining the mean and variation of individual-level attributes.

Actually, I generated the data shown in Figure 7.3 by simulating the scenario described above. I defined a species-level variable – the frequency of major disturbance events – that quantifies how often a habitat experiences a major event of density-independent mortality. The causal effect of this variable is to select for individuals with both larger seeds and lower relative growth rates as the frequency of disturbance events decreases and as the successional age of the vegetation increases. In other words, although relative growth rate increases with increasing seed size within each species, the *average* seed size and the *average* relative growth rates of a given species are both determined by the common cause 'disturbance frequency'. This is shown in Figure 7.4.

As Figure 7.4 makes clear, the relationship between individual seed size and individual relative growth rate consists of causes that operate at two different hierarchical levels. At the level of individual plants there is a *positive* direct effect (seed size→relative growth rate). At an interspecific level there is a *negative* indirect effect between the two variables that is generated by the common cause of selection for habitats experiencing different disturbance frequencies. Whether the overall relationship between seed size and relative growth rate is positive, negative or ambiguous depends on the relative strengths of these different paths. In the data that I have simulated, the species-level effect dominated, which is why the overall trend in Figure 7.3(b) is a downward-sloping cloud of points. How can we incorporate these hierarchical effects into our models? We could ignore the individual level and simply work with the species means. If we did this then not only would we lose a lot of information (by reducing our data set from 50 observations to 10) but we would also ignore the fact that the relationship at the individual level is quite different from the relationship at the species level. We could incorrectly conduct a standard multiple regression of relative growth rate on both seed size and average disturbance frequency. This mistake is like our previous example of correlating eye colour and hair length. The average disturbance frequency is the same for all individuals within a given species; we would be inflating the number of 'pieces' of information that we really possess. Instead, we should take an explicit multilevel approach.

First, we model the individual-level relationship between RGR and seed size: $RGR_{ij} = \beta_{0j} + \beta_1 seedsize + \varepsilon_{ij}$. Here, we are specifying that relative growth rate varies at both the individual (i) level and the species (j) level. The intercept (β_{0j}) varies only at the level of species. The slope of the relationship between RGR and seed size (β_1) is constant across all species. Finally, the residual variation in RGR within each species (ε_{ij}) is assumed to be normally distributed with a constant variance.[10] Since the intercepts of RGR are, themselves, random variables that change from species to species, we next model this species-level variation in these intercepts: $\beta_{0j} = \alpha_{00} + \alpha_{01} disturbance_j + \mu_{0j}$. Now we are specifying that the species-level intercepts are functions of the average disturbance frequency, which is a species-level variable. There is a constant intercept (α_{00}) term, which represents the average seed size across all species in the statistical population. The species-level slope between average disturbance frequency and the intercepts of each species is α_{01}. Finally, the deviations of each species' intercept from that predicted by average disturbance frequency is the random variable μ_{0j}. Putting this all together we get $RGR_{ij} = \alpha_{00} + \beta_1 seedsize_{ij} + \alpha_{01} disturbance_j + \mu_{0j} + \varepsilon_{ij}$. The standard error of the slope of seed size is based on the error variance at the level of individuals within species (ε_{ij}), which has been corrected for the species-level variation. The slope of the average disturbance frequency is based on the error variance at the level of species (μ_{0j}), which has been corrected for the error variance at the individual level.

We next specify a multilevel model for seed size. According to Figure 7.4, seed size is caused only by average disturbance frequency, and this effect occurs only at the species level. Our multilevel model is therefore $seedsize_{ij} = \alpha_{00} + \alpha_{01} disturbance_j + \mu_{0j} + \varepsilon_{ij}$. If I fit these models using the mixed model function lme() in the nlme library of R then I obtain the following results:

seed size = 443.76 − 16.32 disturbance frequency
RGR = −0.284 + 0.0006 seed size + 0.0217 disturbance frequency

The residual variation of the mean seed sizes per species was 55.28 (13 per cent) and the residual variation of individual seed sizes was 361.76 (77 per cent). The residual

---

[10] Both the assumption of normality and the assumption of constant variance can be relaxed in multilevel modelling.

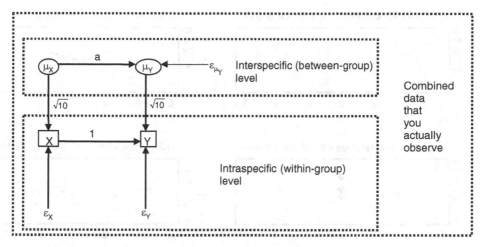

Figure 7.5 A very simple two-level model

Notes: In this scenario there are two attributes (X and Y) of each individual from each species. Each species has a mean value of each variable (μ_{Xi} and μ_{Yi}) and each individual has a value of each variable that varies around its species mean ($X_{ij} = \mu_{Xi} + \varepsilon_{Xij}$ and $Y_{ij} = \mu_{Yi} + 1X_{ij} + \varepsilon_{Yij}$). Variable ε_{Xij} takes values from -0.25 to 0.25.

variation of the mean RGR values per species was 4.96×10^{-5} (64 per cent) and the residual variation of the individual RGR values was 2.79×10^{-5} (46 per cent). The big difference between this multilevel model and an ordinary regression model is best seen in the standard errors of the parameters of the equation for RGR. If we do an ordinary regression then we get an estimate of the standard error of the slope for average disturbance frequency of 1×10^{-4}, while the standard error estimated from the multilevel model was 8.3×10^{-4}. In other words, by ignoring the hierarchical nature of RGR the ordinary regression overestimated the precision of the slope by eight times. Since significance tests of the effects of a variable in a regression are based on these standard errors, the effect of ignoring the partially dependent nature of observations is to produce probability estimates that are much smaller than they should be.

Hierarchically ordered data not only cause problems for parameter estimation and inferential tests of significance. The patterns that can be generated with multilevel data can often be downright counter-intuitive. To give you a feeling for such patterns, I have generated data from a very simple two-level model, shown in Figure 7.5. In this scenario there are two attributes (X and Y) of each individual i (1 to 11) from each species j (1 to 10). Each species has a mean value of each variable (μ_{Xj} and μ_{Yj}) and each individual has a value of each variable that varies around its species mean ($X_{ij} = \mu_{Xj} + \varepsilon_{Xij}$ and $Y_{ij} = \mu_{Yj} + 1X_{ij} + \varepsilon_{Yij}$). Variable ε_{Xij} takes values from -0.25 to 0.25. Since we are interested in comparing the within-species and between-species patterns, I will ignore the random variation of Y at the individual level (ε_{Yij}) and concentrate on the expected value of y ($E[Y|X] = \mu_{Yj} + 1X_{ij}$). Substituting for X_{ij} we get $E[Y|X] = \mu_{Yj} + 1\mu_{Xj} + 1\varepsilon_{Xij}$). This equation represents the intraspecific (i.e. within-group) level. The interspecific (i.e. between-group level) will be generated in different ways,[11] and the combined

---

[11] Ignore for now the fact that the path coefficient from each μ to the observed variable is fixed at the square root of the number of individuals per species. This will be explained later.

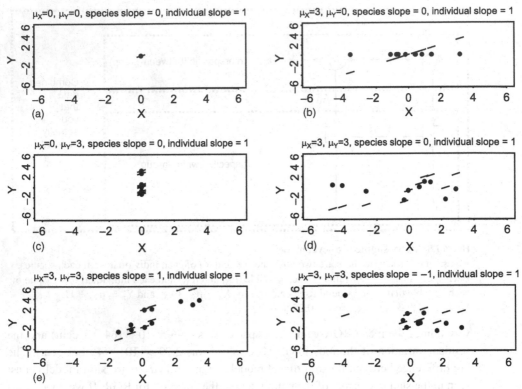

Figure 7.6 Simulated data based on Figure 7.5 based on six different scenarios
Notes: The six scenarios involve the population mean values (μ_X, μ_Y), the species-level slope linking these population values, and the individual-level slope. The lines show the systematic relationships within each species and the solid circles show the measured mean values of X and Y for each species.

data are then shown in Figure 7.6. We imagine that the mean value of variables X and Y (μ_X and μ_Y) differ randomly between the 10 species and that the interspecific relationship between these interspecific means follows the following equation $\mu_Y = a\mu_X + \varepsilon_{\mu X}$.

First, let's see what happens when there is no species-level variation or covariation. To simulate this, I set a in Figure 7.5 equal to zero and make the values of μ_X and μ_Y the same (zero) for all 10 species (thus, the variance of these two variables is zero). If we put these values into our generating equation we obtain $E[Y|X] = (0) + 1(0) + 1\varepsilon_{Xij}$. Figure 7.6(a) shows the resulting pattern. There are actually 10 lines in this figure but they are superimposed on each other since we have exactly the same values for all 10 species (remember that I am plotting the expected values). The solid circles show the mean values of X and Y for each species. Since all 10 species have the same means, these 10 circles are also superimposed.

In Figure 7.6(b) I simulate what happens if each species has a different mean value for X but has the same mean value for Y – that is, I allow μ_X to randomly vary but not μ_Y. All 10 lines – one for each species – appear to line up along the same trend as observed at the intraspecific level. Notice that species whose mean for X (μ_{Xj}) is less than other species have their individual values of both X and Y (their line in the graph) in the lower left. Similarly, those species whose mean for X (μ_{Xj}) is greater than other species have their individual values of both X and Y (their line in the graph) in the upper right. In other words, there is a positive correlation between the mean values of X and Y of these 10 species even though there is no real relationship between the species means μ_{Xj} and μ_{Yj}. To see why, we have only to write the generating equation for this simulation: $E[Y|X] = (0) + 1\mu_{Xj} + 1\varepsilon_{Xij}$. Whenever the mean value of X for a given species (μ_{Xj}) happens, by chance, to be less than average, this decreases the values of Y that individuals of this species will possess. Similarly, whenever the mean value of X for a given species (μ_{Xj}) happens, by chance, to be greater than average, this increases the values of Y that individuals of this species will possess. The observed interspecific correlation that we observe is simply an artefact of mixing together the two levels of variation.

In Figure 7.6(c) I simulate what happens if each species has a different mean value of Y but the same mean value of X – that is, I allow μ_Y to randomly vary but not μ_X. Returning to our generating equation and substituting, we get $E[Y|X] = \mu_{Yj} + 1(0) + 1\varepsilon_{Xij}$. The result is a series of lines stacked on top of each other. The overall correlation between X and Y is severely diluted.

In Figure 7.6(d) I simulate what happens if each species has a different mean value of both X and Y but there is still no true interspecific relationship between these mean values – that is, I allow both μ_X and μ_Y to randomly and independently vary. The result is intermediate between panels (b) and (c).

In Figure 7.6(e) I simulate what happens when each species has a different mean value of both X and Y and there is also a positive interspecific relationship between these mean values. To do this, I allow μ_X and μ_Y to randomly vary but link them: $\mu_{Yj} = 1\mu_{Xj} + \varepsilon_{\mu Xj}$. Substituting this into the generating equation, I get $E[Y|X] = \mu_{Yj} + 1\mu_{Xj} + 1\varepsilon_{Xij} = 2\mu_{Xj} + \varepsilon_{\mu Xj} + 1\varepsilon_{Xij}$. The result is that the slope between the means appears twice as large as it really is.

Finally, in Figure 7.6(f) I simulate what happens when each species has a different mean value of X and Y and there is also a negative interspecific relationship between these mean values. In other words, the interspecific relationship is the opposite of the intraspecific relationship. To do this, I allow μ_X and μ_Y to randomly vary but link them: $\mu_{Yj} = -1\mu_{Xj} + \varepsilon_{\mu Xj}$. Substituting this into the generating equation, I get $E[Y|X] = \mu_{Yj} + 1\mu_{Xj} + 1\varepsilon_{Xij} = 0(\mu_{Xj}) + \varepsilon_{\mu Xj} + 1\varepsilon_{Xij}$. The result is that the correlation between the means disappears even though there are really strong (but opposite) relationships between the variables at both hierarchical levels. The moral of this simple set of simulations is that combining data that have relationships at different levels and analysing them as if the hierarchical structure did not exist can lead to incorrect conclusions.

There is much more to be said about multilevel regression than has been said so far. What I have described is not so much an introduction as an appetiser, and the interested reader should consult the references given earlier in this chapter for more information. Now that we recognise the problem that hierarchically organised data can cause, and have an intuitive understanding of how multilevel regression deals with

Table 7.2 Decomposition of the variance in a one-way analysis of variance

Source of variance	Sum of squares	Degrees of freedom	Variance
Total	$\sum_{i=1}^{G}\sum_{j=1}^{N_i}(Y_{ij}-\bar{Y})^2$	N-1	$\frac{\sum_{i=1}^{G}\sum_{j=1}^{N_i}(Y_{ij}-\bar{Y})^2}{N-1}$
Between groups	$C\sum_{j=1}^{G}(\bar{Y}_{.j}-\bar{Y})^2$	G-1	$\frac{C\sum_{j=1}^{G}(\bar{Y}_{.j}-\bar{Y})^2}{G-1}$
Within groups	$\sum_{j=1}^{G}\sum_{i=1}^{N_j}(Y_{ij}-\bar{Y}_{.j})^2$	N-G	$\frac{\sum_{j=1}^{G}\sum_{i=1}^{N_j}(Y_{ij}-\bar{Y}_{.j})^2}{N-G}$

the problem, let's see how these notions can be incorporated into structural equation modelling.[12]

7.4 Multilevel SEM

Lavaan is not currently written to incorporate multilevel SEM. However, it is possible to 'trick' lavaan into modelling multilevel SEM for random intercepts (but not slopes) by setting up the model as if it was a multigroup analysis with a particular pattern of fixed values for some parameters. There is currently much work on more general versions of multilevel SEM with both random slopes and intercepts, via full-information maximum likelihood, and I hope that they will be included in new releases of lavaan.[13]

Suppose that we have a variable that has been measured on N observations. These observations are organised into G groups with N_1, N_2, ..., N_G observations in each group; for the moment we will assume that there are the same number (C) of observations in each group. I write Y_{ij} to mean the ith observation in group j and I write $\bar{Y}_{.j}$ to mean the mean of the value in the jth group. The deviation of this value from the overall mean is $(Y_{ij}-\bar{Y})$. We can decompose this deviance as follows: $(Y_{ij}-\bar{Y}) = (\bar{Y}_{.j}-\bar{Y})+(Y_{ij}-\bar{Y}_{.j})$. This leads to the one-way ANOVA table that is fondly remembered by everyone who has taken an introductory course in statistics (Table 7.2).

The above decomposition has the useful property that the between-group deviations have zero correlation with the within-group deviations. Remembering that a variance is simply the covariance of a variable with itself, we can do the same trick with covariances. In this way, we can define both a pooled within-group covariance matrix (S_{PW}) and a between-group covariance matrix (S_B) for our data. The pooled within-group covariance matrix is constructed by first centring each variable by its group mean,

---

[12] Although I do not discuss multilevel tests of path models based on the method presented in Chapter 3, the reader should be aware that multilevel regression methods can be used to obtain probability estimates of conditional independence by fitting a series of regressions in accordance with the hypothesised causal graph. These probability estimates can then be combined using Fisher's C statistic.

[13] An R library, called xxM, is currently under development that uses these newer methods. This can be obtained at http://xxm.times.uh.edu.

calculating the sum of squares and cross-products of these group-centred variables and then dividing by N-G. One easy way to obtain this matrix from any statistical program is to simply calculate the covariance matrix of the centred variables (which has a denominator of N − 1) and multiply by (N − 1)/(N − G). The between-group covariance matrix is constructed by calculating the sum of squares and cross-products of the group means and dividing by G − 1. An easy way to obtain this matrix from any statistical program is to simply calculate the covariance matrix of the group means and multiply by C, which will have a denominator of G − 1.

The sample pooled within-group covariance matrix (S_{PW}) is an unbiased estimate of the population within-group covariance (Σ_{PW}) matrix. Unfortunately, the sample between-group covariance matrix (S_B) is not a simple estimator of the population between-group covariance matrix (Σ_B); instead, it estimates $S_B = \Sigma_{PW} + C\Sigma_B$. If you look carefully at this equation then you will notice that it looks suspiciously like the multigroup structural equations formulation that we studied earlier in this chapter, with two 'groups'. We can therefore trick commercial SEM programs into fitting a multilevel SEM by treating it like a multigroup SEM with particular cross-group constraints.

To set up the analysis we trick lavaan into thinking that it is actually conducting a multigroup analysis with two groups. The first 'group' represents the group-centred data, for which there is the pooled sample covariance matrix obtained from the group-centred variables (S_{PW}) based on N − G 'observations'. To get this matrix you will create a new data set by (a) calculating the mean values of each variable for each group and (b) subtracting the mean values of the variables of the corresponding group from each observation. If this new matrix is called temp then the resulting covariance matrix is obtained by `cov(temp)*(N-1)/(N-G)`, where N is the total number of observations in the original data set and G is the total number of groups. This is the level-1 covariance matrix. We specify our within-group causal structure for this 'group'.

Next, we define a second 'group', for which we have the sample between-group matrix (S_B) based on G 'observations', where G is the number of groups in the multilevel model. To get this covariance matrix we create a new data set (call it temp2) consisting of the group means of our variables and then calculate `C*cov(temp2)`, where C is the number of observations per group. For this second 'group' we specify both the within-group causal structure and the between-group causal structure. These two causal structures are linked by latent variables that represent the true values of the group means in the statistical population. Why? Because our calculated group means are only estimates of these underlying (i.e. latent) parameters. Since the variances and covariances of the group means are multiplied by the constant C (the number of individuals within each group), we fix the path coefficients leading from these latent variables to the individual variables by $\sqrt{C}$.[14] Finally, we must constrain all the free parameters in the first 'group' (i.e. the model at the level of the individual) to be equal to the equivalent parameters in the second group (i.e. those parameters in the second group dealing with the model for the individual level). When we fit this model to our

---

[14] If we have an equation $Y = aX + e$ then the variances are $Var(Y) = a^2 Var(x) + Var(e)$. By setting the path coefficient from the latent variable representing the group mean to the individual-level variable at $\sqrt{C}$ (i.e. $L = \sqrt{C}X + e$) then we obtain $Var(L) = C*Var(X) + var(e)$.

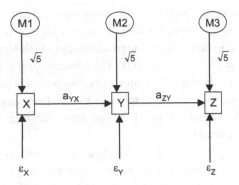

Figure 7.7 A two-level model involving three variables (X, Y and Z) and their population means

data the estimation procedure will correct for the partial non-independence of our data due to their hierarchical nature.

When we have different numbers of observations per group, we can calculate an approximate scaling factor due to Muthén:

$$C = \frac{N^2 - \sum_{i=1}^{G} N_i^2}{N(G-1)}$$

Estimates based on this scaling factor have been shown to be fairly accurate so long as the group sizes are not extremely different (Hox 1993; McDonald 1994; Muthén 1994). Of course, when group sizes are equal this reduces to the common group size. The parameter estimates, standard errors and maximum-likelihood chi-square statistics are still asymptotic values but now the requirement for sufficient sample sizes applies both at the level of individuals and at the level of groups. Note that, at the level of groups, we are considering random samples of means. This means that the central limit theorem applies and the distribution of means will be closer to multivariate normal than the distribution of the actual values is.

In order to better understand how to interpret such multilevel structural equation models, I will analyse simulated data generated by different models. First, let's see what happens in the simplest case, when all the observations are really independent observations generated by the same causal process – that is, when there is really no group-level structure and all variances and covariances exist at the individual level. To do this, I generate 200 independent observations from the following equations:

$$X = N(0, 1)$$
$$Y = 0.5X + N(0, 0.75)$$
$$Z = 0.5Y + N(0, 0.75)$$

Now, I randomly divide these 200 independent observations into 40 groups of five observations each. Since this assignment is completely random, the only variation between the group means is due to sampling variation, and the systematic variation in the group means is zero. Figure 7.7 shows the multilevel model. The variables M1, M2 and M3 are latent variables representing the population means of each variable centred around the overall mean of each variable. Since there are five observations per group, the scaling constant C is 5. Since there are no causal paths linking these latent variables, we are assuming that there is no covariance between them, although we could allow for such covariances; this will be shown in a later example.

I now fit two models, one nested within the other. First, I fit the model shown in Figure 7.7 (remembering to constrain all the free parameters at level 1 to be equal in the model associated with the within-groups covariance matrix and the model associated with the between-groups covariance matrix) while fixing the variance of the latent M1, M2 and M3 to zero. The model in Figure 7.7 is the between-groups model that necessarily contains the within-groups model $(X \rightarrow Y \rightarrow Z)$ nested within it. By fixing the variance of the latents M1, M2 and M3 to zero I am assuming that the variance in the population means between groups are zero for all three variables and that any observed variance at this group level is due only to sampling variation. This assumption is, of course, correct for these data. The resulting model gives a maximum-likelihood chi-square statistic of 7.611 (7 df) for a probability of 0.368.

There were two covariance matrices, the within-groups covariance matrix and the between-groups covariance matrix, and each had $3 \times 4/2 = 6$ non-redundant elements. Therefore, we had 12 non-redundant elements in total. The within-groups model had to estimate five free parameters (a_{YX}, a_{ZY}, ε_X, ε_Y and ε_Z). The between-groups model also had to estimate the same five free parameters, since the variances of the latents M1, M2 and M3 were fixed at zero. I also had to constrain the five free parameters associated with the within-groups model to be equal to these same free parameters in the between-groups model. Therefore, I had $12 - 5 = 7$ degrees of freedom.

Since this model, with the variances of the latents M1, M2 and M3 fixed to zero, provides a non-significant probability level, we could stop there. However, to test the hypothesis that there is really no group-level variance contributing to X, Y or Z, I now refit the model by allowing the variances of the latent M1, M2 and M3 to be freely estimated. This new model gives a maximum-likelihood chi-square statistic of 7.475 (4 df) for a probability level of 0.113. Note that we have reduced the degrees of freedom from seven to four because we are now estimating three parameters that were previously fixed. Since this model is nested within the first, we can calculate the probability that the variances of M1, M2 and M3 really were zero by calculating the difference in the chi-square statistics (0.136) with three degrees of freedom. The resulting probability level (0.987) tells us that the observed variation in the group means was very likely to have been due only to sampling variation. If we go back to our first model and look at the estimated variances of M1, M2 and M3 and the standard variation of these estimates, we find that each is very small and less than one standard error from zero. Here is the lavaan code:

```
mod←"
# This next line tells lavaan that there are different
# causal #models for different groups.
between groups:
#latent variances connect to individual level variables
#(x,y,z)for this between-groups model
# multiply by sqrt(5) since there are 5 obs per group.
LMx =~sqrt(5)*x
```

```
LMy =~sqrt(5)*y
LMz =~sqrt(5)*z
# this next part exists for both levels. I name the free
# parameters to constrain them to be equal between "groups"
y~a*x
z~b*y
# no covariances between latents
LMx~~0*LMy
LMx~~0*LMz
LMy~~0*LMz
LMx~~start(1)*LMx
LMy~~start(1)*LMy
LMz~~start(1)*LMz
# I name the residual variances of x,y,z to constrain them
# to be #equal between the two "groups"
x~~start(1)*Va*x
y~~start(1)*Vb*y
z~~start(1)*Vc*z
# Now enter the within-group model, which has no latents.
within groups:
# Notice that the free parameters have the same names in both
# "groups"
y~a*x
z~b*y
x~~start(1)*Va*x
y~~start(1)*Vb*y
z~~start(1)*Vc*z
"
combined.cov←list(between = between.cov,within = within.cov)
combined.n←list(between = 40,within = 200-40)
fit←sem(model = mod,sample.cov = combined.cov,sample.nobs =
combined.n,fixed.x = FALSE)
```

The first simulation exercise was simply to show you that, if there really is no group-level variation that results in partial non-independence of the observations, the multilevel model will detect this fact. Next, let's look at a case in which there really is group-level variation, but not group-level covariation. This time we generate our 200 observations according to the following equations:

(a) for the 40 groups:

$$\mu_{Xj} = N(0, 1)$$
$$\mu_{Yj} = N(0, 1)$$
$$\mu_{Zj} = N(0, 1)$$

(b) for each of the five observations within each of the 40 groups:

$$X_{ij} = \mu_{Xj} + N(0, 1)$$
$$Y_{ij} = \mu_{Yj} + 0.5X_{ij} + N(0, 1)$$
$$Z_{ij} = \mu_{Zj} + 0.5Y_{ij} + N(0, 1)$$

Note that, in this simulation, each variable receives variation from two sources: the variation at level 1 (the group level) due to the random variation of each of the group means, and the variation at level 2. I again fit two nested models, as in the first example. In the first model I fix all the group-level variation associated with the latent M1, M2 and M3 to zero (see Figure 7.7) – i.e. I include the commands LMx~~0*LMX etc. in the between-group model. This model does poorly (MLX^2 = 197.079, 7 df, p < 10^{-7}), as it should. Since the variance of the group-level latent means was fixed at zero then all the group-level error variance is incorrectly forced down to the within-groups level. The resulting estimates of the three within-groups error variances are 1.647, 1.904 and 1.857 instead of the correct value of 1. Now I refit the model, but allow the variances of the latent population means (M1, M2 and M3) to be freely estimated. This time the model provides an adequate fit (MLX^2 = 6.931, 4 df, p = 0.140), as it should. Since this model is nested within the first, the difference in the maximum-likelihood chi-square statistic tests the hypothesis that there was group-level variance. This difference is highly significant (MLX^2 = 190.148, 3 df, p < 10^{-7}). Both the group-level variances and the individual variances are correctly estimated. If we ignore the multilevel nature of these data and simply put all 200 observations into the same data set and fit the model X→Y→Z, the model fails (MLX^2 = 6.206, 1 df, p = 0.013). This is a general result (Muthén 1997).

Up to now we have seen how ignoring the multilevel nature of data can result in improper parameter estimates and probability levels. The real strength of multilevel SEM is that we can actually model how the group-level variables interact separately from the level 1 variables, and how these two levels interact together. To show this, I will simulate data from the process shown in Figure 7.8.

This model has four observed variables. Three of these variables (the number of seeds produced per plant, the seed size for an individual plant and the relative growth rate of an individual plant) are properties of individual plants, and form the within-group model. The fourth variable (the average disturbance frequency of the habitat) is a property of each species – that is, individuals within a given species tend to be found in habitats with the same average frequency of disturbance. The model proposes that, at the level of individuals, increasing seed size causes an increase in relative growth rate but a decrease in the number of seeds produced. At the level of species, selection for habitats of different disturbance frequencies results in species that are adapted to frequently disturbed habitats (early during secondary succession) producing fewer seeds (because they tend to be smaller plants), as well as smaller seeds on average, and seedlings having faster average relative growth rates. Note that the relationship between seed size and RGR is positive at the level of individuals but negative at an interspecific level. Figure 7.9 shows the simulated data set.

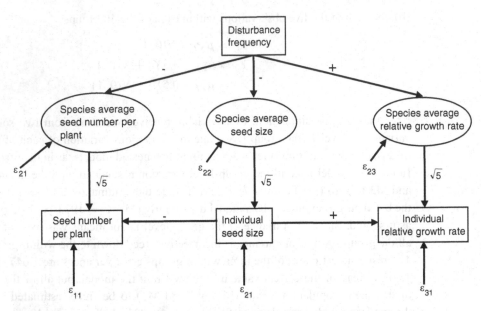

Figure 7.8 A hypothetical causal structure involving four observed variables and three latents
Notes: One of the observed variables ('Disturbance frequency') is a species-level variable that is a common cause of the three latent species' means. The other three observed variables are individual-level variables that are caused by the latent species' means.

Before analysing these simulated data, I want you to notice some interesting trends. First, there is a negative relationship between seed size and RGR. This is because the species-level effect of selection, based on average disturbance frequency, dominates the within-species tendency for larger seeds to increase RGR. Second, notice that there is a positive relationship between seed number and disturbance frequency even though the direct effect of disturbance frequency on these two variables, at an interspecific level, is negative. This is because an increasingly disturbed habitat selects for species with smaller seeds on average and this, in turn, reduces the seed size within such species. However, smaller seeds increase seed number within a given species, resulting in the overall effect along this path being positive. Since this path dominates the direct effect at the species level, we see a positive overall relationship.

Now we fit a series of nested models. First, we specify no variance or covariance at the species level. This model is rejected ($MLX^2 = 746.249$, 7 df, $p < 10^{-10}$). Next we allow variance, but not covariance, at the species level. This nested model is also rejected ($MLX^2 = 51.385$, 4 df, $p < 10^{-10}$), but the change in the maximum-likelihood chi-square statistic is significant ($MLX^2 = 694.864$, 3 df, $p < 10^{-10}$), showing that there is significant species-level variation. Finally, we allow the three latent variables to freely covary amongst themselves. This model, nested within the second, provides an acceptable fit ($MLX^2 = 2.096$, 1 df, $p = 0.148$), and the change in the maximum-likelihood chi-square statistic is significant ($MLX^2 = 49.289$, 3 df, $p = 1.13 \times 10^{-10}$), showing that there is also species-level covariation.

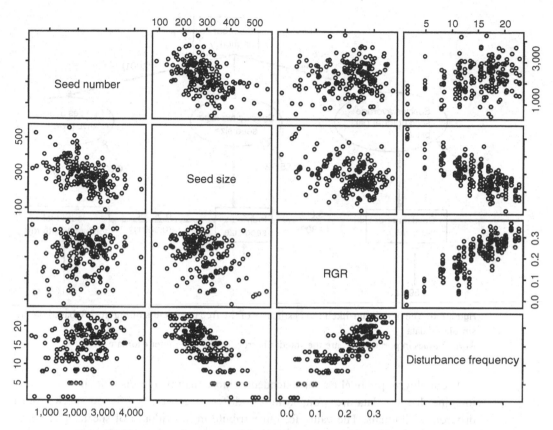

Figure 7.9 The scatterplot matrix of the simulated data generated by the causal structure shown in Figure 7.8

Since there is significant covariation between the latent species-level means, we introduce the species-level variable 'Average disturbance frequency' and specify that this variable is the sole common cause of these three species-level variables. This model (which is the true model that generated these data) also provides an acceptable fit (MLX^2 = 5.123, df = 4, p = 0.275). Figure 7.10 shows the final parameter estimates and their standard errors in parentheses.

The parameter estimates are all close to the true values that I had simulated, and all are within the approximate 95 per cent confidence intervals (two times the standard errors). This model is quite remarkable. Not only have we been able to account for the partial non-independence of the data within each species due to their hierarchical nature but we have also been able to separate the within-species structure from the between-species structure and link the two hierarchical processes together. Although the data were simulated, they are biologically realistic. Natural selection based on some average environmental property determines the average values of attributes shown by different species and the covariation between these average values. These average values then limit the range of values shown by particular individuals within each species but still allow variation between individuals and covariation of the attributes at the level of individuals.

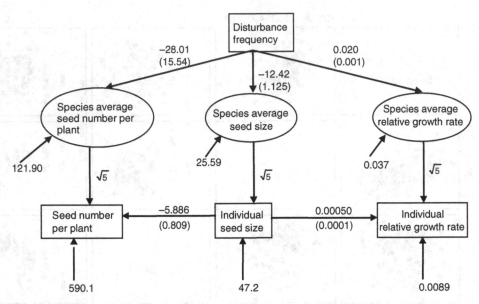

Figure 7.10 The maximum-likelihood estimates of the free parameters of Figure 7.8, based on the simulated data
Note: Values in parentheses are the standard errors of the parameter estimates.

The multigroup model can be extended to more than two levels. For simplicity, let's imagine that our data are grouped into G different genera, S different species and I different individuals. The value for our attribute in individual i of species j of genus k is Y_{ijk}. Now we can write $Y_{ijk} = (\bar{Y}_{..k} - \bar{Y}) + (\bar{Y}_{.jk} - \bar{Y}_{..k}) + (Y_{ijk} - \bar{Y}_{.jk})$, where $\bar{Y}$ is the grand mean over all I observations, $\bar{Y}_{..k}$ is the mean of Y for each genus and $\bar{Y}_{.jk}$ is the mean of Y for each species. It follows that we can decompose the sum of squares of Y_{ijk} into a term representing the deviations of each genus mean from the grand mean $(\bar{Y}_{..k} - \bar{Y})$, a term representing the deviations of each species from its genus mean $(\bar{Y}_{.jk} - \bar{Y}_{..k})$ and a term representing each individual from its species mean $(\bar{Y}_{ijk} - \bar{Y}_{.jk})$. Following exactly the same logic as we used to derive the two-group multilevel model, we can therefore obtain a genus-level sample covariance matrix (S_G), a species-level pooled sample covariance matrix (S_{PS}) and an individual-level pooled sample covariance matrix (S_{PI}):

$$S_{PI} = \frac{\sum_{k=1}^{G} \sum_{j=1}^{S} \sum_{i=1}^{I} (Y_{ijk} - \bar{Y}_{.jk})^2}{N - S}$$

$$S_{PS} = \frac{\sum_{k=1}^{G} \sum_{j=1}^{S} N_{.jk}(\bar{Y}_{.jk} - \bar{Y}_{..k})^2}{S - G}$$

$$S_{G} = \frac{\sum_{k=1}^{G} N_{..k}(\bar{Y}_{..k} - \bar{Y})^2}{G - 1}$$

If our data are completely balanced with N_G species per genus and N_S individuals per species, we can write: $\Sigma_T = N_G \Sigma_G + N_S \Sigma_S + \Sigma_I$. If this is not the case then we have to use Muthén's approximate scaling factor for each level of the hierarchy:

$$C_L = \frac{N^2 - \sum_{i=1}^{N_L} N_{iL}^2}{N(N_L - 1)}$$

Here, C_L is the scaling factor for level L of the hierarchy, N is the total number of observations in the data set, N_L is the total number of units at level L (for instance, the number of species or genera) and N_{iL} is the number of units within unit i of level L. In this way, we can model structural relationships at various levels of organisation and account for partial non-independence due to common ancestry at various taxonomic levels. Of course, these levels do not have to represent traditional taxonomic classifications. For instance, you might have measures at different times for the same individual, defining a within-individual level. Some readers might have noticed that the above description looks a lot like a nested type II ANOVA. Box 7.1 describes the relationship for those who are interested.

Box 7.1

Consider a typical balanced ANOVA table for a nested analysis with three levels. There are 160 observations. These observations are grouped into $N_1 = 40$ level-1 groups, $N_2 = 80$ level-2 groups (i.e. two level-2 groups per level-1 group) and $N_3 = 160$ level-3 observations (two level-3 observations per level-2 group).

Source	Df	SS	MS	Expected MS
Level-1	$N_1 - 1 = 40 - 1$	SS_1	$SS_1/(N_1 - 1)$	$\sigma^2_{L_3 \subset L_2} + C_2\sigma^2_{L_2 \subset L_1} + C_1\sigma^2_{L_1}$
Level-2	$N_2 - N_1 = 80 - 40$	SS_2	$SS_2/(N_2 - N_1)$	$\sigma^2_{L_3 \subset L_2} + C_2\sigma^2_{L_2 \subset L_1}$
Level-3	$N_3 - N_2 = 160 - 80$	SS_3	$SS_3/(N_3 - N_2)$	$\sigma^2_{L_3 \subset L_2}$
Total	$N_3 - 1 = 160 - 1$	SS_T	$SS_T/(N_3 - 1)$	

Here, the notation $\sigma^2_{L_3 \subset L_2}$ means the variation of the level-3 units nested within the level-2 units. C_2 is the number of level-3 units within each level-2 unit and C_1 is the number of level-2 units within each level-1 unit. We see that the higher-level mean squares (MS: they are sample variances if the units are randomly sampled) do not estimate variation unique to that level but a weighted sum of variation at that level and all levels below it.

If we subtract the variance (i.e. MS) at a given level with the variance directly below it in the hierarchy, and divide by the number of observations per unit (i.e. C_i), then we obtain an estimate of the *variance component* at that level. For instance, to obtain the variance component at level 1 we write

$$\frac{[(\sigma^2_{L_3 \subset L_2} + C_2\sigma^2_{L_2 \subset L_1} + C_1\sigma^2_{L_1}) - (\sigma^2_{L_3 \subset L_2} + C_2\sigma^2_{L_2 \subset L_1})]}{C_1} = \sigma^2_{L_1}$$

We estimate this variance component by calculating

$$\frac{[MS_{L_1} - MS_{L_2}]}{C_1}$$

The variance at a given level measures the total amount of variation that is found at that level. However, such variation is due to the combined effect of variation at lower levels and the added variation contributed at that level. The variance components measure the amount of added variation at each level. Usually, one expresses these variance components as percentages of the total variation.

A variance (or a sum of squares) is simply a special type of covariance (or sum of cross-products) – namely the covariance of a variable with itself. We can therefore apply the same logic to each variance and covariance in a covariance matrix. If we measure a whole set of variables on each observational unit instead of just one then we can produce a table that summarises the decomposition of the entire covariance matrix. Rather than sums of squares (SS), we would calculate sums of squares and cross-products (SSCP). Rather than mean squares (variances) we would calculate mean squares and cross-products (MSCP) – i.e. covariances.

Source	Df	SSCP	MSCP	Expected MSCP
Level-1	$N_1 - 1 = 40 - 1$	$SSCP_1$	$SSCP_1/(N_1 - 1)$	$\Sigma^2_{L_3 \subset L_2} + C_2\Sigma^2_{L_2 \subset L_1} + C_1\Sigma^2_{L_1}$
Level-2	$N_2 - N_1 = 80 - 40$	$SSCP_2$	$SSCP_2/(N_2 - N_1)$	$\Sigma^2_{L_3 \subset L_2} + C_2\Sigma^2_{L_2 \subset L_1}$
Level-3	$N_3 - N_2 = 160 - 80$	$SSCP_3$	$SSCP_3/(N_3 - N_2)$	$\Sigma^2_{L_3 \subset L_2}$
Total	$N_3 - 1 = 160 - 1$	$SSCP_T$	$SSCP_T/(N_3 - 1)$	

The variance components can be extracted from the diagonal elements of these covariance matrices.

One important type of multilevel model involves repeated measurements over time on the same set of individuals. In such a sampling design, the first level would be the intra-individual level (i.e. variation over time in the same individual); this is analogous to 'repeated-measures' analyses, with which many biologists are familiar. Unfortunately, I am not aware of any multilevel models that have been used in a biological context, and only very few in any other context (Muthén 1990; 1994; 1997; Muthén and Satorra 1995). Nonetheless, I suspect that multilevel models will become very important in biology, since hierarchies are so ubiquitous.

8 Exploration, discovery and equivalence

8.1 Hypothesis generation

If this were a textbook of statistics then this chapter would not exist. Modern statistics is almost entirely concerned with *testing* hypotheses, not *developing* them. Such a bureaucratic approach views science as a compartmentalised activity in which hypotheses are constructed by one group, data are collected by another group and then the statistician confronts the hypothesis with the data. Since this book is a user's guide to causal modelling, such a compartmentalised approach will not do. One of the main challenges faced by the practising biologist is not in testing causal hypotheses but in developing causal hypotheses worth testing.

If this were a book about the philosophy of science then this chapter might not exist either. The philosophy of science mostly deals with questions such as: how can we know if a scientific hypothesis is true or not? What demarcates a scientific hypothesis from a non-scientific hypothesis? For most philosophers of science, the question of how one looks for a useful scientific hypothesis in the first place is someone else's problem. For instance, Karl Popper, in his influential *Logic of Scientific Discovery* (Popper 1980: 32), says that 'there is no such thing as a logical method of having new ideas, or a logical reconstruction of this process. My view may be expressed by saying that every discovery contains "an irrational element", or "a creative intuition"...' Later, he says that '[scientific laws] can only be reached by intuition, based on something like an intellectual love of the objects of experience'. Again, one gets the impression that science consists to two hermetically sealed compartments. One compartment, labelled *hypothesis generation*, consists of an irrational fog of thoughts and ideas, devoid of method, out of which a few gifted people are able to extract brilliant insights. The other compartment, labelled *hypothesis testing*, is the public face of science. Here, one finds method and logic, in which established rules govern how observations are to be taken, statistically manipulated and interpreted.

At a purely analytic level there is much to be gained by taking this schizophrenic view of the scientific process. After all, how a scientific idea is developed is irrelevant to its truth. For instance, the history of science documents many important ideas whose genesis was bizarre.[1] Archimedes reportedly discovered the laws of hydrostatics after

---

[1] The appendix of *The Art of Scientific Investigation* (Beveridge 1957) lists 19 cases in which the origin of important scientific ideas arose from bizarre or haphazard situations. In fact, Beveridge devotes an entire chapter to the importance of chance in scientific discovery.

jumping into a bathtub full of water. Kukulé discovered the ring structure of benzene after falling asleep before a fire and dreaming of snakes biting their tails. These curious stories are entertaining, but we remember them only because the laws of hydrostatics hold and benzene really does have a ring structure. As a public activity, science is interested in the result of the creation, not in the creative act itself.

The day-to-day world of biology does not exist at such a purely analytic level. Although it is possible to conceptually divide science into distinct hypothesis-generating and hypothesis-testing phases, the two are often intimately intertwined in practice. When the two are not intertwined the science can even suffer. Peters (1991), in his *A Critique For Ecology*, points out that, because empirical and theoretical ecology are often carried out by different people, the result is that much ecological theory is crafted in such a way that it cannot be tested in practice, and much of field ecology cannot be generalised because it is not placed into a proper theoretical perspective. In this context I like the citation, attributed to W. H. George, given at the beginning of Beveridge's (1957) *The Art of Scientific Investigation*: 'Scientific research is not itself a science; it is still an art or craft.' Unlike the assembly-line worker who receives a partly finished object, adds to it and then passes it along to someone else, the craftsman must construct the object from start to finish. In the same way, the craft of causal modelling consists as much of the generation of useful hypotheses as of their testing. Certainly, hypothesis generation is more art than method, and hypothesis testing is more method than art, but this does not mean that we must relegate hypothesis generation to a mystical world of creative intuition in which there are no rules. The purpose of this chapter is to describe reliable methods of generating causal hypotheses.

8.2   Exploring hypothesis space

How does one go about choosing promising hypotheses concerning causal processes? To place the problem in context, imagine that you have collected data on N variables, and at least some of these variables are not amenable to controlled randomised experiments. Why you suspect that these N variables possess interesting or important causal relationships may well be due to the irrational creative intuition to which Popper refers, but you are still left with the problem of forming a multivariate hypothesis specifying the causal connections linking these variables.

To simplify things, let's assume that all the data are generated by the same unknown causal process (i.e. causal homogeneity), that there are no latent variables responsible for some observed associations (i.e. causal sufficiency) and that the data are faithful[2] to the causal process. How many different causal graphs could exist under these conditions? Each pair of variables (X and Y) can have one of four different causal relationships: either X directly causes Y, Y directly causes X, X and Y directly cause each other or the two have no direct causal links. We now have to count up the number of different

---

[2] See Chapter 2 for the definition of faithfulness. In fact, much of this chapter makes use of notions introduced in Chapter 2, and the reader might want to reread that chapter before continuing.

Table 8.1 The number of different cyclic causal graphs without latent variables that can be constructed given N variables

N	Number of graphs
2	4
3	64
4	4,096
5	1,048,576
6	1,073,741,824

pairs of variables, which is just the number of combinations of two objects out of N. The combinatorial formula is therefore $4^{\frac{N!}{2!(N-2)!}}$. Table 8.1 gives the number of different potential causal graphs of this type that can exist given N variables.

If we think of the full set of potential causal graphs having N variables as forming an 'hypothesis space', and your research programme as a search through this space to find the appropriate causal graph, then Table 8.1 is bad news. Even if we could test one potential graph per second it would take us almost 32 years to test every potential graph containing only six variables! If we were to restrict our problem to acyclic graphs then the numbers would be smaller, but still astronomical (Glymour et al. 1987). If it is true that the process of hypothesis generation (in this case, proposing one causal graph out of all those in the hypothesis space) is pure intuition, devoid of method, then it is a wonder that science has made any progress at all. That science *has* made progress shows that efficient methods of hypothesis generation, though perhaps largely unstated, do exist.

So, how should we go about efficiently exploring this hypothesis space? To go back to my previous question: how does one go about generating promising hypotheses concerning causal processes (Shipley 1999)? One way would be to choose a graph at random and then collect data to test it. With five variables there is a bit less than one chance in a million of hitting on the correct structure. There is nothing logically wrong with such a search strategy; we will have proposed a falsifiable hypothesis and tested it. However, no thinking person would ever attempt such a search strategy, because it is incredibly inefficient. We need search strategies that have a good chance of quickly finding those regions of hypothesis space that are likely to contain the correct answer. What would be our chances of hitting on the correct structure if we were to appeal only to 'pre-existing theory', as recommended by many SEM books? Clearly, that would depend on the quality of the pre-existing theory. However, if the theory really was so compelling that the researcher did not feel a need to search for alternatives then the problem would be firmly within the 'hypothesis testing' compartment and no question of a search strategy would be posed.

Very often biologists find themselves in the awkward position of straddling the 'hypothesis generation' and 'hypothesis testing' compartments. Often we have some background knowledge that excludes certain causal relationships and suggests others, but not enough firmly established background knowledge to specify the full causal structure without ambiguity. In such situations the goal is not to test a pre-existing theory – which might not be sufficiently compelling to justify allocating scarce resources and time in testing it – but, rather, to develop a more complete causal hypothesis that would be worth testing with independent data. The real problem is less in testing hypotheses than in finding hypotheses that are worth testing in the first place. We need search strategies that can be proved to be efficient at exploring hypothesis space, at least given explicitly stated assumptions. Until very recently such search strategies, which are described in this chapter, did not exist. You will see that these search strategies rely heavily on the notion of d-separation and on how this notion allows a translation from causal graphs to probability distributions.

8.3   The shadow's cause revisited

I have repeatedly compared the relationship between cause and correlation to the relationship of an object and its shadow. There is something missing in this analogy when applied to actual research projects. When we measure a correlation in a sample of data we are almost never interested in the value as such. Rather, we use the value to infer what the correlation might be in the population from which we randomly chose our sample data. It is as if, in nature's shadow play, not only do the causal processes cast potentially ambiguous correlational shadows, but these shadows are randomly blurred as well. We therefore have two problems. First, we have to find a way of provably deducing causal processes from correlational shadows; and, second, we have to take into account the inaccuracies caused by using sample correlations to infer population correlations. It is important to keep these two problems distinct. The second problem – that of dealing with sampling variation – is a typical problem of mainstream statistics. For this reason, we will first see how to go from correlations to causes when there is no sampling variation. In other words, we will consider asymptotic methods.

The history of the development of these exploratory methods, or 'search' algorithms, is fascinating. The word 'history' has connotations of age but, in fact, all these methods date to less than a few decades from the writing of this book. The mathematical relationships between graphs, d-separation and probability distributions were worked out in the mid-1980s by Judea Pearl and his students at the University of California, Los Angeles (UCLA) (Pearl 1988). This was the translation device between the language of causality and the language of probability distributions that had been missing for so long. As soon as it became possible to convert causal claims into probability distributions the dam was burst, and the conceptual flood came pouring out. It became immediately obvious that one could also convert statements concerning probabilistic independencies into causal claims. Pearl and his team at UCLA developed a series of algorithms to extract causal information from observational data during the period from 1988 to 1992.[3] Interestingly, a group of people at the Philosophy Department at Carnegie Mellon University (Clark Glymour, Peter Spirtes, Richard Scheines and their students) had also been working on

---

[3] This brief history, and the algorithms of Pearl and his students, are given by Pearl (2000: chap. 2).

the same goal. In 1987 they had published a book (Glymour et al. 1987) in which zero partial correlations and vanishing tetrad differences were used to infer causal structure, but without the benefit of d-separation or the mathematical link between causal graphs and probability distributions. As soon as they encountered Pearl's work on d-separation (they didn't know about the discovery algorithms of Pearl) they immediately began to independently derive and prove almost identical search algorithms. These algorithms (and much more) were proved and published by the Carnegie Mellon group in the book *Causation, Prediction, and Search* (Spirtes, Glymour and Scheines 1993) and incorporated into their TETRAD II program. An algorithm called the inductive causation algorithm was proved and published by Verma and Pearl (1990), and it is very similar to the causal inference algorithm of the Carnegie Mellon group that is presented in this chapter. I will leave it to the people involved to sort out questions of priority. I think that it is fair to say that, once the d-separation criterion had been developed and the dam had burst, the various algorithms were 'in the air' and had only to be brought down to earth by those with the knowledge. The philosopher's dream of inferring (partial knowledge of) causation from observational data had been realised.

In Chapter 2 I explained how to translate from the language of causality, with its inherently asymmetric relationships, to the language of probability distributions, with its inherently symmetric relationships. The Rosetta Stone allowing this translation was the notion of d-separation. Using d-separation we could reliably convert the causal statements expressed in a directed acyclic graph into probabilistic statements of dependence or independence that are expressed as (conditional) associations. This translation strategy was used in Chapters 3 to 7 to allow us to test hypothesised causal models using observational data.

When attempting to discover causal relationships the problem has been turned on its head. We now have to start with probabilistic statements of (conditional) dependence or independence and somehow back-translate into the language of causality. As you will see, this back-translation is almost always incomplete. There is almost always more than one acyclic causal graph that implies the same set of probabilistic statements of (conditional) dependence or independence. In other words, there are almost always different acyclic causal graphs that make different causal predictions but exactly the same predictions concerning probabilistic dependence or independence. This gives rise to the topic of equivalent models – a topic that has been recognised in SEM for a long time, and generally ignored for just as long.

The methods that I describe in this chapter are based on the strategy of back-translation that I described above. The first step is to obtain a list of probabilistic statements of (conditional) dependence or independence involving the variables in question. From this list, we construct an *undirected dependency graph*. An undirected dependency graph looks like a causal graph in which all the arrows have been converted into lines without arrowheads. However, the lines in the undirected dependency graph have a very different meaning. Two variables in this graph have a line between them if they are probabilistically dependent conditional on every subset of other variables in the graph. The lines in the undirected dependency graph express symmetric associations, not asymmetric causal relationships. Since we cannot measure associations involving variables that we have not measured, the undirected dependency graph cannot have latent variables. The next step is to convert as many of the symmetric relationships in the undirected dependency graph as possible into asymmetric causal relationships. This is called *orienting* the edges, and uses the notion of d-separation.[4] Generally, not all the undirected lines will be converted into directed arrows, and so we do not end up with a directed graph. Rather, we end up with a partially oriented graph.

8.4   Obtaining the undirected dependency graph

Before explaining how to obtain an undirected dependency graph from observational data, it is useful to explain how to convert a directed acyclic graph into an undirected dependency graph involving only measured variables. Doing this will help to underline the difference between the undirected dependency graph and the causal graphs with which you are now familiar. In acyclic graphs without latent variables, the undirected dependency graph is simply a directed acyclic graph in which all the arrows are replaced with lines lacking arrowheads. However, to be useful in discovery, we can work only with those variables that we have actually measured. If the directed graph contains latent variables then the resulting undirected dependency graph, involving only observed variables, will usually require modifications, and these modifications help to illustrate the proper interpretation of such graphs. To get the undirected graph from a directed acyclic graph[5] (Figure 8.1), or from a typical acyclic path diagram if it contains correlated errors, do the following things.

(1) If there is not already an arrow or curved double-headed arrow between any two observed variables, but d-separation of the pair requires conditioning on latent variables, then draw a line (not an arrow) between the pair.
(2) If there are curved double-headed arrows between any pairs of variables (i.e. correlated errors) then replace these with a line (not an arrow).
(3) Remove the latent variables, and also any arrows going into, or out of, these latent variables.
(4) Change all remaining arrows to lines.

Figure 8.1(a) shows a path diagram with both latent variables and correlated errors. Figure 8.1(b) shows the undirected dependency graph that results when considering only the observed variables. There are lines between {B,C}, {B,D} and {C,D} in the undirected dependency graph even though these pairs of variables were not adjacent in the original path diagram. This is because, following the first rule, d-separation of each of these pairs required conditioning on a latent variable (A). Since we have not measured variable A we can't condition on it, and so the three pairs of observed variables remain probabilistically associated even after conditioning on any set of other observed variables. Similarly, there is a line between {F,G} in the undirected dependency graph

---

[4] Vanishing tetrad differences (Chapter 5) are also used but these zero-tetrad equations can be reduced to statements concerning d-separation involving latent variables.
[5] The case of cyclic directed graphs will be dealt with later.

Path diagram

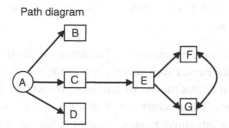

(a) A path diagram involving six observed variables and one latent variable

Undirected graph

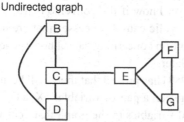

(b) The undirected dependency graph corresponding to the path diagram

Figure 8.1 A path diagram and the corresponding undirected dependency graph

even though the two variables were not adjacent in the original path diagram. This is because, following the second rule, this pair of variables has correlated errors. After changing all remaining arrows in the path diagram to lines, we end up with the undirected graph.

A *direct* cause between two variables in a causal graph is a causal relationship between them that cannot be blocked by other variables or sets of variables involved in the causal explanation. Similarly, we can define a *direct association* between two variables in an undirected dependency graph as an association that cannot be removed upon conditioning on any other observed variable or set of variables.[6] The undirected dependency graph is therefore a graph that shows these direct associations. Of course, if we are attempting to discover causal relationships then we will not already have the directed acyclic graph. Our first task, therefore, is to discover the undirected dependency graph from the data alone (remembering that, for the moment, we are assuming that our sample size is so large that we can ignore sampling variation) when we don't know what the true directed acyclic graph looks like.

Let's begin with the following assumptions:

(1) every unit in the population is governed by the same causal process (i.e. causal homogeneity);
(2) the probability distribution of the observed variables measured on each unit is faithful to some (possibly unknown) cyclic[7] or acyclic causal graph; and
(3) for each possible association, or partial association, among the measured variables, we can definitely know if the association or partial association exists (is different

---

[6] This will be more formally defined as an inducing path later on.
[7] The subsequent orientation phases will differ depending on whether or not we assume an acyclic structure.

from zero) or doesn't exist (is equal to zero); this is simply the assumption that there is no sampling variation.

We don't have to assume that there are no unmeasured variables generating some associations (this assumption is called *causal sufficiency*), or that the variables follow any particular probability distribution or that the causal relationships between the variables take any particular functional form. No assumptions of an acyclic structure are needed, although the algorithms for cyclic structures require linearity in the functional relationships between variables. The method uses d-separation, and we know the d-separation implies zero (partial) associations (Spirtes 1995; Pearl and Dechter 1996) under such conditions. Unfortunately, we don't yet know if the converse is true – that is, if there can be independencies generated by cyclic causal processes that are not implied by d-separation (Spirtes 1995). The assumption concerning causal homogeneity can be partly relaxed as well, as will be described later.

Given these assumptions, Pearl (1988) has proved that there will be an edge (a line) in our undirected dependency graph between a pair of variables (X and Y) if X and Y are dependent conditional on every set of variables in the graph that does not include X or Y. We can therefore discover the undirected graph of the causal process that generated our data by applying the algorithm below. Following the definition of the order of a partial correlation, let's define the conditioning *order* of an association as the number of variables in the conditioning set. So, a zero-order association is an association between two variables without conditioning, a first-order association is an association between two variables conditioned on one other variable, and so on. How one measures these associations will depend on the nature of the data; the various methods described in Chapter 3 can be used for different types of data.

8.5   The undirected dependency graph algorithm[8]

The first step is to form the *complete* undirected graph involving the V observed variables – in other words, adding a line between each variable and every other variable. Since latent variables are, by definition, unmeasured, we cannot include them in our complete undirected graph. Now, for each unique pair of observed variables (X, Y) that have a line between them in the undirected dependency graph at any stage during the implementation of the algorithm, do the following.

(1) Let the order of the association be zero.
(2.1) Form every possible set of conditioning variables, containing the number of variables specified by the order, out of the remaining observed variables in the graph.
(2.2) If the association between the pair of variables (X, Y) is zero when conditioned on any of these sets then remove the line between X and Y from the

---

[8] This algorithm is included in the SGS algorithm of Spirtes, Glymour and Scheines (Spirtes, Glymour and Scheines 1990).

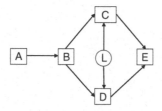

Figure 8.2 A directed graph, including one latent variable, used to illustrate the undirected dependency graph algorithm
Note: This causal graph is unknown to the observer.

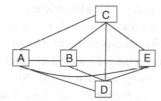

Figure 8.3 Step 1 in the construction of the undirected dependency graph
Note: There is an undirected edge between each pair of observed variables.

undirected dependency graph, move on to a new pair of variables and then go to step 1.

(2.3) If the association between the pair of variables (X, Y) is not zero when conditioned on all these sets then increase the order of the association by one, and go to step 2.1. If you cannot increase the conditioning order then the line between your two variables is kept. Move on to a new pair of variables.

Once you have applied this algorithm on every set of observed variables, the result is the undirected dependency graph. Given the assumptions listed above, you are guaranteed to obtain the correct undirected dependency graph of the causal process that generated the data if the algorithm is properly implemented.

To illustrate this algorithm, let's imagine that we have been given data (lots of it, so that we do not have to worry about sampling variation) that, unknown to us, were generated by the causal graph shown in Figure 8.2.

Now, we don't know about Figure 8.2; this causal structure is hidden behind the screen of nature's shadow play. In fact, we might not even know of the existence of the latent variable (L), since, had we known about it, we probably would have measured it. All that we have is a (very large) data set containing observations on the variables A to E and a series of measures of association and partial association between them; these are the shadows that we can observe on the screen. Our task is to infer as much about the structure of Figure 8.2 as we can. To begin, we create the complete undirected dependency graph of these five variables (Figure 8.3).

Notice that the latent variable (L) doesn't appear in Figure 8.3, because we are dealing only with observed variables at this point. Let's begin with the pair (A, B) and apply the algorithm. Since A and B are adjacent in the true causal structure (Figure 8.2) then these two variables are not unconditionally d-separated. We will therefore find that the pair is

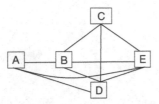

Figure 8.4 The undirected dependency graph with an undirected edge removed
Note: The undirected edge between A and C has been removed because we have found a subset of observed variables – {B} – that makes A and C independent upon conditioning.

associated in our data when we test for a zero association (independence) without conditioning (zero-order conditioning). Therefore, the line between A and B in Figure 8.3 remains after the zero-order step. We increase the conditioning order to one and see if A and B become independent upon conditioning on the following first-order sets: {C}, {D}, {E}. These are the only first-order conditioning sets that we can form from five variables while excluding variables A and B. From Figure 8.2 we know that A and B are not d-separated given any of these sets. Therefore, they will not be independent in our data upon first-order conditioning, and the line between them in Figure 8.3 remains after this step. We continue by increasing the conditioning order to two and test for a zero association relative to the following sets: {C,D}, {C,E}, {D,E}. These are the only second-order conditioning sets that we can form. Given the true causal structure in Figure 8.2 we will find that the second-order association between A and B remains. We increase the conditioning set to three and test for a zero association relative to the conditioning set {C,D,E} but still the association between A and B will remain. Since we cannot increase the conditioning order any more, we conclude that there is a line between A and B in the final undirected dependency graph.

We then go on to a new set of variables – in this case, A and C. When we apply the algorithm to the pair (A, C) we will find that A and C are still zero-order associated since they are d-connected in Figure 8.2. When we increase to order 1 and form the sets {B}, {D} and {E} we will find that A and C become independent upon conditioning on B. This is because, in Figure 8.2 (the true graph), A and C are d-separated given B, and d-separation implies probabilistic independence. We therefore remove the line between A and C in Figure 8.3, giving Figure 8.4. Since we have removed the line we don't have to go any further with this pair.

If we apply the algorithm to the pair (A, D) we find that A and D also become independent upon conditioning on B, and so we remove the line from A and D in Figure 8.4. A and E also become independent either upon conditioning on B or on the sets {B,C}, {B,D} or {B,C,D}. This is because A and E are d-separated by any of these conditioning sets. The pair (C, D) would never become independent since, in Figure 8.2, they are both caused by a latent variable that will never therefore appear in any of the conditioning sets. Similarly, the pair (C, E) will always remain associated, as will the pair (D, E). The undirected dependency graph that results after applying the algorithm to every possible pair is shown in Figure 8.5; this is the correct undirected dependency graph given the causal process shown in Figure 8.2.

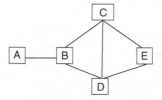

Figure 8.5 The completed undirected dependency graph
Notes: The undirected edges between A and D, between A and E and between B and E have been removed because we have found a subset of observed variables that renders each pair independent upon conditioning.

8.6   Interpreting the undirected dependency graph

The undirected dependency graph informs us of the pattern of direct associations in our data. It doesn't inform us of the pattern of direct *causes* in our data. For instance, there is a line between C and D in Figure 8.5 even though, peeking at the causal process that generated the data (Figure 8.2), we know that the association between C and D is due only to the effect of the latent variable (L). Just as the term 'direct' cause can have meaning only in relation to the other variables in the causal explanation, a 'direct' association can have meaning only in relation to the other variables that have been measured. However, we can infer from the undirected dependency graph that if two variables have a line between them then either:

(a) there is a direct causal relationship between the two; and/or
(b) there is a latent variable that is a common cause of the two; and/or
(c) there is a more complicated type of path between the two, called an inducing path (this will be explained in more detail later).

At the same time, we can exclude other types of latent variables. For instance, we know that there is no latent variable that is a common cause of A, B and C in Figure 8.5. If there were, A and C would not be d-separated given any set of other observed variables, and there would therefore be a line between A and C in the undirected dependency graph.

The first two explanations for a direct association in an undirected dependency graph should be understandable by now. The third possibility is less obvious but can be illustrated by an example given by Spirtes, Glymour and Scheines (1993). Consider Figure 8.6. On the left (Figure 8.6(a)) is the directed acyclic graph with a latent variable F. Neither variables A nor B have any direct causal link with variable D. On the right (Figure 8.6(b)) is the undirected dependency graph. Notice that there is a line between A and D, and also between B and D, in the undirected dependency graph even though there are neither direct causal links between the pairs nor latent variables that are causes common to both. This is because A and D would never be d-separated given any subset of variables B and C, and thus would always be probabilistically associated. Similarly, B and D would never be d-separated given any subset of variables A and C, and so would always be probabilistically associated. For instance, if we look at the pair (A, D)

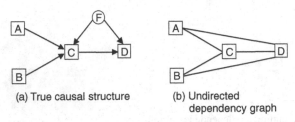

(a) True causal structure   (b) Undirected
dependency graph

Figure 8.6 A less obvious translation from a DAG to an undirected dependency graph
Notes: The true causal structure is shown in panel (a) and the resulting undirected dependency graph is shown in panel (b). Notice that there are edges between A and D and between B and D in the undirected dependency graph even though no such directed edges exist in the directed graph. This is due to the presence of the latent variable F generating an induring path between these pairs of variables.

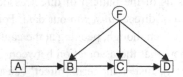

Figure 8.7 A directed graph with one latent variable (F)

in Figure 8.6(a) then A and D would be unconditionally associated through the path A→C→D and associated conditional on B, associated conditional on C through the path A→C←F→D and also associated conditional on both B and C for the same reason.

More definitions are required before the explanation as to how this third possibility can arise in general can be understood.

Directed versus undirected paths

Look at the directed acyclic graph in Figure 8.7. Imagine that this DAG is a road map consisting only of one-way streets whose direction is shown by the arrows. Normally, to get from one variable to another we have to respect the traffic rules and follow the arrows. If we can get from one variable to the other by following these rules then we will call our route a *directed path*. For instance, we can go from A to D by following the directed path A→B→C→D. A→B←F→D is *not* a directed path, because we have gone the wrong way on a one-way road (B←F) when going from B to F. However, if we ignore the rules of the road and drive in whatever direction we want, irrespective of the direction of the arrows, then we can go from A to F along the path A→B←F→D. Such a path, in which the direction of the arrows is ignored, is called an *undirected path*. We haven't erased the arrows, we have simply decided to ignore them.[9] Of course, a directed path must also be an undirected path, but an undirected path might not also be a directed path. For instance, the following are undirected paths in Figure 8.7 but they are not directed paths: A→B←F→D, A→B→C←F→D.

---

[9] Pretend you are a diplomat who is working at the United Nations in New York City. You can't change the laws about one-way streets but you can safely ignore them because of diplomatic immunity.

Inducing paths[10]

List all the variables in the DAG and call it the set **V**. In Figure 8.7 the set **V** is {A,B,C,F,D}. We can call this complete DAG the graph G. Now chose some subset of variables in the DAG and call it **O**. For instance, you might choose the set **O** = {A,B,C,D}, thus leaving out the variable F. By doing this you will have a new graph (call it G') in which the variable F is latent; in other words, the variable F still has the same causal relationships to the other (**O**) variables as before, but variable F doesn't appear in G'. Because G' doesn't show the variable F, it is not a complete description of the full causal process. Now choose two variables in your chosen set (**O**) of variables and find an undirected path between them in the complete graph (G). For instance, if we choose A and D in Figure 8.7 then we can find the undirected paths A→B←F→D, A→B→C←F→D, A→B←F→C→D and A→B→C→D. Some of these undirected paths might be a special type of path called an *inducing path relative to* **O**. To determine if a given undirected path is an inducing path relative to **O**, look at those variables in the undirected path in G that are also in your chosen set **O**. If (a) every variable in **O** along the undirected path except for the end points (here, A and D) is a collider along the path, and if (b) every collider along this undirected path is an ancestor of either of the end points, then the path is an inducing path between the end points relative to **O**. Such an inducing path has the property that the end points will never be d-separated given any subset of other variables from the set **O**.

Let's look at the first undirected path between A and D (A→B←F→D) and choose **O** = {A,B,C,D}. The only other variable along this undirected path, except for A and D (the end points), that is in **O** is B, since F has been left out (i.e. is latent). Since B is a collider along this path and is an ancestor (because of the path B→C→D) of D (one of the end points), the path A→B←F→D is an inducing path relative to **O** = {A,B,C,D}. To see that this inducing path results in A and D never being d-separated given any subset of variables from **O**, we have only to look at each possible conditioning set. The empty set (i.e. unconditional conditioning) allows d-connection though the path A→B→C→D. The set {B} allows d-connection through the undirected path A→B←F→D. The set {C} allows d-connection through the undirected path A→B→C←F→D. The set {B,C} allows d-connection through either of these last two undirected paths.

None of the other undirected paths between A and D are inducing paths relative to **O** = {A,B,C,D}. For instance, the undirected path A→B→C←F→D has the variable B that is in **O** but is not a collider along the path. Therefore, conditioning on B will d-separate A and D. Similar reasons exclude the paths A→B←F→C→D and A→B→C→D.

Notice that the variables at the ends of such an inducing path will never be d-separated given the variables in **O** because one will always be conditioning on a collider and thus opening a path through some variable not in **O**. Therefore, the undirected dependency graph involving the variables in **O** will always have a line between two variables if there is an inducing path between them. Noting that **O** will usually consist of the set of 'observed' variables, you might start to see the usefulness of the notion of an inducing

[10] The properties of inducing paths were described by Verma and Pearl (1991) but the name for such paths was introduced by Spirtes, Glymour and Scheines (1993).

path. If you see a line between two variables in the undirected dependency graph then you will know that there is an inducing path between them.

One practical problem with the algorithm that I have presented for obtaining the undirected dependency graph is that, as the number of observed variables increases, the number of sets of conditioning variables increases geometrically. When faced with large numbers (say 50) observed variables, even fast personal computers might take a long time to construct the undirected graph if the topology of the true causal graph is uncooperative. A slightly modified version of the algorithm[11] is presented by Spirtes, Glymour and Scheines (1993) that is more efficient when dealing with many observed variables. The two algorithms are equivalent given population measures of association, but the more efficient algorithm can make more mistakes in small data sets.

We sometimes have independent information about some of the causal relationships governing our data. In such cases it is straightforward to modify the algorithm for the undirected dependency graph to incorporate such information. If we know that the association between two observed variables is due only to the fact that another measured variable, or set of measured variables, is a common cause of both then we simply remove that edge before applying the algorithm. Similarly, if we know that two observed variables either have a direct causal relationship, or share at least one common latent cause, then we simply forbid the algorithm from considering that pair. Note that it is not enough to know (say from a randomised experiment) that one measured variable is a cause of another; we must know that it is (or is not) a *direct* cause. A randomised experiment will not be able to tell us this if some of the observed variables are attributes of the experimental units, as explained in Chapter 1.

8.7   Orienting edges in the undirected dependency graph using unshielded colliders assuming an acyclic causal structure

In Chapter 2 I discussed how d-separation predicts some counter-intuitive results concerning statistical conditioning. Consider a simple causal graph of the form X→Z←Y. X and Y are causally independent and, since they are unconditionally d-separated, they are also probabilistically independent. However, if we condition on Z (the common causal descendant of both X and Y) then X and Y become conditionally dependent. This is because X and Y are not d-separated conditional on Z. In general,[12] if we have two variables (X and Y) and condition on some set of variables $\mathbf{Q}$ that contains at least one common causal descendant of both X and Y, X and Y will not be d-separated. Because of this X and Y will not be probabilistically independent upon conditioning on $\mathbf{Q}$ even if X and Y are causally independent.

This fact allows us to determine the causal direction of some lines in the undirected dependency graph. In Chapter 2 I defined an *unshielded collider* as a causal relationship

---

[11] This modified algorithm is incorporated in their PC algorithm. The algorithm that I have described forms part of their SGS algorithm.

[12] This is true for acyclic causal structures but not for cyclic causal structures. This is discussed in more detail later.

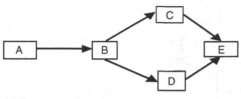

(a) True causal graph

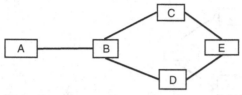

(b) Undirected dependency graph

Figure 8.8 The true (unknown) causal graph and the resulting undirected dependency graph

between three variables (X, Y and Z) such that both X and Y are direct causes of Z (X→Z←Y) but there is no direct causal relationship between X and Y (i.e. there is no arrow going from one to the other).[13] Let's now define an *unshielded pattern* in an undirected dependency graph as one in which we have three variables (X, Y and Z) such that there is a line between X and Z, a line between Y and Z (X–Z–Y) but no line between X and Y. Since there is no line between X and Y we know that X and Y are d-separated given some subset of other variables in the undirected dependency graph. Given such an unshielded pattern we can decide if there are arrowheads pointing into Z from both directions or not in the causal graph that generated the data. If there were arrowheads pointing into Z from both directions in the actual causal process generating the data then X and Y would never by probabilistically independent conditional on any set of other observed variables that includes Z.

To illustrate this method of orienting our undirected paths in the undirected dependency graph, imagine that the unknown causal process generating our observed data is as shown in Figure 8.8. Even though the causal process is hidden from us, we will obtain the undirected dependency graph shown in Figure 8.8(b) once we apply the algorithm to our data. Now, since we don't know what the actual causal process looks like, we don't know if there are latent variables generating some of the direct associations.

Before going on, let's introduce some more conventions for modifying our undirected dependency graph. A graph in which only some of the edges are oriented is called a *partially directed* graph or a *partially oriented* graph. Since we don't yet know whether or not there are arrowheads at the ends of any of the lines in our undirected dependency graph (i.e. we don't yet know the directions of the causal relationships shown in the causal graph in Figure 8.8(a)), let's admit this fact by adding an open circle (Xo–oY) at the end of each line (Figure 8.9(b)). By doing this we are no longer dealing with an

---

[13] If we were dealing with a path diagram rather than a directed acyclic graph then there must not be any edge at all, either an arrow or any double-headed arrows, between X and Y.

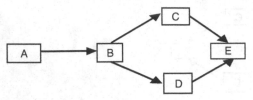

(a) True causal graph

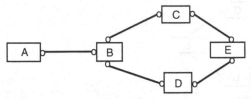

(b) Partially oriented graph

Figure 8.9 The true causal graph and a corresponding partially oriented graph with no orientations of the edges specified

undirected graph; rather, we are dealing with a partially oriented graph whose directions are not yet known. An open circle simply means that we don't know whether there should be an arrowhead or not. Therefore, given Xo–oY, the oriented edge in the true causal graph might be X→Y, X←Y or X←→Y. The final oriented edge (X←→Y) doesn't mean a feedback relationship between X and Y (remember our assumptions). Rather, it means that there is an unmeasured (latent) common cause generating the direct association between X and Y. It doesn't necessarily mean that there is a common latent cause of X and Y either, as Figure 8.6 makes clear.

The partially oriented graph in Figure 8.9 has six unshielded patterns, as given in Table 8.2. To orient some of the edges by detecting an unshielded collider, apply the following algorithm to each unshielded pattern (Xo–oZo–oY).

8.8   The orientation algorithm[14] using unshielded colliders

Let the conditioning number (i) be 1.

(1) Form all possible conditioning sets of i observed variables consisting of the variable in the middle of the unshielded pattern (Z) plus any observed variables other than the variables at the ends of the unshielded pattern (X and Y). Call each such conditioning set **Q**.

(2.1) If the partial association between X and Y, conditioned on any set **Q** of other variables, is zero then stop and conclude that the three variables forming the

---

[14] This algorithm is used in Pearl's IC (inductive causation) algorithm. The related algorithm by Spirtes, Glymour and Scheines (1993) uses a set called **Sepset**(X,Y) that reduces the computational burden. The output is identical in acyclic causal structures but can be different in cyclic causal structures.

Table 8.2 Applying the orientation algorithm using unshielded colliders to the partially oriented graph in Figure 8.9

Unshielded pattern	Partially oriented pattern	Explanation
Ao–oBo–oC	Ao–o<u>Bo</u>–oC	B must be in **Q**. A and C are always d-separated given B and any other observed variable.
Ao–oBo–oD	Ao–o<u>Bo</u>–oD	B must be in **Q**. A and D are d-separated given B and any other observed variable.
Co–oBo–oD	Co–o<u>Bo</u>–oD	B must be in **Q**. C and D are d-separated given B and any other observed variable.
Bo–oCo–oE	Bo–o<u>Co</u>–oE	C must be in **Q**. B and E are d-separated given {C,D} and any other observed variable.
Bo–oDo–oE	Bo–o<u>Do</u>–oE	D must be in **Q**. B and E are d-separated given {D,C} and any other observed variable.
Co–Eo–oD	Co→E←oD	E must be in **Q**. C and D are never d-separated given E plus any other observed variable.

unshielded pattern do not form an unshielded collider in the true causal graph (i.e. not Xo→Z←oY). We can call such a pattern a *definite non-collider*.

(2.2) If the partial association between X and Y, conditioned on every set **Q** of other variables, is not zero then increase the conditioning number (i) by one and go to step 1.

After cycling through all possible orders of i, if we have not declared the unshielded pattern to be a definite non-collider then it is a collider. Orient the pattern as Xo→Z←oY.

Since, in this example, we can peek at the true causal graph (Figure 8.9(a)), we can use d-separation to predict what would happen if we applied the above algorithm to each of the six unshielded patterns that we found in our partially oriented graph. For instance, when we test the unshielded pattern Ao–oBo–oC we begin the algorithm by testing for a zero association (probabilistic independence) between A and C given B. The order of the conditioning set is initially one, and we already have one variable (B). Since A and C are d-separated by B we will find A and C to be probabilistically independent given B and therefore stop right away, concluding that the causal effects do not collide at B. This unshielded pattern is a definite non-collider. The full results, and their explanations, are given in Table 8.2. You will notice one more new notation in Table 8.2. If we have concluded that an unshielded pattern is a definite non-collider (i.e. that there definitely are not arrowheads pointing into the middle variable) then underline the middle variable. Thus, the notation Xo–o<u>Yo</u>–oZ means that we still don't know what the actual orientation is, but it is definitely not Xo→Z←oY.

The final partially oriented graph that results is shown in Figure 8.10. In fact, since the partially oriented edges indicate inducing paths, Spirtes, Glymour and Scheines (1993) call these *partially oriented inducing path graphs*, or POIPGs.

At this point, other information about some of these partially oriented relationships might help us. If, for instance, we knew from previous work that the direct association

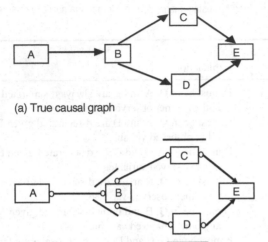

(a) True causal graph

(b) Partially oriented graph

Figure 8.10 The true but unknown causal graph and the final partially oriented graph, with orientation using unshielded colliders

between A and B was due at least in part to a common latent cause then we could orient this as A←→B. This would immediately restrict the orientations of the other lines, since we know that the three unshielded patterns of which B is the middle variable are all definite non-colliders. Therefore, we can exclude A←→B←oC and A←→B←oD.

Because a randomised experiment, when it can be done, can give us information about causal direction, the combination of prior information from randomised experiments and these search algorithms can often be very useful. For instance, imagine that the five variables in Figure 8.10 represent five attributes of a plant and that we cannot perform randomised experiments to untangle the causal relationships between them. However, we can introduce a new variable that is a property of the external environment, such as light intensity. It is possible to randomly allocate plants to the different treatment groups representing light intensity, and so we can tell, for each of the five plant attributes, if changes in light intensity cause changes in the attribute. For the reasons given in Chapter 1 we can't say that light intensity is a *direct* cause but we can say that, if the values of the attribute differ between treatments, the causal signal (direct or indirect) goes from light intensity to the attribute and not the other way around. Now, if, in an observational study, we measure the five attributes plus light intensity, and find that there is an edge in the partially oriented graph between light intensity and some of the plant attributes, then we can use this information to orient such edges. Once some edges are oriented this will usually help to orient others.

If we are willing to assume that there are no latent variables responsible for some of the lines in an undirected graph (i.e. causal sufficiency) then we can further restrict the number of alternative graphs. For instance, if we take the partially oriented graph in Figure 8.10 and assume causal sufficiency then there are only four different directed acyclic graphs that are compatible with the partially oriented graph (Figure 8.11). There

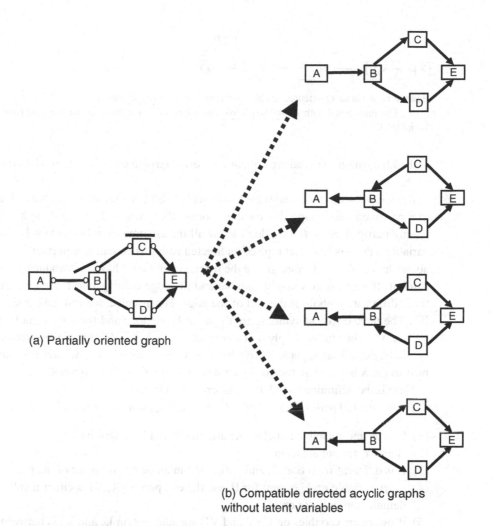

(a) Partially oriented graph

(b) Compatible directed acyclic graphs
without latent variables

Figure 8.11 Alternative DAGs without latent variables consistent with a single partially oriented graph
Notes: Panel (a) shows the partially oriented graph of Figure 8.10. Panel (b) shows the four possible completely directed acyclic graphs without latent variables that are consistent with the partially oriented graph.

were a huge number of potential acyclic causal graphs involving five variables in our initial hypothesis space, and we have reduced this number to four.

8.9   Orienting edges in the undirected dependency graph using definite discriminating paths

In order to orient edges using unshielded colliders it is necessary for the pattern to be unshielded. When we have a shielded pattern (three variables with lines between each of the three pairs forming a 'triangle') we can sometimes still orient the pattern if it is

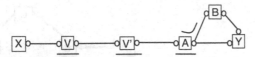

Figure 8.12 A partially oriented graph involving six observed variables

Note: The undirected path Xo–oVo–oV'o–oAo–oBo–oY is a definite discriminating path for the variable B.

embedded within a special type of partially oriented path called a *definite discriminating path*.

Let's start with some undirected path (call it U) between two variables (X and Y) in a partly oriented graph that contains some other variable B. Even though the graph is only partly oriented (so we don't know all the asymmetric relationships between the variables) there is a special type of undirected path that contains important information about the variable B. Before giving the formal definition, I have to introduce yet another symbol. If we look at a single variable and the edge coming into it then we can have three different symbols at the end of the edge. For instance, we can have X←, Xo– or X–. The three different symbols are the arrowhead, the o and the empty mark. Now, if I write X*–, the star is simply a placeholder that can refer to any of the three different symbols. So, if I say replace X*–oY by X*→Y then I mean 'keep whatever symbol was next to the X but change the "o" symbol next to the Y to a ">" symbol'.

Here is the definition of a definite discriminating path.

An undirected path U is a *definite discriminating path* for variable B if and only if:

(1) U is an undirected path between variables X and Y containing B;
(2) X and Y are not adjacent;
(3) B is different from both X and Y (i.e. B cannot be an end point of the path);
(4) every variable on U except for B and the end points (X, Y) is either a collider or a definite non-collider on U;
(5) if two other variables on U (V and V') are adjacent on U, and V' is between V and B on U, then the orientation must be V*→V' on U;
(6) if V is between X and B on U and V is a collider on U then the orientation must be either V→Y or else V←*X; and
(7) if V is between Y and B on U and V is a collider on U then the orientation must be either V→X or else V←*Y.

To see the usefulness of such definite discriminating paths, consider Figure 8.12. Each of the four unshielded patterns in the undirected dependency graph (X–V–V', V–V'–A, V'–A–B and V'–A–Y) derived from this partially oriented graph allows us to apply the algorithm for unshielded colliders; in this case, all were determined to be definite non-colliders. Unfortunately, the shielded pattern involving A, B and Y cannot be oriented this way. However, the undirected path Xo–oVo–oV'o–oAo–oBo–oY is also a definite discriminating path for the variable B. What happens if, in the underlying causal graph, X and Y are d-separated given A and B, but not given A alone? The only way that this could occur is if B were a definite non-collider along the undirected path (Ao–oBo–oY) since, if the orientation was really Ao→B←oY, conditioning on A and B

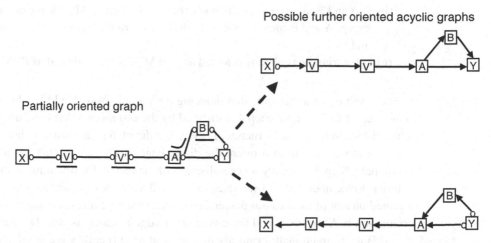

Figure 8.13 The partially oriented graph on the left implies only two alternative partially oriented acyclic graphs on the right

would not d-separate X and Y. So we can definitely state that the partial orientation is Ao–o Bo–oY. Yet this is not all. Since we have assumed that the unknown causal graph is acyclic, there are only two different partially oriented acyclic causal graphs that accord with this information (Figure 8.13).

Now we can put all the pieces together and state the causal inference algorithm of Spirtes, Glymour and Scheines (1993).

8.10   The causal inference algorithm[15]

(1) Apply the algorithm to obtain the undirected dependency graph.

(2) Orient each edge in the undirected dependency graph as o–o.

(3) Apply the orientation algorithm using unshielded colliders. For each unshielded pattern (A–B–C), orient unshielded colliders as Ao→B←oC and orient each definite non-collider as Ao–oBo–oC.

(4) Repeat steps 4 to 6 until no further changes can be made.

(5) If there is a directed path from A to B, and an edge A*–*B, orient A*–*B as A*→B.

(6) If B is a collider along a path A*→B←*C, B is also adjacent to another variable D (i.e. B*–*D), and A and C are conditionally independent[16] given D, then orient B*–*D as B←*D.

(7) If there is an undirected path U that is a definite discriminating path between variables A and B for variable M, and variables P and R are adjacent to M along U, and P*–oMo–*R forms a triangle, then

   (a) if A and B are conditionally independent given M plus any other variable except A and B then P*–oMo–*R along U is oriented as a non-collider: P*–oMo–*R;

---

[15] My description of the causal inference algorithm differs from the original formulation only in replacing **Sepset** sets with the actual d-separation claim.

[16] There was an error by Spirtes, Glymour and Scheines (1993) at this point, which was corrected in a subsequent errata.

(b) if A and B are never conditionally independent given M plus any other variable except A and B then P*–oMo–*R along U is oriented as a collider: P*→M←*R; and

(c) if the triangle is already oriented as P*→M̲*–*R then orient it as P*→M→R.

The result is a partially oriented inducing path graph. You should be able to understand steps 1 to 3 by now. Step 4 is justified by the assumption that there are no cyclic relationships in the causal structure. If there is a directed path from variable A to variable B and we were to also orient the direct edge as B→A then this would create a cyclic path. Step 5 is simply a generalisation of the reason for orienting an unshielded collider. Remember that two variables (X, Y) will never be d-separated if they are conditioned on any of their causal descendants. Since we have already established that the orientation is A*→B←*C and there is another edge oriented as Bo–*D, and that both A and B are (conditionally) causally independent of D (i.e. they are not d-connected), then d-separation predicts that A and D would become probabilistically dependent when conditioned on D if the orientation was B–*D and remain probabilistically independent if the orientation was B←*D. Steps 6(a) and 6(b) derive from the notion of a definite discriminating path, as described before. In step 6(c) we have already established that M is a non-collider along P*→M̲–*R. Therefore, there cannot be arrowheads pointing into M from both directions, and we can orient the triplet as P*→M̲–*R. There is now only one orientation possible, namely P*→M→R.

8.11   Equivalent models

The inferential testing of structural equation models, described in Chapters 3 to 7, consisted of deriving the observational predictions of the hypothesised causal process (the correlational shadows) and then comparing the observed and predicted patterns of correlation or covariation. I have emphasised that failing to reject such an hypothesised model provides support for it but does not allow us to accept it without other (non-statistical) evidence. One reason might be that the sample size was too small to permit us to detect a real (but small) deviation between the observed and predicted patterns. However, the search algorithms in this chapter should alert us to another reason: different causal processes can cast the same observational shadows.

This leads to the topic (rarely discussed in the SEM literature) of observationally *equivalent models* – that is, different causal models that cannot be distinguished based on observational data. Such equivalent models will produce exactly the same chi-square values, and exactly the same probability levels, when tested against the same data set. This is true no matter how big your data set is. In fact, it will be true even if you have the population values rather than sample values. When we test a structural equation model we are really testing the entire set of observationally equivalent models against all non-equivalent models. In one sense, this might be disappointing: we cannot distinguish between some competing causal explanations. In another sense, this is useful: when we reject a particular model we are also simultaneously rejecting all the observationally equivalent models as well.

The search algorithms in this chapter can allow us to find all the causal models that are observationally equivalent[17] to our hypothesised one. Given your path diagram, here are the steps.

(1) Change all arrows (even double-headed ones)[18] to lines.
(2) Draw the 'o' symbol at either end of each line.
(3) Redraw each unshielded collider that was in the original path diagram; in other words, if there was an unshielded collider in the path diagram (X→Z←Y) then replace Xo–oZo–oY with Xo→Z←oY.
(4) For each non-collider triplet that was in the original path diagram, add an underline; in other words, if there was either X→Z→Y, X←Z←Y or X←Y→Z in the path diagram then replace Xo–oZo–oY with Xo–oZo–oY.

Figure 8.14 summarises these steps.

At this point you can permute the different possible orientations, so long as you never introduce an unshielded collider that was not in your original path diagram and never remove an unshielded collider that was in your original path diagram.

8.12   Detecting latent variables

One practical problem with the causal inference algorithm is that it can be quite uninformative when many observed variables are all caused by a small number of latent variables. In such cases the application of the causal inference algorithm will not be very informative. Consider the simple measurement model (panel (a)) shown in Figure 8.15 and the resulting output from the causal inference algorithm.

The output of the causal inference algorithm tells us that each of the observed variables (A to D) is probabilistically associated with each of the others. Since there are no unshielded patterns among the observed variables in either of the two output graphs, we cannot orient any of the edges. Whenever you see a set of observed variables that form such a pattern (I will can this a *saturated* pattern) you should suspect latent variables. However, it is possible for such saturated patterns to arise even without latent variables, as causal process B (panel (b) of Figure 8.15) shows. Is there any way of differentiating between the two? Yes. For this, we need to look again at vanishing tetrad equations, which we briefly studied in Chapter 5.

You will recall that Spearman (1904) derived a set of equations, called vanishing tetrads, which must be true given the type of structure shown in causal process A (Figure 8.15(a)). He argued that if such vanishing tetrad equations held then this was

---

[17] The algorithm for observational equivalence in acyclic models was first published by Verma and Pearl (1990).

[18] A model having two variables sharing correlated errors (i.e. a double-headed arrow between them) is equivalent in its d-separation consequences to a model having a latent variable that is a common cause of both (Spirtes et al. 1998).

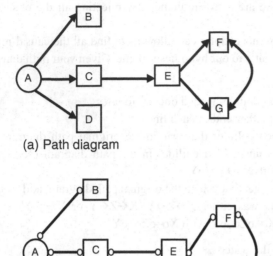

(a) Path diagram

(b) Steps 1 and 2

(c) Steps 3 and 4

Figure 8.14 A path diagram and the steps in obtaining all models that are equivalent to it

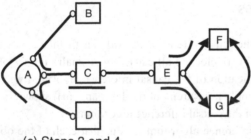

(a) Causal process A   (b) Causal process B

(c) Output of the causal inference algorithm

Figure 8.15 Directed acyclic graphs of two different causal processes that both imply the same partially oriented graph

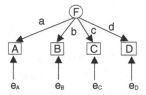

Figure 8.16 A directed graph involving four observed variables and one latent variable (F)

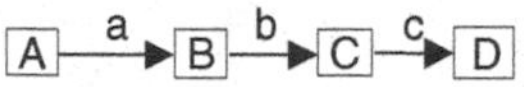

$$A \xrightarrow{a} B \xrightarrow{b} C \xrightarrow{c} D$$

Figure 8.17 A vanishing tetrad is implied even without latent variables
Note: This directed acyclic graph implies the vanishing tetrad $\rho_{AC} \cdot \rho_{BD} - \rho_{AD} \cdot \rho_{BC} =$ (ab)(bc) − (abc)(b) = 0 even though there are no latent variables.

evidence for the presence of a common latent cause of the observed variables. As will be explained below, this claim is not true, but a modification of it can indeed be used to detect such a common latent cause if the relationships between the latent variable and the observed variables are linear.

A vanishing tetrad equation is a function of four correlation (or covariance) coefficients. Because of the causal structure of models such as those in Figure 8.16, and because of the rules of path analysis, such a vanishing tetrad equation must be zero in the population regardless of the (non-zero) values of the path coefficients. For instance, the population correlation coefficient (ρ_{AB}) between the observed variables A and B is a*b. The population correlation coefficient (ρ_{CD}) between the observed variables C and D is c*d. Therefore, $\rho_{AB}\rho_{CD}$ = a*b*c*d. However, we also know that ρ_{AC} = a*c and that ρ_{BD} = b*d. Therefore, $\rho_{AC}\rho_{BD}$ = a*c*b*d. It follows that $\rho_{AB}\rho_{CD}$-$\rho_{AC}\rho_{BD}$ = 0, since a*b*c*d–a*c*b*d = 0. The tetrad equation ($\rho_{AB}\rho_{CD}$–$\rho_{AC}\rho_{BD}$) becomes zero, or *vanishes*, because of the way the observed variables relate to the latent variable. The causal process shown in Figure 8.16 implies three different vanishing tetrad equations (of which only two are independent). In fact, every set of four variables can have three possible tetrad equations regardless of the true causal process, though they don't have to be zero.

$$\rho_{AB} \cdot \rho_{CD} - \rho_{AD} \cdot \rho_{BC} = 0$$
$$\rho_{AC} \cdot \rho_{BD} - \rho_{AD} \cdot \rho_{BC} = 0$$
$$\rho_{AC} \cdot \rho_{BD} - \rho_{AB} \cdot \rho_{CD} = 0$$

Unfortunately, a causal structure such as the one in Figure 8.17 also implies vanishing tetrad equations. For instance, using the rules of path analysis we find that $\rho_{AC} \cdot \rho_{BD} -$ $\rho_{AD} \cdot \rho_{BC}$ = (ab)(bc) − (abc)(b) = 0. Clearly, simply showing that a vanishing tetrad equation holds is not evidence for the presence of a common latent cause of the observed variables. Although it may not seem immediately obvious, there is a close relationship between d-separation and vanishing tetrad equations.[19]

---

[19] Theorem 6.11 of Spirtes, Glymour and Scheines (1993) states that a vanishing tetrad equation of the type $\rho_{IJ}\rho_{KL}-\rho_{IL}\rho_{JK}$ = 0 is linearly implied by an directed acyclic graph only if either ρ_{IJ} or ρ_{KL} equals zero

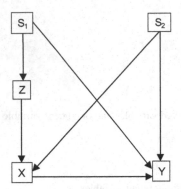

Figure 8.18 A directed acyclic graph to illustrate the concept of a trek

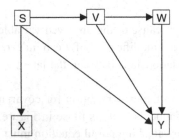

Figure 8.19 A directed acyclic graph to illustrate the concept of a *choke point* for a set of treks

A vanishing tetrad equation can be given a graphical interpretation. Let's define a *trek* between two variables (X, Y) as a pair of directed paths; one directed path goes from a *source* variable (S) to X and the other directed path goes from the same source variable to Y. One of the two directed paths can be of length zero (i.e. S = X or S = Y). For instance, in Figure 8.18 there are three treks between X and Y. One is from the source variable S_1 (X←Z←S_1→Y), one is from the source variable S_2 (X←S_2→Y) and one is from the 'source' variable X in which one directed path is of length zero (X→Y). I will write T(X,Y) to mean 'a trek between X and Y', **T**(X,Y) to mean 'the set of all treks between X and Y' and X(T(X,Y)) to mean 'the directed path in a trek between X and Y that goes into X'.

In Figure 8.19 there are three different treks between X and Y: X←S→Y, X←S→V→Y and X←S→V→W→Y. Notice that all the directed paths in all these treks leading into X pass through S. When this occurs we say that S is a *choke point* for X(**T**(X,Y)). There was no choke point for X(**T**(X,Y)) in Figure 8.18.

To see what all this has to do with vanishing tetrads, let's consider a set of four variables (I, J, K and L). If we have a set of treks **T**(I,J) between two variables (I, J) and a set of treks **T**(K, L) between two other variables (K, L), and all the directed paths in **T**(I, J) that are into J (i.e. J(**T**(I, J))) and all the directed paths in **T**(K, L) that are into

and if either ρ_{IL} or ρ_{JK} equals zero or there is a (possibly empty) set **Q** of variables in the directed acyclic graph such that $\rho_{IJ.\mathbf{Q}} = \rho_{KL.\mathbf{Q}} = \rho_{IL.\mathbf{Q}} = \rho_{JK.\mathbf{Q}} = 0$.

L (i.e. L(T(K, L))) intersect at the same variable Q, then Q is called a *JL choke point*. The tetrad representation theorem (Spirtes, Glymour and Scheines 1993) states that if we see a vanishing tetrad in the statistical population ($\rho_{IJ}\rho_{KL} - \rho_{IL}\rho_{JK} = 0$) then this means that there is either a JL choke point or an IK choke point.

How can vanishing tetrads help to detect the presence of latent variables? If you see a saturated pattern in your undirected dependency graph involving four variables[20] then test to see if there are any vanishing tetrads between these variables. If vanishing tetrads exist then this is evidence for a latent variable. To see why, consider that if the choke point implied by this vanishing tetrad was an observed variable then the two variables (J, L) or (I, K) would be d-separated by this choke point, and therefore could not form part of a saturated pattern.[21]

This fact provides a simple algorithm to test whether the observed correlations among a set of four observed variables is due to a common latent cause. These are the steps.

8.13   Vanishing tetrad algorithm

Given a set **O** of observed variables and a set **T** of four observed variables from **O** that form a saturated pattern in the undirected dependency graph, assume that there is no reason to invoke a common latent cause for these four variables in **T** and then do the following.

(1) Choose one of the three tetrad equations that are possible given the four chosen variables in **T**. If you have tried all three then stop.
(2) If the tetrad equation doesn't equal zero, go to step 1.
(3) If the tetrad equation does equal zero then there is a latent variable that forms the IK choke point of IK(**T**(I,J),**T**(K,L),**T**(I,L),**T**(JK)) or the JL choke point of JL(**T**(I,J),**T**(K,L),**T**(I,L),**T**(J,K)).

To illustrate this algorithm, let's go back to Figure 8.15. The undirected dependency graph will contain a saturated pattern for variables A, B, C and D. Here again are the three tetrad equations:

$$\rho_{AB} \cdot \rho_{CD} - \rho_{AD} \cdot \rho_{BC} = 0$$
$$\rho_{AC} \cdot \rho_{BD} - \rho_{AD} \cdot \rho_{BC} = 0$$
$$\rho_{AC} \cdot \rho_{BD} - \rho_{AB} \cdot \rho_{CD} = 0$$

All three tetrad equations vanish in panel (a) (causal process A) of Figure 8.15. Because the first equation vanishes we know that there is either an AC and/or a BD choke point. Because the second equation vanishes we know that there is either an AB and/or a CD choke point. Because the third equation vanishes we know that there is either an AD or

---

[20] If there are more than four variables forming a saturated pattern then take each unique set of four variables.
[21] In Figure 8.17 there was a vanishing tetrad but no latent variable. The treks between each of the four pairs of variables (AC, BD, AD and BC) all had directed paths of zero length (A→B→C, B→C→D, A→B→C→D and B→C). The choke point for these four treks was the variable B, which was an observed variable. This is part of the reason why these four variables do not form a saturated pattern.

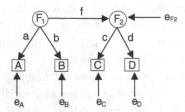

Figure 8.20 A path diagram involving four observed variables and two latent variables
Note: a, b, c, d and f are path coefficients.

a BC choke point. In fact, all these choke points exist in this graph, and all are the same variable (F). If we do the same thing to panel (b) (causal process B) of Figure 8.15 we will see that no tetrad equation vanishes.[22]

Let's go on and apply the vanishing tetrad algorithm to a causal graph involving two latent variables (Figure 8.20). Here are the three tetrad equations:

$$\rho_{AB}\rho_{CD} - \rho_{AD}\rho_{BC} = (ab)(cd) - (afd)(bfc) = abcd(1 - f^2) \neq 0$$
$$\rho_{AC}\rho_{BD} - \rho_{AD}\rho_{BC} = (afc)(bfd) - (afd)(bfc) = 0$$
$$\rho_{AC}\rho_{BD} - \rho_{AB}\rho_{CD} = (afc)(bfd) - (ab)(cd) = f^2(1 - abcd) \neq 0$$

Notice that the only tetrad equation that vanishes is the one with either an AB and/or a CD choke point. In fact, both choke points exist. All the directed paths leading into A and B of all treks between the four pairs of variables (AC, BD, AD and BC) pass through F_1. All the directed paths leading into C and D of all treks between these four pairs of variables also pass through F_2.

8.14   Separating the message from the noise

The ancients knew how to discover causal relationships. Things happened in the world because the gods willed them. One had only to ask and, if the gods were willing and the diviner gifted, the causes would be revealed. Unfortunately, the gods were capricious and their words couched in allegory. A good seer had to be able to separate the message from the noise; to know when a bump in a goat's intestine foretold war and when it was simply undigested grass.[23] If the methods presented in this chapter are the modern version of the diviner's art then we still need to separate the causal message from sampling noise.

The various algorithms all require that we know if sets of random variables are independent or not. We are constantly being asked 'Is the statistical association zero or different from zero?'. So far I have assumed that we can always answer such a question

---

[22] Unless the graph is unfaithful. It is always possible to choose path coefficients in such a way as to make a particular tetrad equation vanish, but the vanishing tetrad equation is not implied by the topology of the graph.

[23] The ancient Greek philosophers were the first to conceive of a world governed by natural causes rather than by divine will. They then confronted the subject of this chapter. Democritos (460–370 BC) is reported to have said: 'I would rather discover one causal law than be King of Persia' (Pearl 2000).

Table 8.3 Possible combinations of decisions to a null hypothesis and its alternative, giving rise to type I and type II errors

		True value in the statistical population	
		$\rho_{XY} = 0$	$\rho_{XY} \neq 0$
Your answer after looking at r_{XY} and	$\rho_{XY} = 0$	Right choice	Type II error
calculating its probability	$\rho_{XY} \neq 0$	Type I error	Right choice

unambiguously, because I have assumed that we have access to the entire statistical population. If correlations are the shadows cast by causes then I have assumed that these shadows are always crisp and well defined. Given such an assumption, we can extract an amazing amount of causal information from purely observational data; certainly, much more than is intimated by the old mantra *correlation does not imply causation*.

Let's get back to reality. We almost never have access to the entire statistical population. Rather, we collect observations from random samples of the statistical population, and these random samples are not perfect replicas of the entire population. If correlations are the shadows cast by causes then *sample correlations* are randomly blurred correlational shadows. We have to find a way of dealing with the imperfect information contained in these blurred correlational shadows. Inferring population values from sample values is the goal of inferential statistics, and inferential statistics is the art of drawing conclusions based on imperfect information. In practice, we can never unambiguously know if the statistical association is zero or different from zero. How can we deal with this problem when applying the various discovery algorithms, and what sort of errors might creep into our results? To see this, we first need to review some basic notions of hypothesis testing.

Consider the problem of determining if the population value of a Pearson correlation coefficient (ρ_{XY}) between two random variables, X and Y, is zero based on the measured sample value (r_{XY}). There are only two possible choices: either it really is or it really isn't. Similarly, we can give only one of two answers based on our sample measure: either we think it is or we think it isn't. These define four different outcomes (Table 8.3).

Normally, the types of biological hypotheses that interest us are ones in which variables are associated, not independent. Because we want evidence beyond reasonable doubt before accepting this interesting hypothesis (see Chapter 2), we usually begin by assuming the contrary – that there is no association – and then look for strong evidence against this assumption before rejecting it and therefore accepting our biologically interesting one. In other words, we want to see a value of r_{XY} that is sufficiently large that there is very little probability that it would have come from a statistical population in which $\rho_{XY} = 0$ is true.

So, what is a small enough probability that we would be willing to declare that ρ_{XY} really is different from zero? This is a somewhat subjective decision, as described in Chapter 2, but Table 8.3 shows that part of our decision will depend on how important it is for us to avoid either a type I error (incorrectly declaring that $\rho_{XY} \neq 0$ when, in reality, $\rho_{XY} = 0$) or a type II error (incorrectly declaring that $\rho_{XY} = 0$ when, in reality, $\rho_{XY} \neq 0$). Because the presence of a real association usually (but not always) gives us useful biological information, and because we know that our ever-present sampling variation can sometimes fool us into observing a large value of r_{XY} even when the variables are independent, we usually place more importance in reducing our type I errors than in reducing our type II errors. Therefore, we usually choose a small probability before we are willing to declare our value of r_{XY} as being 'significantly' different from zero. For instance, choosing a significance level of 0.05 means that we are only willing to accept a 5 per cent chance of making a type I error. However, notice that by decreasing our significance level to the low value of 0.05 we are simultaneously willing to accept a larger chance of making a type II error. This is usually acceptable, because we have already decided that it is more important to be quite sure that ρ_{XY} *is not* zero than to be quite sure that ρ_{XY} *is* zero.

In each of the algorithms described in this chapter you are repeatedly asked to decide if the measure of association is zero or not. You have to make this choice based on a random sample of data and, therefore, you have to conduct statistical tests and choose how important it is to minimise either type I or type II errors. Here's the rub: by definition, these are discovery algorithms. You don't already have a preferred causal hypothesis that you wish to test. You can't have any a priori preference for either $\rho_{XY} = 0$ or $\rho_{XY} \neq 0$, and neither outcome provides more information than the other to the various algorithms.[24]

This brings us to the notion of statistical power. The hypothesis that $\rho_{XY} \neq 0$ is not really a single hypothesis at all; rather, it is a composite hypothesis that includes $\rho_{XY} = 0.01$, $\rho_{XY} = 0.1$, $\rho_{XY} = 0.9$ and an infinite number of other individual hypotheses. Intuitively, it is obvious that it would be much more difficult to distinguish between 0.0 and 0.01 than between 0.0 and 0.9 in any sample of data. If we had a huge data set (say 1,000 observations) and the population value was 0.0 then our sample correlation would almost always be extremely close to zero (Figure 8.21). Sampling variation would only very rarely result, by chance, in a value greater than even a low number such as 0.1. Therefore, if the population value was even slightly different from 0.0, we would almost always find a very small probability value for our measured r_{XY} and would almost never conclude that $\rho_{XY} = 0$ when, in reality, $\rho_{XY} \neq 0$. Our test would be very powerful in detecting even very slight associations between X and Y. If, on the other hand, we had only a small data set (say 10 observations) and our population value was 0.0 then our sample correlation would fluctuate quite widely around zero simply due to sampling variation (Figure 8.21). Therefore, even if the population value was quite different from 0.0 (say $\rho_{XY} = 0.4$) we would often observe sample values close to zero simply due to sampling variation. Because of this, the probability of incorrectly concluding that $\rho_{XY} = 0$ would not be negligible. Our test would not be very powerful in detecting even moderate associations between X and Y and we would make type II errors more often.

The *power* of a statistical test is the probability of rejecting the null hypothesis when, in fact, it is false. It is defined as $1-\beta$, where β is the probability of a sample statistic, taken from a statistical population in which the null hypothesis is false, falling within

---

[24] One exception might be in the orientation phase of the algorithms. It might be better to plead ignorance, and leave edges unoriented, than to definitely make a choice about declaring an unshielded pattern to be a collider or a definite non-collider.

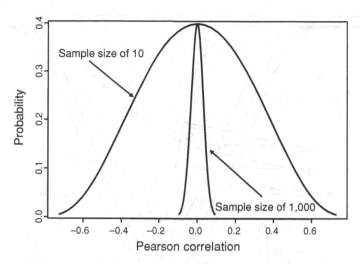

Figure 8.21 The probability of observing a Pearson correlation coefficient of various values when the population value is zero at two different sample sizes

the acceptance region of the null hypothesis. The power is affected by sample size, by the significance level chosen for rejecting the null hypothesis and by the difference (the 'effect size') between the true value of the test statistic and the value assumed by the null hypothesis. Figure 8.22 plots the statistical power to reject the null hypothesis that $\rho = 0$ when the true value varies from –0.9 to 0.9 at two sample sizes (30 and 300 observations) and three different significance levels ($\alpha = 0.05, 0.10$ and 0.20).

Figure 8.22 clearly shows the compromise that must be made. If the null hypothesis ($\rho - 0$) is true then increasing the significance level from $\alpha = 0.05$ to $\alpha = 0.2$ increases our chances of incorrectly rejecting the null hypothesis (type I error). We will be incorrectly declaring associations to exist more often than they really do. On the other hand, increasing the significance level from $\alpha = 0.05$ to $\alpha = 0.2$ increases our power to reject the null hypothesis when associations really do exist but are weak. As the sample size increases then power increases irrespective of the chosen significance level. This is a good thing, because now we can increase our power to detect real but weak associations without increasing our significance level and therefore without increasing our changes of falsely accepting associations that don't really exist. At large sample sizes we are best to set a low significance level (say $\alpha = 0.05$, or even $\alpha = 0.01$), since at such large sample sizes we will keep both type I and type II error rates low. At small sample sizes we are best to increase our significance level if, in fact, we don't have any preference for the presence or absence of a real association. This is because, at a low significance level, only very large values of the correlation coefficient would have a reasonable chance of being detected. As the sample size increases that power approaches 1.0 even as α approaches 0, meaning that the chances of committing both type I and type II errors approach zero.

You will see that significance levels of $\alpha = 0.2, 0.4$ or even higher might be used with very small sample sizes. Clearly, applying these algorithms to small samples means accepting more and more errors due to sampling fluctuations. Remember that these are

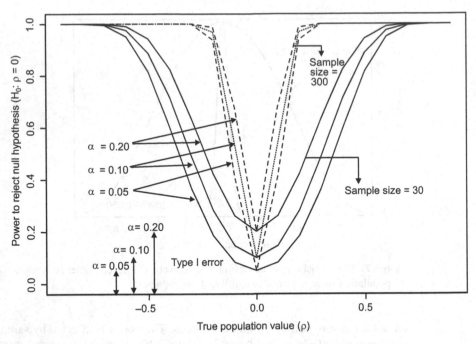

Figure 8.22 The statistical power to detect a non-significant Pearson correlation
Note: This power is affected by the sample size and the chosen significance level (α) used to reject the null hypothesis.

exploratory methods, not methods designed to test a preconceived hypothesis. If you set out to hike through an unfamiliar area then you will probably take a map. The search algorithms are like imperfect maps to a causal landscape. At large sample sizes these maps will give you all the detail that can be obtained, even though no one map might be able to provide all the information that is wanted. At small sample sizes these maps will give you only the major hiking trails, may quite possibly miss some of the smaller trails and may even include some incorrect paths. Independent tests using new data are always important after applying the search algorithms but this is especially important as sample size decreases. Nonetheless, you will see that the error rates are not that bad, especially for constructing the undirected dependency graph, even at small sample sizes. With these points in mind, let's look at the causal inference algorithm in the presence of sampling error.

8.15   The causal inference algorithm and sampling error

At this stage it is useful to look at a numerical example. Figure 8.23 shows a path model from which I will generate sample data and apply the causal inference algorithm. I will generate two different data sets, one with 30 observations and one with 300 observations. The coefficient α equals 0.4. Let's begin with the larger sample size (N = 300).

Table 8.4 Variances (diagonal), covariances (lower subdiagonal) and correlations (upper subdiagonal) of 300 simulated data from a multivariate normal distribution generated according to the causal structure shown in Figure 8.23

	A	B	C	D	E
A	0.93	0.35	0.14	0.08	0.10
B	0.35	1.03	0.42	0.38	0.35
C	0.15	0.45	1.11	0.22	0.52
D	0.07	0.37	0.22	0.94	0.52
E	0.11	0.38	0.59	0.54	1.12

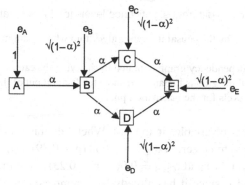

Figure 8.23 Path model used to generate sample data

Table 8.4 shows the variances on the diagonal, the covariances below the diagonal, and the correlations above the diagonal for a simulated data set with N = 300.

The first step is to obtain the undirected dependency graph. I will choose a significance level[25] of 0.05. After constructing the saturated undirected graph I then remove any lines between variables whose zero-order correlations are not significantly different from zero at a rejection level of 0.05 – that is, if $|r| < 0.113$. There is one sample correlation coefficient in Table 8.4, between A and E, that is judged to be zero in the population. Note that this decision is actually a type II error, since the population correlation coefficient between A and E is not zero, though it is weak, with a true value of 0.128. However, this does not introduce any errors in the undirected dependency graph, since this graph is concerned only with direct associations. In other words, the algorithm is robust to these types of errors.

I next look at each of the 10 pairs of variables ({A,B}, {A,C},..., {D,E}) and, for each, test for zero first-order partial correlations. In other words, for each pair I calculate the partial correlation conditional on each of the remaining three variables in turn. With each test I see if the absolute value of the partial correlation is less than 0.113 and, if it

---

[25] This significance level refers to each individual statistical test, not the final partially oriented graph. From Figure 8.17 I know that I will have almost 100 per cent power to detect correlations whose population values are greater than 0.2 in absolute value, although, in an empirical study, I would not know what the population values were.

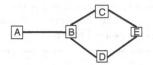

(a) Undirected dependency graph

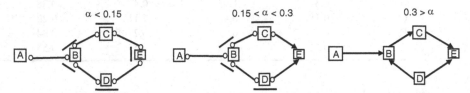

(b) Final output of the algorithm using different significance levels for the orientation phase

Figure 8.24 Different DAGs proposed by the causal inference algorithm when changing the significance level
Notes: Panel (a) is the undirected dependency graph obtained by applying the causal inference algorithm to the sample data. The three graphs in panel (b) show the final output of the algorithm when using different significance levels for the orientation phase.

is, I remove the line joining the two variables in the pair. When I do this I find only four first-order partials that are judged to be zero: $r_{AC|B} = -0.008$ (p = 0.89), $r_{AD|B} = -0.062$ (p = 0.28), $r_{AE|B} = -0.022$ (p = 0.70) and $r_{CD|B} = 0.069$ (p = 0.23). Note that we can't remove the line between A and E since it has already been removed (by error) when looking at the zero-order correlations.[26] This is why I said that the algorithm is robust.

I next look at each of the remaining pairs of variables that are still adjacent and, for each, test for a zero second-order partial correlation. In other words, for each pair I calculate the partial correlation conditional on each possible pair of the remaining three variables in turn and remove the line between any pair whose absolute value of the second-order partial is less than 0.113. There is only one such zero second-order partial: $r_{BE|\{CD\}} = 0.009$ (p = 0.88). I then go on to test for zero third-order partials, but I do not find any. The result is the correct undirected dependency graph (Figure 8.24(a)).

I then go on to the orientation phase. If I maintain the same significance level (0.05) then the algorithm makes a mistake. The path $Co{\rightarrow}E{\leftarrow}oD$ collides at E but, in order to detect this, I must find that the partial correlation between C and D, conditioned on every other possible subset of the other variables that includes E, is never zero. However, even at a sample size of 300 we don't have a lot of power to detect small non-zero associations. What's even worse, we have to conduct four different tests ($r_{CD|\{E\}}$, $r_{CD|\{EA\}}$, $r_{CD|\{EB\}}$, $r_{CD|\{EAB\}}$). If these were independent tests that our actual rejection level for all four tests together would be $0.05^4 = 6.25 \times 10^{-6}$. If these four tests were perfectly correlated (i.e. the probability level for each were the same) then we would have maintained an overall significance level of 0.05. In other words, by setting the significance level of each test at 0.05 we were really demanding very strong evidence – perhaps as low as six chances in 1 million, and certainly fewer than five chances in

---

[26] Actually the algorithm would not even calculate this partial correlation, since the two variables are not adjacent at this stage.

100 – before we were willing to recognise a collider triplet. Of course, we don't know the degree to which these four tests are correlated, but if they were independent then we should choose a significance level of about 0.47 for each test in order to maintain an overall level of 0.05. This is because $0.47^4 = 0.049$.

These considerations emphasise the exploratory nature of the method. The best approach is to try different significance levels and see how the undirected graph and the partially directed graph change. In any real study you could also include any prior information about the data, such as if some variables occurred earlier in time than others, in order to choose between the different results. Figure 8.24(a) shows the undirected graph, and Figure 8.24(b) shows the partially directed graphs that result from our data using different significance levels. The structure of this undirected dependency graph is very stable; varying the significance level from 0.001 to 0.6 always gives the same result. The structure of the partially oriented graphs is less stable. Below a significance level of around 0.15 all unshielded patterns are declared unshielded colliders, including the one that is really a collider. Between 0.15 and 0.3 the correct partially oriented graph is obtained. Beyond 0.3 a second (incorrect) unshielded collider is detected.

At this stage I can use my d-sep test to evaluate each of the three partially oriented graphs.[27] When I do this I find that the equivalent graphs that result when the significance level used in the orientation stage is 0.3 or larger are clearly rejected. This is because these graphs all predict that A and C are unconditionally independent. In fact, $r_{AC} = 0.35$, and, with a sample size of 300, this would occur much less than once in 1 million times if the data were really generated according to this graph. Therefore, we have to choose between the first two partially oriented graphs. When we look at the first partially oriented graph we see that it is impossible for all the unshielded patterns to be definite non-colliders and for there to also be no cycles in any equivalent DAG that is consistent with it. We are led to accept the middle partially oriented graph as the most consistent with our data.

Figure 8.25 shows the results when the undirected dependency graph algorithm is applied to a sample data set, generated from Figure 8.23, but with a small sample size of 30 observations. Now we see lots of errors. No equivalent model from the undirected dependency graph, obtained using $\alpha = 0.01$, provides an acceptable fit to the data. However, all the remaining undirected graphs have equivalent models that do produce an acceptable fit, based on the d-sep test and a significance level of 0.05. To go any further requires information beyond that which exists in this little data set. For instance, the undirected graphs at $\alpha = 0.05$ to 0.2 all predict that A and B are independent of C, D and E.[28] If, in other studies, either A or B was found to be correlated with either C, D or E then this undirected graph could be rejected.[29] Once you decide upon a particular

---

[27] Every possible partially oriented graph can be tested in this way. Simply choose one of the equivalent DAGs consistent with the partially oriented graph and apply the d-sep test. Since every equivalent graph will give the same probability level under the null hypothesis, it doesn't matter which one you choose.

[28] Since there are no undirected paths between these two sets of variables, there can be no directed paths either. Therefore, they must be independent.

[29] In Shipley (1997) I describe how embedding the algorithm for the undirected dependency graph inside a bootstrap loop helps to reduce the effects of small sample sizes.

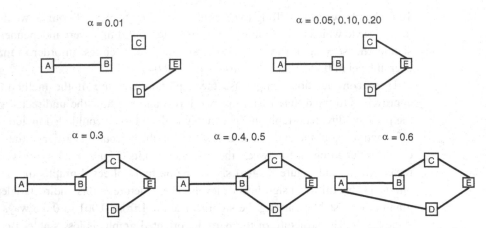

Figure 8.25 Output of the undirected dependency graph algorithm when applied to a sample data set, generated from Figure 8.23, with a small sample size of 30 observations

undirected graph as being most consistent with all the information you can begin to explore the orientation phase.

The best way to see what types of errors these search algorithms will make at different sample sizes is to generate data with different characteristics and count the types of errors that occur. Spirtes, Glymour and Scheines (1993) have conducted such simulation studies for the case of causal sufficiency and using not just the algorithms described here but others that are more efficient with large numbers of variables but that commit more errors at small sample sizes. I have also explored the error rates of the causal inference algorithm with small sample sizes, and using bootstrap techniques (Shipley 1997). The general results that come from these simulation studies are as follows.

(1) Error rates are lower for constructing the undirected dependency graph than for orienting the edges.
(2) The error rates for adding a line in the undirected dependency graph when there shouldn't be one are quite low. Even at very small sample sizes (say 30 observations), if a line appears then it probably exists unless the rejection level is very high (say 0.5 or more).
(3) The error rates for missing a line in the undirected dependency graph when there should be one are higher. As the strength of the direct causal relationships decreases this error rate increases. As the number of other variables to which a given variable is a direct cause increases this error rate increases. As the sample size increases this error rate decreases.
(4) The error rates for orienting edges are higher than those related to the undirected dependency graph. This is to be expected, since the orientation phase depends on the number and type of unshielded patterns; therefore, any errors in the undirected dependency graph will be propagated into the orientation phase.
(5) The rejection level used in constructing the undirected dependency graph should increase as the sample size decreases. At very small sample sizes, values of 0.2 or higher should be used. At sample sizes of around 100 to 300, a rejection level of

0.1 should be used. At higher sample sizes, a value of 0.05 is fine since statistical power does not have to be traded off with the ability to avoid type I errors.

8.16   The vanishing tetrad algorithm and sampling variation

The vanishing tetrad algorithm has been much less studied than the other algorithms that I have described. In part, this is because the assumption of a linear relationship between the latents and the observed variables limits its application. Another reason is perhaps because it is less informative than the other algorithms; it can alert us to the presence of latent variables but cannot tell us exactly how these latents connect to the observed variables. Another reason is that, unlike the tests for (conditional) independence that are used in the other algorithms, the test for a zero-tetrad equation is only asymptotic. In other words, in order to get accurate probability estimates based on the null hypothesis that a tetrad equation is zero, you need a certain minimum sample size. No one (to my knowledge) has formally studied the asymptotic requirements for this test, and so I will present some Monte Carlo results to give you some rules of thumb when interpreting the asymptotic probability levels of the test statistic.

The test statistic is $\tau = \rho_{IJ}\rho_{KL} - \rho_{IL}\rho_{JK}$, where I, J, K and L are the four variables involved in the tetrad; remember that there are always three tetrad equations for each set of four variables. Under the null hypothesis this value will be zero in the population. Wishart (1928) derived the asymptotic sampling variance of this statistic in the first part of the twentieth century but no one (to my knowledge) has ever derived the exact sampling variance. The asymptotic sampling variance[30] is

$$\left(\frac{D_{IK} D_{JL} (N + 1)}{(N - 1)} - D \right) \left(\frac{1}{N - 2} \right)$$

where D is the determinant of the population correlation matrix of the four variables, D_{IK} is the determinant of the 2×2 matrix consisting of the population correlation matrix of variables I and K, D_{JL} is the determinant of the 2×2 matrix consisting of the population correlation matrix of variables J and L, N is the sample size and the four variables follow a multivariate normal distribution. There are six possible pairs of four variables; four of these pairs define a tetrad equation and the other two pairs define the 2×2 submatrices whose determinants are used in calculating the asymptotic variance. If the null hypothesis is true then the test statistic is asymptotically distributed as a normal variate with a zero mean and the given variance. Therefore, the value $\tau/\sqrt{\mathrm{var}(\tau)}$ asymptotically follows a standard normal distribution.

To conduct the statistical test you replace the population values by the sample values. In doing this you are only approximating the true probability level, and so it is important to know how good (or bad) this approximation is. Table 8.5 shows some results of Monte Carlo simulations in which a four-variable measurement model, of the sort shown in

---

[30] The formula given by Spirtes, Glymour and Scheines (1993) is incorrect, but these authors give the correct formula in their earlier book (Glymour et al. 1987).

Table 8.5 Statistical bias in the vanishing tetrad test

						Quantiles	
α	ρ	N	$\sigma(\tau)$	SD(τ)	0.20	0.10	0.05
0.1	0.01	25	0.0603	0.0974	0.09	0.05	0.02
		50	0.0293	0.0483	0.11	0.06	0.02
		1,000	0.0015	0.0024	0.12	0.05	0.02
0.3	0.09	25	0.0603	0.0974	0.09	0.05	0.02
		50	0.0374	0.0514	0.13	0.06	0.03
		100	0.0218	0.0285	0.16	0.07	0.03
		500	0.0079	0.0085	0.21	0.10	0.05
0.4	0.16	25	0.0811	0.1061	0.12	0.06	0.03
		50	0.0481	0.0582	0.17	0.09	0.05
0.5	0.25	25	0.0964	0.112	0.20	0.09	0.04
0.6	0.36	25	0.1094	0.1175	0.23	0.11	0.05

Notes: Each line summarises the results of 500 independent data sets generated from a model like that of Figure 8.16 with path coefficients α between the latent and each measured variable. ρ gives the population correlation between each measured variable, N is the sample size, $\sigma(\tau)$ is the asymptotic standard deviation of the tetrad equation and SD(τ) is the average standard deviation of the 500 data sets. Also shown are the 20 per cent, 10 per cent and 5 per cent quantiles of the standardised tetrad equations $(\tau/\sqrt{\text{VAR}(\tau)})$.

Figure 8.16, was used to generate 500 independent data sets. Each such simulation used a different sample size per data set and a different value for the path coefficients (α) between the latent variable and the observed variables. Remember that, according to the rules of path analysis, the population correlation coefficient (ρ) between each pair of observed variables in such a model is α^2.

It is clear that, when the correlations between the four observed variables are very low, the approximation is not good. When the population correlation between them is only 0.01 then even a sample size of 1,000 produces conservative probability levels, and you would reject the null hypothesis that the tetrad is zero too often. By the time that the population correlation between the four observed variables is about 0.25 then even small sample sizes cause no problems.

In using the vanishing tetrad algorithm, you would first construct the undirected dependency graph. If no set of four (or more) variables in the undirected dependency graph is saturated (i.e. in which each variable has a line to each other variable in the set) then there is no need to apply the algorithm. If you do see such a pattern then apply the vanishing tetrad algorithm to this set of variables, keeping in mind the approximate nature of the calculated probabilities. If the correlations between the variables are very weak then you should increase the significance level.

8.17   Empirical examples

In Chapter 3 I used data from Jordano (1995) consisting of five variables measured on 60 trees of St Lucie cherry to test a path model. In fact, the path model was derived from the causal inference algorithm. The five variables were (1) the area of the tree canopy projection (a measure of the photosynthetic biomass), (2) the total number of ripe fruit produced per tree during the year, (3) the average fruit diameter, (4) the average seed weight and (5) the number of seeds dispersed from the tree by birds. The primary variable of interest was the number of seeds dispersed from the tree. This is because seeds that fall directly beneath the tree and germinate will generally die due to shading from the parent, and so the evolutionary fitness of the tree will be more closely related to the number of seeds that are dispersed away from the parent. However, it is reasonable to suppose that the other measured variables will interact to affect the number of seeds dispersed, and so we wish to understand how these variables relate to one another.

There are approximately 59,000 potential acyclic graphs to consider given our five measured variables. I could propose a specific causal model, and even provide reasonable biological arguments to back it up. I am quite sure that most readers could also propose a specific causal model and provide reasonable biological arguments. I am also quite confident that the proposed causal models will not be the same.[31] In fact, I could alternatively propose different causal models and come up with equally good reasons for each one! The problem is that my biological knowledge of the phenomenon is not sufficiently detailed to strongly favour one model over other reasonable models. This is a situation in which it does not make much sense to waste a lot of effort in applying an inferential test to a specific model. When applying an inferential test (either the d-sep test or that based on structural equation modelling) the question is: are the data consistent with *this* model? However, since we can easily come up with a number of different reasonable models, and have no idea if there might be others, we are really at a stage in which we want to ask: *which* models are consistent with the data and *which* aren't? It is analogous to the difference between testing if a regression coefficient is zero and obtaining a confidence limit for the regression coefficient.

After transforming the original variables to their natural logarithms, I gave these data to the causal inference algorithm. Figure 8.26 shows the resulting undirected dependency graph for various rejection levels. The undirected graph that results (Figure 8.26(a)) when the rejection level is low (0.05 or 0.10) can be rejected without even orienting it. This is because it predicts that each of {Seed weight, Fruit diameter} is unconditionally d-separated from – and therefore independent of – each of {Canopy projection, Number of fruits produced, Number of seeds dispersed}. Applying the d-sep test to only these independent statements yields a χ^2 value of 27.18 with 12 degrees of freedom (p = 0.007). If we then go to the second undirected graph (Figure 8.26(b)), obtained using rejection levels of 0.2 to 0.5, we can apply the orientation phase of the algorithm. Until we get to a very high rejection level for this phase (0.4) we always find that each unshielded pattern is a definite non-collider. At a rejection level of 0.4 we are informed that fruit diameter is a collider. Figure 8.27 shows the partially oriented acyclic graph based on the middle undirected graph.

Despite the small sample size (60 observations) we have already discovered quite a lot of information about the possible causal relationships between these variables. There

---

[31] In fact, the model that I proposed in Chapter 3 differed from the model proposed in Jordano (1995); see Chapter 3.

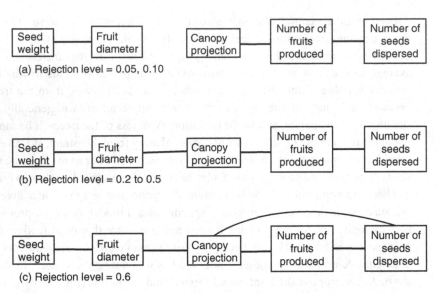

(a) Rejection level = 0.05, 0.10

(b) Rejection level = 0.2 to 0.5

(c) Rejection level = 0.6

Figure 8.26 Output of the undirected dependency graph algorithm when applied to the empirical data of Jordano (1995) at various significance levels

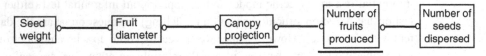

Figure 8.27 The final partially oriented graph that is produced based on the middle undirected graph of Figure 8.26

is no evidence that there are latent variables that are common causes of more than two observed variables; if there were, we would see three or more variables with a saturated pattern between them. Remember that we are in an exploratory mode. We are looking for possible models that accord with our available evidence about the correlational shadows but we also want our model to accord with any previous biological knowledge that we might possess. For instance, consider the relationship between the number of cherry fruits produced and the number of seeds dispersed by the birds (Number of fruits produced o–o Number of seeds dispersed). Since seeds cannot be dispersed by birds before the fruit has been produced, we can exclude the orientation (Number of fruits produced ← Number of seeds dispersed). It is possible that the orientation is (Number of fruits produced ←→ Number of seeds dispersed), though it is difficult to conceive of a latent variable that both determines how many fruits the tree will produce and also how many of these fruits will be eaten by the birds. However, if we accept this orientation involving a latent variable then we must also exclude the orientation (Canopy projection o→ Number of fruits produced), since this would produce a collider. This would therefore force us to accept the following orientation: (Canopy projection←Number of fruits produced). Such an orientation disagrees not only with much empirical evidence but also with the time ordering of the phenomenon, since the canopy is produced before the

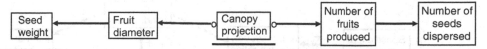

Figure 8.28 The partially oriented graph that is retained as most biologically plausible

fruits are made. If we begin with the biologically reasonable hypothesis that the total photosynthetic capital of the tree, of which the canopy projection area is a measure, determines both how many fruit will be produced and the average size of each fruit, we are immediately led to the partially oriented directed acyclic graph in Figure 8.28.

Such a result, to me, is incredible. With five observed variables we had a little over 59,000 possible directed acyclic models. This algorithm, combined with a few reasonable biological observations, reduced this huge number to a few reasonable models. Since I know that the statistical power to detect small non-zero correlations is not great with only 60 observations I would not bet my salary on the accuracy of Figure 8.28, but I would feel much more confident about proposing a model derived from Figure 8.28 as a useful biological hypothesis to be tested with independent data.

This is the real strength of these discovery algorithms. Unless pre-existing theory is already quite solid, proposing a complete causal model from such theory often degenerates into asking: 'If I were God, and the world was a machine, then how would I construct it?' Since few of us are gods and the world is not really a machine, such 'hypothesis generation' can easily mask unbridled speculation. The discovery algorithms first show us the correlational shadows that our data contain, which causal processes might reasonably have cast them and which causal processes were unlikely to have cast these correlational shadows. This constrains our speculation and forces us to consider different alternative models, and also forces us to explicitly justify any causal process that appears to contradict what the data seem to say.

The next empirical example shows that we shouldn't accept the output of these discovery algorithms blindly. Their purpose is to help us develop useful causal hypotheses, not to replace the scientist with a computer algorithm. In Chapters 3 and 4 I presented a path model relating specific leaf area, leaf nitrogen content, stomatal conductance, net photosynthetic rate and the CO_2 concentration within the leaf. This model (Figure 8.29(a)) was based on the pre-existing model of stomatal regulation produced by Cowan and Farquhar (1977). When I apply the causal inference algorithm to the empirical data[32] the resulting partially oriented graphs make no biological sense. At low rejection levels (0.2 and lower) none of the suggested graphs fit the data. At higher rejection levels (0.2 to 0.5) a graph (Figure 8.29(b)) is suggested that does produce a path model with a non-significant χ^2 value, but this graph contradicts some well-established biological knowledge of leaf gas exchange. Note that in this graph the net photosynthetic rate is causally independent of the CO_2 concentration within the leaf even though this is physically impossible. The amount of CO_2 within the leaf is determined by the rate at

---

[32] I use only the 35 species that have a C_3 photosynthetic system, and each variable is transformed to its natural logarithm to insure multivariate normality.

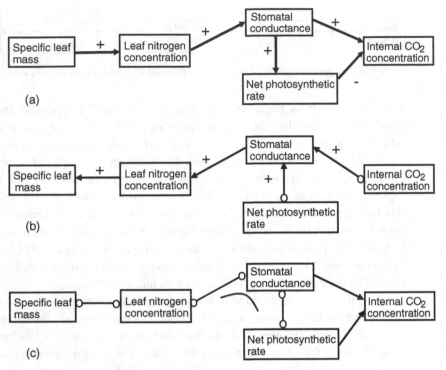

Figure 8.29 The causal inference algorithm applied to empirical data
Notes: Panel (a) is the model proposed in Chapter 3 based on biological arguments. The partially oriented graph in panel (b) is the output of the causal inference algorithm when no constraints were placed on it. The partially oriented graph in panel (c) is the output of the causal inference algorithm when simple constraints are placed on it based on well-known physical laws.

which it is diffusing into the leaf across the concentration gradient from the higher outside concentration to the lower concentration within the leaf (i.e. stomatal conductance) and the rate at which it is being removed from the intercellular air by photosynthesis. The concentration within the leaf is therefore determined by the net rate at which it is being fixed by photosynthesis (by definition, the net photosynthetic rate) and the rate at which gases are diffusing across the stomates (measured by the stomatal conductance).

Why would the causal inference algorithm produce such an erroneous output? The reason is almost surely because this biological process violates one of the assumptions of the algorithm, namely that the probability distribution of these variables is faithful to the causal process that generated it.[33] This means that independence or partial independence relationships are assumed to reflect the way in which the variables are causally linked together, rather than special numerical values of the strengths of the direct causal relationships that manage to cancel each other out. Imagine that you glance out of the window and see a single person walking down the lane. One hypothesis for this observation might be that there are really two people but that one is positioned behind the

---

[33] Unfaithfulness only properly applies to the population probability distribution.

other in such a way that he or she is perfectly hidden by the person in front. A simpler, and more parsimonious, hypothesis is that there is really only one person coming down the lane. Both hypotheses are possible but the first requires that you are witnessing a very special juxtaposition of distances, shapes and sizes of people. The illusion of a single person would disappear as soon as any of those conditions change. We could say that such special conditions are *unfaithful* to our general expectations, and so we would reject the hypothesis unless we had very good independent reasons to believe that someone might be hiding due to such special conditions. In the same way, these discovery algorithms assume that if we observe an observational independence between two variables then this means that the two are causally independent. It is always possible that the two variables only *appear* to be independent because positive and negative direct and indirect relationships cancel each other out, but this would require a very special balancing of causal effects. Just as with the example of two people appearing to be a single person, unless we have good reasons for suspecting such a curious observation, we would choose the more parsimonious explanation.

In fact, the correlation coefficient between the Ln-transformed net photosynthetic rates and the Ln-transformed internal CO_2 concentrations in these data was only 0.051 ($p = 0.77$). Therefore, the line between these two variables would be immediately removed when constructing the undirected dependency graph. We know the net photosynthetic rate must be a cause of the amount of CO_2 within the leaf, and yet there appears to be no relationship between the two variables in these data. The reason for this apparent contradiction can be found in the Cowan–Farquhar (1977) model of stomatal regulation upon which the path model in Figure 8.24(a) is based. According to this theory, the stomates are regulated in order to maintain the internal CO_2 concentration at the 'break point' – the point at which carbon fixation is equally limited by the regeneration of RuBisCO due to ATP production from the light reaction of photosynthesis and the amount of RuBisCO available in the dark reaction of photosynthesis. In other words, the overall correlation between the net photosynthetic rate and the internal CO_2 concentration is determined by two different causal paths. One path is the direct effect of net photosynthetic rate in reducing the internal CO_2 concentration (net photosynthesis→internal CO_2). The other path is the trek from stomatal conductance that is a common cause of both net photosynthetic rate and internal CO_2 concentration (net photosynthesis←stomatal conductance→internal CO_2). Increasing the stomatal conductance increases the amount of CO_2 that enters the leaf, thus increasing both the photosynthetic rate and the internal CO_2 concentration. Furthermore, in order to maintain a constant internal CO_2 concentration, the stomates must ensure that the increase in internal CO_2 due to diffusion through the stomates is just enough to counter the decrease in CO_2 that is caused by the resulting increase in the net photosynthetic rate. By balancing these positive and negative effects, such a homeostatic control maintains a constant internal CO_2 concentration but also produces an unfaithful probability distribution.

Since the operational definition of the net photosynthetic rate is the rate at which CO_2 is being removed from the air within the leaf, we have good independent reasons to suspect that the net photosynthetic rate will exert a direct negative effect on the internal CO_2 concentration. We can now apply the causal inference algorithm again but add the constraint that stomatal conductance and net photosynthetic rates must each remain as direct causes of the internal CO_2 concentration. This constraint is justified by simple physical laws of passive diffusion of gases across a concentration gradient. The resulting graph is shown in Figure 8.29(c). The causal inference algorithm has suggested a partially oriented graph that is statistically equivalent to the path model that was proposed on the basis of the Cowan–Farquhar theory of stomatal regulation. We have only to note that it is biologically more reasonable to suppose that the leaf nitrogen concentration is caused by specific leaf mass rather than the inverse[34] and we recover the path model in its entirety.

The assumption of faithfulness is really based on a parsimony argument. It says that, if the only causal information available is that obtained from the observational data at hand and we have different possible causal structures that are exactly equivalent in their predictions of (partial) independence, it is preferable to assume a causal structure whose independence predictions are robust rather than to assume a causal structure whose independence predictions require a special balancing of direct and indirect causal effects. When we have good causal information exterior to the data at hand then such information should be used. With small samples this is especially important, because low statistical power means that real, but weak, effects might be incorrectly interpreted as independence.

8.18 Orienting edges in the undirected dependency graph without assuming an acyclic causal structure

A recurring theme in science fiction stories is the Universal Translator; a device that can infallibly translate back and forth between any set of languages. In the case of acyclic causal structures we had an imperfect, but still quite serviceable, translation device. D-separation, applied to the directed acyclic graph, could infallibly translate from the language of causality to the language of probability distributions. We could not use it to infallibly translate backwards from a probability distribution to the causal graph because different causal structures can generate the same joint probability distribution. This is why the discovery algorithms generate as output a partially oriented acyclic graph rather than a single directed acyclic graph. Even so, this is quite useful in reducing the hypothesis space down to a manageable set of possible DAGs. When we move on to search algorithms for (possibly) cyclic causal processes then the problem gets even more difficult, because our translation device, d-separation, cannot be generally applied to non-linear cyclic causal processes. Nonetheless, Richardson (1996b) has produced an algorithm that is provably correct for cyclic causal structures (given population measures of association and faithfulness) under the assumption that the functional relationships between the variables are linear and that there are no latent variables generating associations between more than two observed variables.

---

[34] The concentration of nitrogen does not vary strongly through the depth of the leaf. Therefore, if there is more leaf biomass per leaf area (i.e. the leaf is thicker) then there will be more nitrogen per unit leaf area.

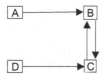

Figure 8.30 A directed cyclic graph with a feedback relationship between variables B and C

As you might have already feared, this algorithm is not just more complicated; it also requires some new definitions and notational conventions. Taken one at a time, each part of the algorithm is still intuitively comprehensible. The algorithm is based on the notion of a *partial ancestral graph* (PAG). A PAG is an extension of the partially oriented inducing path graph of acyclic models that was used in the causal inference algorithm. Here are the conditions for a graph to be a PAG.

(1) There is an edge between two variables, A and B, if and only if A and B are d-connected given any subset of other observed variables in the graph – i.e. if and only if there is an inducing path between A and B. This is the same as for the graphs generated as output from the causal inference algorithm.
(2) If there is an edge between A and B that is out of A with the notation A–*B (but not necessarily into B) then A is an ancestor of B.
(3) If there is an edge between A and B that is into B with the notation A*→B (but not necessarily out of A) then B is *not* an ancestor of A.
(4) If there is an underlining at the middle variable of a triplet with notation A*–*B*–*C then the edges *do not* collide at B. Therefore, B is an ancestor of either A or B, but not both.
(5) If there is an edge from A to B and from C to B (A→B←C), but B is *not* a descendant of a common child of A and B, then B is doubly underlined[35]; thus A → B ← C. This is the first big difference from the POIPG graphs generated as output from the causal inference algorithm. In fact, such a condition is impossible in an acyclic causal structure, as will be explained in more detail below.
(6) Any edge end point not marked in one of the above ways is left with a small circle; thus, A○– means that the mark after the A is unknown and could be any of A←, A– or A*–.

If you go over these points slowly then you will see that the only major difference is in point (5). Let's look more closely at it. To begin, let's look at what would happen if we applied the causal inference algorithm to the causal structure, shown in Figure 8.30, containing a feedback loop.[36]

---

[35] Richardson (1996b) uses a dotted underline rather than a double underline.
[36] In Chapter 2 I described how such cyclic graphs can result from a dynamic process in which the same variables are measured at different times and included in the same data set without any explicit indexing of the time dimension. Richardson (1996b) presents an extended discussion of the interpretation of cyclic patterns in causal graphs.

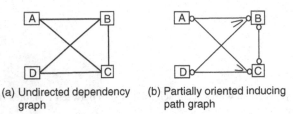

(a) Undirected dependency
graph

(b) Partially oriented inducing
path graph

Figure 8.31 Development of the partially oriented inducing path graph
Notes: Panel (a) is the undirected dependency graph that results from the directed cyclic graph in
Figure 8.30. Panel (b) is the partially oriented inducing path graph.

Figure 8.31 shows the development of the partially oriented graph. The undirected dependency graph has no line between A and D because, applying d-separation to the true graph in Figure 8.30, A and D are unconditionally d-separated. However, there is a line between A and C and another between D and B even though none appear in the true graph. This is not a mistake. Remember that a line in the undirected dependency graph simply means that the two variables joined by the line are d-connected given any subset of other variables (i.e. that there is an inducing path between them). A is always d-connected to C because there are two undirected paths given the feedback relationship (A→B→C and A→B←C), and so conditioning on B (or {B,D}) will always leave one path open. The line between D and B occurs for the same reason. Now, when we apply the orientation phase of the causal inference algorithm to the undirected dependency graph, we find that the two unshielded patterns (A–C–D and D–B–A) are non-colliders. To see this, write out the two undirected paths between A and D: A→B→C←D and A→B←C←D. The unshielded pattern A–C–D is a collider (according to the causal inference algorithm) if A and D are never d-separated given C plus any possible subset of other observed variables (i.e. {C} and {B,C}). Now, A and C are never d-separated given {C} because of the undirected path A→B→C←D. However, given {B,C} then both undirected paths are d-separated, and so the causal inference algorithm would declare the unshielded pattern (A–C–D) to be a definite non-collider. The same result would occur for the unshielded pattern (D–B–A).

This mistake made by the causal inference algorithm occurs because of the feedback relationship between B and C. It could never occur in an acyclic causal process. Yet this mistake is actually very informative, because it suggests a way of detecting feedback relationships. To see this we have to go back to the algorithm used to construct the undirected dependency graph.

In the algorithm for the undirected dependency graph we chose two variables (X and Y) and then look for a conditioning subset of other observed variables that renders X and Y independent (thus, d-separated). We begin with the smallest such conditioning subset – the null subset containing no other variables – and test for independence of X and Y given this null subset (i.e. unconditional independence). If this does not occur then we test the first-order conditioning subsets, and so on. As soon as we find a conditioning subset that renders X and Y independent we remove the line between them and stop. Let's call the conditioning subset that renders X and Y independent during the algorithm the *Separation set of X and Y*, or **Sepset**(X,Y). Now, in an acyclic causal process in which the directed acyclic graph has a collider, such as $X^* \to Z \leftarrow^* Y$, Z will *never* be a member of **Sepset**(X,Y), because any conditioning set that contains Z will make X and Y d-connected. Furthermore, in an acyclic causal process of this type in which a given variable is a non-collider, such as $X^*-oZo-Y$, Z will *always* be a member of **Sepset**(X,Y). Therefore, given an unshielded pattern (X–Z–Y) and the assumption of no cyclic causal relationships, we can orient this unshielded pattern simply by determining if Z is a member of **Sepset**(X,Y). If Z is a member of **Sepset**(X,Y) then Z is a non-collider in the unshielded pattern; if not, Z is a collider. Since **Sepset**(X,Y) is the smallest subset that d-separates X and Y there can be other subsets that also d-separate X and Y. However, if Z is a collider along X–Z–Y then neither **Sepset**(X,Y) nor any other subset that d-separates X and Y will ever contain Z in an acyclic graph. However, in a cyclic causal process this is not true. Because the causal influence goes both ways in a cyclic graph it is possible for Z to be absent from **Sepset**(X,Y), given the unshielded pattern X–Z–Y, even though Z plus some other variables can d-separate X and Y.

For instance, let's look again at Figure 8.31. There is an unshielded pattern (A–B–D), and the conditioning sets that d-separate A and D are {null} – since A and D are unconditionally d-separated – and {B,C}.[37] **Sepset**(A,D) is {null} because this is the smallest conditioning set. Since B is not in **Sepset**(A,D) this would normally mean that B is a collider if there were no cycles. If there were no cycles then this would also mean that no other set that d-separates A and D could contain B, yet {B,C} does d-separate A and D. The discovery algorithm for cyclic causal structures uses this fact. If, given an unshielded pattern X–Z–Y, Z appears in some subsets of other variables that d-separates X and Y but not in others then there is a cyclic orientation.

Before presenting the full algorithm I should head off some potential confusion that would result if people compared my description of the algorithms for the undirected dependency graph and the causal inference algorithm with those published by Spirtes, Glymour and Scheines (1993). In my descriptions I did not refer to Sepsets, but these play an important role in the original descriptions. I did this because, for acyclic causal processes, the result is the same in the population even though the use of Sepsets reduces the computational cost by not having to retest for d-separation during the orientation phases. If they are applied to causal processes having cyclic structures, the result can be different. Figure 8.32 shows the resulting partially oriented inducing path graphs that are obtained from the causal process shown in Figure 8.30 when (panel (a)) applying the causal inference algorithm as I have presented it and (panel (b)) as Spirtes, Glymour and Scheines (1993) originally presented it.

Notice that the unshielded patterns (A–B–D and A–C–D) are oriented as non-colliders in my formulation and colliders in the original formulation. This is because **Sepset**(A,D) is empty (a null set). In the unshielded pattern Ao–oBo–oD, B is not in **Sepset**(A,D), and so B is therefore considered a collider in the original formulation. In my formulation, given Ao–oBo–oD, one must find that A and D are conditionally associated given B plus any other subset of observed variables, not just **Subset**(A,D).

---

[37] There are two undirected paths between A and D: $A \to B \to C \leftarrow D$ and $A \to B \leftarrow C \leftarrow D$. The first undirected path is d-separated given {B,C} because B is blocked. The second undirected path is d-separated given {B,C} because C is blocked.

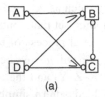   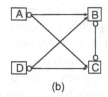

(a)   (b)

Figure 8.32 The partially oriented inducing path graphs that result from the two different formulations of the causal inference algorithm
Notes: Panel (a) is a partially oriented inducing path graph using the formulation presented in this book. Panel (b) is a partially oriented inducing path graph using the original formulation of Spirtes, Glymour and Scheines (1993).

Since A and D are d-separated given {B,C} then Ao–oBo–oD is oriented as Ao–<u>oBo</u>–oD. It is precisely when these two formulations disagree that we have evidence for a feedback relationship. Now, I will give the cyclic causal discovery (CCD) algorithm of Richardson.

8.19   The cyclic causal discovery algorithm

(1) Form the undirected dependency graph. As soon as a pair of variables (X,Y) are found to be independent given a conditioning set **Q**, let **Sepset**(X,Y) = **Sepset**(Y,X) = **Q** and go on to another pair of variables. Orient each edge (X–Y) between two variables in the undirected dependency graph as (Xo–oY).

(2) For each unshielded pattern (Xo–oZo–oY), orient as (Xo→Z←oY) if Z is not in **Sepset**(X,Y) and orient as (Xo–<u>oZo</u>–oY) if Z is in **Sepset**(X,Y).

(3) For each triplet of variables (A,X,Y) such that X and Y are adjacent (i.e. Xo–oY) and A is not adjacent to either X or Y, (a) if **Sepset**(A,Y) is not a subset[38] of **Sepset**(A,X) then orient Xo–*Y as X←Y, otherwise (b) if **Sepset**(A,X) is not a subset of **Sepset**(A,Y) but A and X are d-connected given **Sepset**(A,Y) then orient Xo–*Y as X←Y.

(4) Find each triplet of variables that are now oriented as (A→B←C). Now find those variables V that are *local* to A; the set of all such variables is called **Local**(A). A variable V is *local* to A if either A*–*V or A→Y←V. Next, form a new set (**T**) consisting of all variables in **Local**(A) except for B, C or those also in **Sepset**(A,C). If any subset of **T** d-separates A and C then orient as (A → <u>B</u> ← C) and record T plus **Sepset**(A,C) plus B in new sets called **SupSepset**(A,B,C) and **SupSepset**(C,B,A). The double underline means that X and Y do not both collide at Z.[39]

---

[38] A set is a subset of another set if every element in the first set is also in the second set. For instance, **A** = {x,y,z} is a subset of **B** = {w,x,y,z}.

[39] The original formulation of this algorithm uses a dotted underline rather that a double underline. Although this section appears quite complicated, the complications arise in an effort to save computational time. Basically, this step simply looks for a pattern (A*→B←*C) such that B is not in **Sepset**(A,C) but there is at least one subset of other variables that includes B and that d-separates A and C. This is the clue needed to detect a feedback relationship.

(5) Find a quadruple of variables (A,B,C,D) such that B and D are adjacent and the following patterns exist: A → B̲ ← C and either A→D←C or A → D̲ ← C. If such patterns exist then orient o–oD or B–oD as B→D if D is not in **SupSepset**(A,B,C) and orient as B*–D if D is in **SupSepset**(A,B,C).

(6) Find a quadruple of variables (A,B,C,D) such that B and D are adjacent, D is not adjacent to both A and C, and A → B̲ ← C. If such conditions exist then orient Bo–oD or B–oD as B→D if A and C are not d-separated given **SupSepset**(A,B,C) plus D.

This algorithm probably seems overwhelming. However, since it is incorporated into the TETRAD III program you don't have to understand it sufficiently to actually program it, only well enough to have an intuitive knowledge of what it does. The most important part is to be able to interpret it. I'll go over each section of the algorithm and provide an intuitive explanation. However, note that this algorithm is the most general algorithm of all those presented so far.[40] If the causal process is acyclic then this algorithm will give the same output as the causal inference algorithm even if the functional relationships are non-linear.

> **Section (1)**: This is simply the algorithm for constructing the undirected dependency graph. If there is an edge between two variables (X, Y) then there is an inducing path between them and no other variable, or set of these other variables, can d-separate X and Y. The reason for constructing **Sepset**(X,Y) and **Sepset**(Y,X) is simply so that we don't have to keep conducting the independence tests in the other sections of the algorithm. We could have done this in the causal inference algorithm as well. In fact, the original formulation of the causal inference algorithm, as implemented in TETRAD II and TETRAD III, does use separation sets. The separation sets provide useful information because every variable in **Sepset**(X,Y) is an ancestor of either X or Y.
>
> **Section (2)**: This section is simply the algorithm for determining if variable Y in an unshielded pattern (Xo–oZo–oY) is a collider (thus, Xo→Z←oY) or a non-collider (thus, Xo–oZo–oY). Now that we have **Sepset**(X,Y) we don't have to redo all the (conditional) independence tests. If we see that Z is in **Sepset**(X,Y) then Z is a non-collider and if Z isn't in **Sepset**(X,Y) then it is a collider. This uses the fact (above) that if Z is in **Sepset**(X,Y) then Z is an ancestor of either X or Y. Therefore, the orientation can't be Xo→Z←oY, because this would imply that Z was a descendant of both X and Y. This also explains why causal processes having feedback relationships, such as that shown in Figure 8.25, produce different results when applying my version of the causal inference algorithm and the original version that uses separation sets. In such feedback processes, a variable can be *both* an ancestor and a descendant at the same time. This is not possible in an acyclic causal structure.

---

[40] There are a few 'propagation' rules that can be added after the algorithm is finished. For instance, Xo→Yo–oZ implies Xo→Y–oZ.

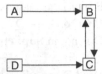

Figure 8.33 A directed cyclic graph with a feedback relationship between variables B and C

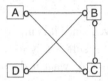

Figure 8.34 The undirected dependency graph, obtained after section (1) of the cyclic causal discovery algorithm

Section (3): We know that X and Y are adjacent (Xo–*Y) and that A is not adjacent to either X or Y by looking at the partially oriented graph. We also know that those variables that d-separate A and Y are not a subset of those variables that d-separate A and X – i.e. that A and X are still d-connected given **Sepset**(A,Y). Therefore, X is not an ancestor of Y and we can orient Xo–*Y as X←Y.

Section (4): This section begins by looking for a triplet of variables that has already been oriented as A→B←C in section B of the algorithm. However, we have already seen that if there are feedback loops then a variable can be both an ancestor and a descendant of another variable. Therefore, this section tries to find some set of variables that d-separates A and C while including B. Remember that, if we see A→B←C, this means that, in section (2), we had found A, B and C to form an unshielded pattern and that B *wasn't* a member of the separation set that d-separated A and C. We will therefore have found two separation sets, one with B and one without B, that d-separate A and C. This is the signal for a feedback loop. Since this section looks for the smallest set that includes B and **Sepset**(A,C) – i.e. **SupSepset**(A,B,C) – this means that every variable in **SupSepset**(A,B,C) is an ancestor of A, B or C. The double underline that is added to B means that both A and C can't both collide at B; some equivalent graphs have A→B and C is not adjacent to B while other equivalent graphs have C→B and A is not adjacent to B.

Sections (5) and (6): Since every member of **SupSepset**(A,B,C) is an ancestor of A, B or C we can now use this information to orient Bo–oD.

The proof of the correctness of each section of this algorithm, given the assumptions, is provided in Richardson (1996b; 1996a). Let's apply this algorithm to the causal structure shown in Figure 8.30, reproduced as Figure 8.33.

Assuming that we have a very large sample size, so that we can ignore errors in determining probabilistic independence due to sampling variations, the undirected dependency graph obtained after section (1) is shown in Figure 8.34.

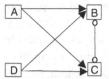

Figure 8.35 The partially oriented graph that is obtained after orienting based on the unshielded colliders

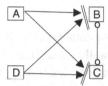

Figure 8.36 The partially oriented graph obtained after section (4) of the cyclic causal discovery algorithm

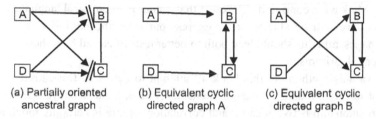

(a) Partially oriented   (b) Equivalent cyclic   (c) Equivalent cyclic
    ancestral graph   directed graph A   directed graph B

Figure 8.37 The partially oriented ancestral graph that results from the cyclic causal discovery algorithm, followed by the two equivalent cyclic graphs

There are only two unshielded patterns (Ao–oBo–oD and Ao–oCo–oD). Since A and D are unconditionally d-separated the **Sepset**(A,D) is empty – that is, **Sepset**(A,D) = {null}. Therefore, we orient these two unshielded patterns as A→B←D and A→C←D. Figure 8.35 shows the partially oriented ancestral graph after this step.

No changes are made after applying section (3) because the necessary patterns don't exist. When we apply section (4) we find that **Sepset**(A,D) = {null} but that A and D are d-separated given {B,C}. Therefore, we reorient A→B←D to be A → B̲ ← D and reorient A→C←D to be A → C̲ ← D. Next we construct **SupSepset**(A,B,D̄) = {B,C} and **SupSepset**(A,C,D) = {B,C̄}. The result is in Figure 8.36.

The double underlining means that A and D can't both be ancestors of either B or C at the same time. Upon arriving at section (5) we see that A → B̲ ← D and A → C̲ ← D and that B and C are adjacent. We can therefore go on and apply this section. Now **SupSepset**(A,B,D) = {B,C} includes C, and so the edge Bo–oC is oriented as B–oC, and **SupSepset**(A,C,D) also includes B, and so the edge B–oC is oriented as B–C, meaning that B and C are each the other's ancestor. The necessary conditions are not met in section (6), and so no further changes are made. The final partially oriented ancestral graph is shown in Figure 8.37, along with the two directed cyclic models that are equivalent to it.

The last step is to devise a general discovery algorithm that is applicable to acyclic or cyclic causal processes with linear or non-linear functional relationships between the variables. In fact Richardson's (1996b) PhD thesis provides just such an algorithm, but, unfortunately, it is based on an unproven (as yet!) conjecture of Spirtes (1995). The study of causal models with feedback is an active area of research, and, with luck, this chapter will soon be out of date.

8.20   In conclusion . . .

Given the dim view of causality that is adopted by most empiricists, it is ironic that the approach to causality taken in this book is almost, well . . . *empirical*. Rather than defining causality, one looks for those properties of relationships that scientists have deemed to call 'causal', and then develops a mathematical language that possesses such properties. In time, perhaps, this will lead to a comprehensive definition that can be accepted by everyone. For myself, I view 'causality' as a relationship between events or classes of events (i.e. variables) that possesses the properties of asymmetry, transitivity and the Markovian condition.[41] I expect that, as our mathematical language of causality improves, we will be able to better express our scientific notions of causality using mathematics, and this should lead both to better tests of causal hypotheses and to better discovery algorithms.

The various methods in this book all attempt to detect or test causal relationships using observational data. I have not intended my book to be an encyclopaedic treatment of the relationship between cause and correlation – there is certainly much more to be said – but I hope that it will be useful as you watch the correlational shadows dance across the screen of nature's shadow play.

Enjoy the show.

---

[41] I know that in Chapter 1 I promised not to give a definition of causality. I couldn't resist the temptation; I'm an academic, and academics are drawn to definitions like young boys are drawn to puddles. We like to jump in and stir up the mud.

Appendix
A cheat-sheet of useful R functions

The purpose of the Appendix is to summarise the usage of the two main R libraries that are discussed in this book: ggm and lavaan.

The ggm (graphical Gaussian model) library

There is no R library that is dedicated to the d-sep test and its generalisations. However, the ggm library of R does have a number of useful functions that can help you. First, the basiSet() function can always be used to obtain the union basis set of d-separation claims that lies at the heart of the d-sep test. This function requires that you specify the DAG, and this is done using the DAG() function. The dSep() function is not required but is a very useful little function for helping you test your understanding of this important concept. The shipley.test() function implements the d-sep test for the special case in which your DAG involves only normally distributed variables and linear relationships.

```
DAG(..., order = FALSE)
```

The R function DAG() is used for imputing a directed acyclic graph. The output of this function is the adjacency matrix of the DAG – i.e. a square Boolean matrix of order equal to the number of nodes of the graph and a 1 in position *(i,j)* if there is an arrow from *i* to *j* and a zero otherwise. The row names of the adjacency matrix are the nodes of the DAG.

Arguments

... = a sequence of model formulae, using the regression (~) operator. For each formula, the right-hand response defines an effect node (a child in the DAG) and the left-hand explanatory variables the parents of that node. If the regressions are not recursive (i.e. if there is a feedback loop in the DAG) then the function returns an error message.

order = logical, defaulting to FALSE. If TRUE then nodes of the DAG are permuted according to the topological order. If FALSE then nodes are in the order they first appear in the model formulae (from left to right). This argument is used for purely aesthetic reasons.

Consider this simple causal chain model: X→Y→Z. To input this model as a DAG you would specify My.DAG<-DAG(Y~X, Z~Y). Note that this uses the same syntax as when specifying linear models in R. For each child node (dependent variable, endogenous variable) you specify dependent variable ~ parent variable1 + parent variable2, etc.

Some examples

```
DAG(y ~ x + z + u, x ~ u, z ~ u)
## A DAG with an isolated node (v):
DAG(v ~ v, y ~ x + z, z ~ w + u)
## There can be repetitions of the same dependent variable
DAG(y ~ x + u + v, y ~ z, u ~ v + z)
## Interactions are ignored
DAG(y ~ x*z + z*v, x ~ z)
## A cyclic graph returns an error!
## Not run: DAG(y ~ x, x ~ z, z ~ y)
## The order can be changed
DAG(y ~ z, y ~ x + u + v, u ~ v + z)
## If you want to order the nodes (topological sort of the
DAG)
DAG(y ~ z, y ~ x + u + v, u ~ v + z, order = TRUE)
```

```
basiSet(amat)
```

This function, basiSet(), generates as output the union basis set for the conditional independencies implied by a directed acyclic graph – i.e. a minimal set of independencies whose null probabilities are themselves independent and that imply all the other ones in the DAG. This function is useful if you have to apply the d-sep test manually rather than via the shipley.test() function.[1] The output is a list of vectors representing the conditional independence statements forming the basis set. Each output vector contains the names of two non-adjacent nodes followed by the names of nodes in the conditioning set (which may be empty).

Arguments

amat = a square matrix with dimnames representing the adjacency matrix of a DAG. This is normally created via the DAG() function.

Example

```
> basiSet(amat=DAG(B~A,C~B,D~B,E~C+D))
[[1]]
[1] "A" "D" "B" # means A is d-sep D given B
[[2]]
```

---

[1] I did not create this library and have nothing to do with the embarrassing naming of this function. I call this test a d-sep test.

```
[1] "A" "C" "B"
[[3]]
[1] "A" "E" "D" "C"
[[4]]
[1] "B" "E" "A" "D" "C" # means B is d-sep E given {A,D,C}
[[5]]
[1] "D" "C" "B"
```

```
dSep(amat, first, second, cond)
```

This function, dSep(), takes, as input, an adjacency graph created using the DAG() function and a query as to whether some set of nodes in the DAG are d-separated from some other set of nodes, given a third set of conditioning nodes. The output is the answer in the form of a logical variable (TRUE/FALSE).

Arguments

amat = a Boolean matrix with dimnames, representing the adjacency matrix of a directed acyclic graph. The function does not check if this is the case. See the function isAcyclic.

first = a vector representing a subset of nodes of the DAG. The vector is normally a character vector of the names of the variables matching the names of the nodes in row names(amat). It can be also a numeric vector of indices.

second = a vector representing another subset of nodes of the DAG. The set second must be disjoint from first. The mode of second must match the mode of first.

cond = a vector representing a conditioning subset of nodes. The set cond must be disjoint from the other two sets and must share the same mode. If there are no conditioning nodes then cond = NULL.

Examples

```
dSep(DAG(y ~ x, x ~ z), first="y", second="z", cond = "x")
dSep(DAG(y ~ x, y ~ z), first="x", second="z", cond = NULL)
## Example using the DAG in Figure 3.1 :
> my.dag←DAG(B~A,C~B,D~B,E~C+D)
> dSep(my.dag,first="A",second="C",cond="B")
[1] TRUE
> dSep(my.dag,first="A",second="C",cond=NULL)
[1] FALSE
> dSep(my.dag,first="A",second="C",cond="D")
[1] FALSE
> dSep(my.dag,first="B",second="E",cond=c("C","D"))
[1] TRUE
```

```
shipley.test(amat, S, n)
```

This function, shipley.test(), computes a simultaneous test of all independence relationships implied by a given Gaussian model (i.e. assuming normally distributed variables and linear relationships) defined according to a directed acyclic graph, based on the sample covariance matrix. The output consists of the Fisher C-statistic, the degrees of freedom and the null probability. Note that this function does not generate as output the (partial) correlation coefficients and associated null probabilities for each element of the basis set. However, you can obtain the basis set via the basiSet() function and you can obtain the partial correlations associated with the elements of the basis set via the pcor() function, described below.

Arguments

amat = a square Boolean matrix, of the same dimension as S, representing the adjacency matrix of a DAG. This is normally created via the DAG() function.

S = the sample covariance matrix, which must be a symmetric positive definite matrix. This is normally created via the cov() function of R.

n = the sample size, which is a positive integer.

Example

This example uses data containing 100 lines that are found in a data frame called my.dat (note that the variable names of this data frame must match those in the DAG):

```
> my.dag←DAG(B~A,C~B,D~B,E~C+D)
> shipley.test(amat = my.dag,S = cov(my.dat),n = 100)
$ctest
[1] 14.37551
$df
[1] 10
$pvalue
[1] 0.1565421
```

pcor(u,**S**)

This function, pcor(), computes the Pearson partial correlation between the first two variables listed in the vector u, conditional on the remaining variables listed in u. This function does not compute the null probability of this partial correlation, but this can be done separately, as shown in the example below. The function returns a scalar numerical value, the Pearson partial correlation between u[1] and u[2], conditional on u[-c(1,2)].

Arguments

u = a vector containing either variable names (which must be the same names as used in the covariance matrix **S**) or the column numbers of these variables in the covariance matrix.

S = a symmetric positive definite matrix, a sample covariance matrix.

Examples

Assume that we have a data frame called my.dat and containing 100 observations on three variables called (in order) x, y and z.

```
# the partial correlation between x and z, conditional on y
pcor(u=c("x","z","y"),cov(my.dat))
# the same thing
pcor(c(1,3,2),cov(my.dat))
# the absolute correlation between x and z
pcor(u=c("x","z",),cov(my.dat))
# here is how to calculate the null probability. First, save
# the partial correlation
p.r<-pcor(u=c("x","z"),cov(temp))
# next, convert the partial correlation to a Student's t
# value
t.r←p.r*sqrt(100-2-1)/sqrt(1-p.r^2)
# finally, get the 2-tailed null probability
>  2*(1-pt(t.r,100-2-1))
[1] 0.5302312
```

The lavaan library

As I write, the lavaan library is still under development, and so the information here is based on version 3.1.2. The following information is likely to change in future versions (hopefully not too much), and so you should always check the help files. The key reference is the article by Rosseel (2012). There are three main functions in this package: lavaan(), cfa() and sem(). The cfa function is optimised for confirmatory factor analysis, the sem function is optimised for structural equation modelling and the lavaan function is the most general function, upon which the other two are special cases.

The general structure of an R session using the lavaan library, as presented in this book, has three parts.

(1) The creation of a model object that specifies the model structure. The model structure is enclosed in quotes.
(2) A call to the sem() function. The sem() function inputs the model object, the data object and other arguments. Using this information, it fits the model to the data and calculates the various output statistics.
(3) A call to one or several extractor functions in order to obtain various types of information about the model fit.

A simple example of this sequence is the following:

```
#input this simple model: X→Y→Z and save it as an object
# called "my.model"
my.model←"Y~X
```

```
z~Y"
# fit "my.model" to a data set called "input.data" using the
# sem() function
model.fit ←sem(model = my.model,data = input.data)
# obtain a summary of the model fit
summary(model.fit)
```

The next section summarises the various details in lavaan related to specifying the model (i.e. the model syntax), choosing values for the arguments of the sem() function that control the fitting of the model to the data, and the various extractor functions that allow you to see various details of the resulting fit.

Specifying the model structure: model syntax

Formula type	Operator	Example	Causal graph	Meaning
Latent variable definition	=~	L = ~ x+y	L→y L→y	'latent cause, is measured by'. Latent variable L *causes* and *is measured by* observed variables x and y.
Regression	~	y ~ x	x→y	'is caused by, is regressed on'. Observed variable y *is caused by*, and *is regressed on*, x.
(Residual) (co)variance	~~	x ~~ y, x~~x	x←→y, x←→x	'free covariance, free variance'.
Intercept		x ~ 1		'estimate the intercept of' x.
New parameter	:=	Total: = a+b		'create a new free parameter'. *Create a new free parameter* called 'Total', which is constrained to be the sum of old free parameters a and b.

Fixing a parameter value

Whenever you specify a structural equation via the model syntax of lavaan you implicitly define free parameters. One exception is when you use the = ~ operator, since the path coefficient of the first observed variable on the right-hand side of the operator is fixed to unity by default (see the subsection 'Allowing the first indicator of a latent variable to be free' to change this default choice). Thus, a model statement such as y~x+z implicitly defines two free parameters that are the path coefficients associated with x and z. In fact, depending on what other command lines you include in the model object, this statement could also implicitly define free error variances and covariances.

In order to force a parameter to take a specific value ('fixing it') rather than allowing it to be estimated from the data, you 'multiply' the variable by the desired fixed value. Thus, y~x+6*z means that the path coefficient associated with the variable z is fixed to a value of 6 and can't be changed during the process of parameter estimation.

Examples

y~1.5*x → the path coefficient associated with x is fixed at a value of 1.5.

y~~1*y → the (residual) variance of y is fixed at a value of 1.

Y~~0*x → the covariance between (the residuals of) y and x is fixed at zero.

Specifying starting values

The iterative process of estimating the values of free parameters requires specifying the initial (starting) values of these free parameters. By default, lavaan sets all starting values to unity. However, more complicated models can fail to converge, and one reason for this is that the starting values were simply too far away from the final values. In such cases, one must supply better initial values. This is done by via the start() argument, which is 'multiplied' to the variable.

Examples

y~start(0.1)*x → the path coefficient associated with the variable x is free but its starting value during the iterative fitting process is equal to 0.1.

y~~start(10)*y → the starting value of the free (residual) variance of y is equal to 10.

L = ~start(0.001)*x+y → the starting value of the free path coefficient associated with x is 0.001 (see also the subsection 'Allowing the first indicator of a latent to be free').

Specifying starting values in multigroup models

In order to fit a model involving more than one group, you need to have a grouping variable in the data frame. If you want the starting value of a parameter to be the same in all groups then simply give a single start value. To specify different start values for each group, you specify these as a vector: start(c(0.1,0.2,...))*x

Preventing exogenous variables from freely covarying

By default, all exogenous variables in lavaan are assumed to have non-zero covariances. The default occurs because the default value of the argument fixed.x in the sem() function is fixed.x = TRUE. This is a poor default choice, and so you should always explicitly specify the state of these exogenous covariances based on your conception of the causal process. To do this, you must specify sem(...,fixed.x=FALSE) in the sem() function and then explicitly specify these covariances in the model syntax as either fixed (to zero or some other value) or free (see the subsection 'Fixing a parameter value').

Example

```
My.model←"z~x+y
# the next line fixes the covariance between the two
# exogenous variables to zero
x~~0*y "
sem(model.syntax = My.model, data = , fixed.x = FALSE)
```

Specifying parameter labels

Every parameter in your model syntax has a name. For example, these are the parameter names that you see when you use the summary() function. The default name for a parameter is simply a concatenation of variable name 1 + operator + variable name 2. In other words if, in your model syntax, you have a line such as y~x+z then the first path coefficient is named y~x and the second path coefficient is named y~z.

However, you can specify your own names for these parameters. To do this, simply 'multiply' the variable name, using the usual naming conventions of R, by the label that you want to use for its associated parameter.

Examples

y ~ x + z → the (free) path coefficient associated with variable x is called y~x and the (free) path coefficient associated with variable z is called y~z.

y ~ a*x + b*z → the (free) path coefficient associated with variable x is called a and the (free) path coefficient associated with variable z is called b.

You can combine these 'multiplication' conventions. For instance, y ~ start(2)*a*x both specifies a starting value and a label name for the free path coefficient associated with x.

Specifying parameter labels in multiple groups

In order to fit a multigroup model, you need to have a grouping variable in the data frame. If you want different labels for each group then specify these different labels as a vector whose length is equal to the number of groups in the model: c(ag1,ag2,...)*x. However, it is necessary to be careful here; if you give the same label to more than one group then this will force the parameter estimation to be equal across these groups sharing the same label name.

Specifying simple equality constraints

You can force combinations of free parameters to be equal during model fitting. In such cases, the fitted values of the parameters are still chosen so as to minimise the difference between the observed and model covariance matrices, but the chosen values are constrained to be equal to each other. This is most often done when fitting multigroup models, but it can be done whenever your causal hypothesis requires it. There are different but equivalent ways of doing this.

(1) Simply give the same parameter label to the parameters whose values are to be equal. For example, y ~ a*x + a*z will force the values of the estimated path coefficients associated with both variables x and z to be equal, since the labels of both parameters are equal.

(2) Use the equal() function. For example, y ~ x + equal("y~x")*z forces the fitted value of the path coefficient associated with variable z to equal the value associated with variable x.

(3) Use the = = operator:

```
y ~ a*x + b*z
a = = b
```

Specifying non-linear equality or inequality constraints

(1) Give explicit labels to the parameters in question.
(2) Specify the desired constraint using the = =, < or > operators and the parameter labels. For example:

```
# give names (a1, a2, a3) to the three free path coefficients
y ~ a1*x + a2*z +a3*e
# here is a nonlinear equality constraint
a1 = = (a2 + a3)^2
# here is a nonlinear inequality constraint
a1 > exp(a2 + a3)
```

Preventing a free variance from being negative

By definition, a variance cannot be negative, but the various algorithms used by lavaan to estimate parameter values don't know this fact. As a result, it sometimes happens that the estimated value of free residual variance is negative. This is usually a sign of a poorly fitting model or some problem that has occurred during the iterative process of estimation. However, it is possible that the model fits the data well but that the true value of the residual variance is very close to zero. If this happens then the estimate can become negative because of sampling fluctuations. If you think that this is the case then you force lavaan to maintain non-negative variance estimates by specifying a non-linear constraint on this residual variance, as follows.

```
# name the parameter label for the variance
x ~~ varx*x
# force this parameter value to remain non-negative
varx > = 0
```

Specifying more than one causal model in multigroup or multilevel models

Inside the model object, you must name each group and then enter the model specifics for each group. For instance, to have x→y→z in group 1 and x→y←z in group 2, with no cross-group equality constraints (thus, different labels for the free parameters), you would specify

```
"
Group 1:
y~a1*x
z~b1*y
```

```
x~~vx1*x
y~~vy1*y
z~~vz1*z
Group 2:
y~a2*x +b2*z
x~~vx2*x
y~~vy2*y
z~~vz2*z
"
```

Calculating compound (indirect, total) effects

To calculate compound effects, such as indirect or total effects, and their standard errors, you must first give labels to the coefficients in question and then use the : = operator to calculate the desired compound effects. For instance, consider the simple path model x–(a)→y, y–(b)→z, x–(c)→z, where a, b and c are the label names for the three direct effects. There is both a direct effect of x on z (x→z) and an indirect effect (x→y→z). Only the direct effect is calculated by default in lavaan (i.e. the value of the path coefficient associated with variable z). To calculate the direct, indirect and total effects of x on z you would do the following.

```
My.model←"
# give label names for the path coefficients
y ~ a*x
z ~c*x + b*y
# define the parameter measuring the indirect effect (a*b)
indirect.effect : = a*b
# define the parameter measuring the total effect
total.effect : = c + (a*b)"
```

When you fit this model then the values and standard errors of the two new defined parameters (indirect.effect and total.effect) will be generated as output.

Allowing the first indicator of a latent variable to be free

When specifying a latent variable via the = ~ operator, the default in lavaan is to fix the path coefficient of the first indicator variable on the right-hand side of the operator to unity in order to fix the scale of the latent. If you want to fix the scale of the latent in this way, you don't have to do anything except to make sure that the first observed variable on the right-hand side is the variable whose scale you want to use. However, if you don't want to do this (for instance, if you want to identify the latent by fixing its variance to unity) then the usual (and easiest) way is to explicitly fix the latent variance to unity via the std.lv argument in the sem() function: sem(..., std.lv=TRUE). This fixes the standard deviation of all the latent variables in the model to unity.

However, there are times when this method is not appropriate. For instance, you might have more than one latent in your model whose scales you want to fix in different ways. Alternatively, you might want to fix the scale of the latent using some value of an observed scale other than unity. In such instances, you can explicitly force the path coefficient of the first observed variable on the right-hand side of the $= \sim$ operator to be free by 'multiplying' it by NA (to free it) or by a number (to fix its scale to a value other than unity). Thus, $L = \sim NA^*x +...$ tells lavaan that the path coefficient associated with x (the first observed variable on the right-hand side) is *not* fixed. Similarly, $L = \sim 2.5x+...$ tells lavaan that the path coefficient associated with x is fixed at 2.5.

Arguments used when fitting the model via sem()

The sem() function is actually a wrapper for another, more general function, called lavaan(). There are two other wrapper functions, called cfa() and growth(), that I won't discuss here. The sem() function contains very many arguments, and several of them deal with advanced topics that are not discussed in this book. Most of these arguments have default values, and there are complicated interactions between these default values affecting things such as the method of parameter estimation, the types of test statistics that are calculated, and so on. Which of these arguments can be safely kept at their default values and which need to be specified will depend on the complexity of your model. I have indicated those arguments that refer to topics that are discussed in this book with an asterisk.

Here is the full function. You will see most of this if you use type help(sem) in R, but I have added some further details.

```
sem(model = NULL, data = NULL,
    meanstructure = "default", fixed.x = "default",
    orthogonal = FALSE, std.lv = FALSE,
    parameterization = "default", std.ov = FALSE,
    missing = "default", ordered = NULL,
    sample.cov = NULL, sample.cov.rescale = "default",
    sample.mean = NULL, sample.nobs = NULL,
    ridge = 1e-05, group = NULL,
    group.label = NULL, group.equal = "", group.partial = "",
    group.w.free = FALSE, cluster = NULL, constraints = ",
    estimator = "default", likelihood = "default", link =
    "default",
    information = "default", se = "default", test =
    "default",
    bootstrap = 1000L, mimic = "default", representation =
    "default",
    do.fit = TRUE, control = list(), WLS.V = NULL, NACOV =
    NULL,
    zero.add = "default", zero.keep.margins = "default",
```

```
zero.cell.warn = TRUE,
start = "default", verbose = FALSE, warn = TRUE, debug =
FALSE)
```

model* = the name of the object holding the model description. This must always be specified.

data* = the name of the data frame holding the observations, including (if applicable) the grouping structure. You must always either provide this data frame or else provide (a) the covariance matrix via the sample.cov argument, (b) the vector of sample means for each variable via the sample.mean argument and (c) the number of observations used to calculate the sample covariance matrix via the sample.obs argument. These last three arguments are described below.

meanstructure* = FALSE by default; if TRUE then the means (intercepts) are also modelled.

fixed.x* = if TRUE, the exogenous variables are not considered random variables, and so the means, variances and covariances of these variables are not estimated but, rather, are fixed to their sample values. This is different from the way these variables are treated in this book. If FALSE (which is the choice for the way they are treated in this book), they are considered random, and the means, variances and covariances are free parameters. If 'default', the value is set depending on the mimic option (see below).

orthogonal* = if TRUE, the exogenous latent variables are assumed to be uncorre-lated – i.e. the covariances between them are fixed at zero.

parameterisation = an argument used to treat categorical data, and not discussed in this book.

std.lv* = if FALSE, the variances of the latent variables are estimated; if TRUE then the variances of the latent variables are fixed at unity.

std.ov* = (the default is FALSE) only if you want all observed variables to be stan-dardised (unit variance, zero mean) before the analysis. This would give standardised coefficients.

missing = if 'listwise', cases with missing values are removed listwise from the data frame before analysis. If 'direct' or 'ml' or 'fiml' and the estimator (see below) is maximum likelihood, full-information maximum-likelihood (FIML) estimation is used using all available data in the data frame. This is valid only if the data are *missing completely at random* (MCAR) or *missing at random* (MAR). If 'default', the value is set depending on the estimator and the mimic option (see below). This is justified only if the missing values are given the precise definitions of the terms *missing at random* or *completely at random* as explained by Rubin (1996) and Schafer (1997). A value is missing *completely at random* if its probability of being missing is unrelated to any other variable – observed or unobserved. For example, if you missed certain values because your measuring device broke down one day then the pattern of missed values is probably missing completely at random. Missing values of a variable are said to be *missing at random* if the values of the other variables in the data set can predict the pattern of missingness. If you are missing data on seed output because some plants had already shed their seeds before you started measurements, but you also have information on, say, the date of flowering, then this would probably be a case of *missing at random*. However, if you have no other information in your data set that is related to the phenology of reproduction then your missing values would not accord with this necessary assumption.

ordered = character vector and used only if the data are in a data frame. Treat these variables as ordered (ordinal) variables if they are endogenous in the model. Importantly, all other variables will be treated as numeric (unless they are declared as ordered in the original data frame).

sample.cov = a sample covariance if you want to input this instead of the actual data set.

sample.cov.rescale = if TRUE, the sample covariance matrix provided by the user is internally rescaled by multiplying it with a factor (N-1)/N. If 'default', the value is set depending on the estimator and the likelihood option; it is set to TRUE if maximum-likelihood estimation is used and likelihood = 'normal', and FALSE otherwise.

sample.mean = a sample mean vector if you want to model means and you have input the sample covariance instead of the actual data set.

sample.nobs = the number of observations used to calculate the sample covariance matrix if you have input this instead of the actual data set.

ridge = a small numeric constant used for ridging. This is used only if the sample covariance matrix is non-positive definite.

group* = the name of the variable in your data frame that codes the group names (only if you are doing a multigroup model). If you do not specify this argument then lavaan assumes that you have only one group.

group.label* = a character vector. You can use this to specify which group (or factor) levels need to be selected from the grouping variable, and in which order. If NULL (the default), all grouping levels are selected, in the order as they appear in the data.

group.equal* = a vector of character strings. This is used only in a multigroup analysis and is used to specify the pattern of equality constraints across multiple groups. Choices can be one or more of the following:

= 'loadings' means that the path coefficients from latents to indicators are equal across groups, as specified by the = ~ operator in the model syntax;

= 'regressions' means that all regression coefficients are equal across groups, as specified by the ~ operator in the model syntax;

= 'residuals' means that the residual variances of the observed variables are equal across groups;

= 'residual.covariances' means that the covariances of the observed variables are equal;

= 'lv.variances' means that the (residual) variances of the latents are equal;

= 'lv.covariances' means that the (residual) covariances of the latent variables are equal;

= 'means' means that the intercepts/means of the latent variables are equal;

= 'intercepts' means that the intercepts of the observed variables are equal; and

= 'threshholds' refers to categorical variables, and these are not discussed in this book.

group.partial* = a vector of character strings containing the labels of the parameters that should be free in all groups; this is used to override the group.equal argument for some specific parameters.

group.w.free = if TRUE, the group frequencies are considered to be free parameters in the model. In this case, a Poisson model is fitted to estimate the group frequencies. If FALSE (the default), the group frequencies are fixed to their observed values. This is not discussed in this book.

cluster = an argument that has not yet been implemented.

constraints* = additional (in)equality constraints not yet included in the model syntax. This is an alternative to including such constraints in the model syntax.

estimator* = the estimator to be used; the default is ML, for maximum likelihood. There are many choices here, only some of which are discussed in this book. The options are ML for maximum likelihood, GLS for generalised least squares, WLS for weighted least squares (sometimes called ADF estimation), ULS for unweighted least squares and DWLS for diagonally weighted least squares. These are the main options that affect the estimation. For convenience, the ML option can be extended as MLM, MLMV, MLMVS, MLF and MLR. The estimation will still be plain ML, but now with robust standard errors and a robust (scaled) test statistic. For MLM, MLMV and MLMVS, classic robust standard errors are used (se = robust.sem); for MLF, standard errors are based on first-order derivatives (se = first.order); for MLR, 'Huber–White' robust standard errors are used (se = robust.huber.white). In addition, MLM will compute a Satorra–Bentler scaled (mean-adjusted) test statistic (test = satorra.bentler), MLMVS will compute a mean- and variance-adjusted test statistic (Satterthwaite-style) (test = mean.var.adjusted), MLMV will compute a mean- and variance-adjusted test statistic (scaled and shifted) (test = scaled.shifted) and MLR will compute a test statistic that is asymptotically equivalent to the Yuan–Bentler T2-star test statistic. Analogously, the estimators WLSM and WLSMV imply the DWLS estimator (not the WLS estimator) with robust standard errors and a mean- or mean- and variance-adjusted test statistic. Estimators ULSM and ULSMV imply the ULS estimator with robust standard errors and a mean- or mean- and variance-adjusted test statistic.

likelihood = relevant only for ML estimation. If 'wishart', the wishart likelihood approach is used. In this approach, the covariance matrix has been divided by N–1, and both standard errors and test statistics are based on N–1; this is the approach described in this book. If 'normal', the normal likelihood approach is used. Here, the covariance matrix has been divided by N, and both standard errors and test statistics are based on N. If 'default', it depends on the mimic option: if mimic = 'lavaan' or mimic = 'Mplus, normal likelihood is used; otherwise, wishart likelihood is used.

link = used currently only if the chosen estimator is MML and you have binary or ordered observed endogenous variables; this topic has not been discussed in this book. If 'logit', a logit link is used for binary and ordered observed variables. If 'probit', a probit link is used. If 'default', it is currently set to 'probit'.

information = If 'expected', the expected information matrix is used (to compute the standard errors). If 'observed', the observed information matrix is used. If 'default', the value is set depending on the estimator and the mimic option.

se = specifies how the standard errors of the parameter estimates are to be calculated. If 'standard' (the default), conventional standard errors are computed based on inverting the (expected or observed) information matrix. If 'first.order', standard errors are computed based on first-order derivatives. If 'robust.sem', conventional robust standard errors are computed. If 'robust.huber.white', standard errors are computed based on the 'mlr' (aka pseudo ML, Huber–White) approach. If 'robust', either 'robust.sem' or 'robust.huber.white' is used depending on the estimator, the mimic option and whether the data are complete or not. If 'boot' or 'bootstrap', bootstrap standard errors are computed using standard bootstrapping (unless Bollen–Stine bootstrapping is requested for the test statistic; in this case, bootstrap standard errors are computed using model-based bootstrapping). If 'none', no standard errors are computed.

test* = if 'standard', a conventional chi-square test is computed. If 'Satorra.Bentler', a Satorra–Bentler scaled test statistic is computed. If 'Yuan.Bentler', a Yuan–Bentler scaled test statistic is computed. If 'mean.var.adjusted' or 'Satterthwaite', a mean- and variance-adjusted test statistic is computed. If 'scaled.shifted', an alternative mean- and variance-adjusted test statistic is computed (as in Mplus version 6 or higher). If 'boot' or 'bootstrap' or 'Bollen.Stine', the Bollen–Stine bootstrap is used to compute the bootstrap probability value of the test statistic. If 'default', the value depends on the values of other arguments.

bootstrap* = number of bootstrap draws, if bootstrapping is used.

mimic = if 'Mplus', an attempt is made to mimic the Mplus program. If EQS, an attempt is made to mimic the EQS program. If 'default', the value is (currently) set to 'lavaan', which is very close to 'Mplus'.

representation = if 'LISREL', the classical LISREL matrix representation is used to represent the model (using the all-y variant).

do.fit = if FALSE, the model is not fit, and the current starting values of the model parameters are preserved. Defaults to TRUE.

control = a list containing control parameters passed to the optimiser. By default, lavaan uses 'nlminb'. See the R help file of nlminb for an overview of the control parameters. A different optimiser can be chosen by setting the value of optim.method. For unconstrained optimisation (the model syntax does not include any ==, > or < operators), the available options are 'nlminb' (the default), BFGS and L-BFGS-B. See the documentation of the optim function for the control parameters of the latter two options. For constrained optimisation, the only available option is 'nlminb.constr'.

WLS.V = a user-provided weight matrix to be used by estimator WLS; if the estimator is DWLS, only the diagonal of this matrix will be used. For a multiple group analysis, a list with a weight matrix for each group. The elements of the weight matrix should be in the following order (if all data are continuous): first the means (if a meanstructure is involved), then the lower triangular elements of the covariance matrix including the diagonal, ordered column by column. In the categorical case: first the thresholds (including the means for continuous variables), then the slopes (if any), the variances of continuous variables (if any) and finally the lower triangular elements of the correlation/covariance matrix excluding the diagonal, ordered column by column.

NACOV = a user-provided matrix containing the elements of (N times) the asymptotic variance–covariance matrix of the sample statistics. For a multiple group analysis, a list with an asymptotic variance–covariance matrix for each group. See the WLS.V argument for information about the order of the elements.

zero.add = a numeric vector containing two values. These values affect the calculation of polychoric correlations when some frequencies in the bivariate table are zero. The first value applies only for 2×2 tables, the second value for larger tables. This value is added to the zero frequency in the bivariate table. If 'default', the value is set depending on the mimic option. By default, lavaan uses zero.add = c(0.5. 0.0).

zero.keep.margins = this logical argument affects only the computation of polychoric correlations for 2×2 tables with an empty cell, and where a value is added to the empty cell. If TRUE, the other values of the frequency table are adjusted so that all margins are unaffected. If 'default', the value is set depending on the 'mimic'. The default is TRUE.

zero.cell.warn = used only if some observed endogenous variables are categorical. If TRUE, give a warning if one or more cells of a bivariate frequency table are empty.

start* = if it is a character string, the two options are currently 'simple' and Mplus. In the first case, all parameter values are set to zero, except for the factor loadings (set to one), the variances of latent variables (set to 0.05) and the residual variances of observed variables (set to half the observed variance). If Mplus, we use a similar scheme, but the factor loadings are estimated using the fabin3 estimator (tsls) per factor. If start is a fitted object of class lavaan, the estimated values of the corresponding parameters will be extracted. If it is a model list, for example the output of the parameterEstimates() function, the values of the est or start or ustart column (whichever is found first) will be extracted.

verbose = if TRUE, the function value is printed out during each iteration.

warn = if TRUE, some (possibly harmless) warnings are printed out during the iterations.

debug = if TRUE, debugging information is printed out.

Extractor functions

Once you have fitted the model to the data using the sem() function and saved it as an object (we'll call it 'fit'), you can extract different types of information about the resulting fit using different extractor functions.

summary(fit, standardized = FALSE, fit.measures = FALSE, rsquare = FALSE, modindices = FALSE). This is the basic extractor function that outputs (as defaults) information on convergence, the basic test statistics (which depend on which estimator you specified in the sem() function) and the parameter estimates and their standard errors. The additional arguments allow you to also obtain standardised parameter estimates, additional fit statistics, the proportion of the total variance (R^2) of the endogenous variables of the model that are explained, and modification indices (Lagrange multipliers).

coef (fit). This generates as output the fitted coefficients only.

parameterEstimates (fit). This generates as output the parameter estimates and their SE and confidence intervals.

standardizedSolution (fit). This generates as output the standardised estimates of the parameters.

residuals (fit, type = "standardized"). This generates as output the standardised differences between the observed and predicted covariance matrices.

AIC (fit), BIC (fit). These generate as output the AIC or BIC values of the model.

fitMeasures (fit). This generates as output all the various fit measures.

inspect (fit, "r2"). This generates as output the R^2 (proportion of variance explained) associated with each endogenous variable.

References

Aldrich, J. (1995). Correlations genuine and spurious in Pearson and Yule. *Statistical Science* 10: 364–76.

Bentler, P. M. (1995). *EQS Structural Equations Program Manual, Version 3.0*. Los Angeles, BMDP Statistical Software.

Bentler, P. M., and Bonnett, D. G. (1980). Significance tests and goodness of fit in the analysis of covariance structures. *Psychological Bulletin* 88: 588–606.

Bernard, C. (1865). *Introduction à l'étude de la médicine expérimentale*. Paris, J. B. Baillière.

Beveridge, W. I. B. (1957). *The Art of Scientific Investigation*. New York, Random House.

Blalock, H. M. (1961). Correlation and causality: the multivariate case. *Social Forces* 39: 246–51. (1964). *Causal Inferences in Nonexperimental Research*. Chapel Hill, University of North Carolina Press.

Blomberg, S. P., Lefevre, J. G., Wells, J. A., and Waterhouse, M. (2012). Independent contrasts and PGLS regression estimators are equivalent. *Systematic Biology* 61: 382–91.

Bollen, K. A. (1989). *Structural Equations with Latent Variables*. New York, Wiley.

Bollen, K. A., and Long, J. S. (1993). *Testing Structural Equation Models*. Newbury Park, CA, Sage.

Bollen, K. A., and Stine, R. A. (1993). Bootstrapping goodness-of-fit measures in structural equation models, in Bollen, K. A., and Long, J. S. (eds.), *Testing Structural Equation Models*: 111–34. Newbury Park, CA, Sage.

Browne, M. W. (1984). Asymptotically distribution-free methods for the analysis of covariance structures. *British Journal of Mathematical and Statistical Psychology* 37: 62–83.

Browne, M. W., and Cudeck, R. (1993). Alternative ways of assessing model fit, in Bollen, K. A., and Long, J. S. (eds.), *Testing Structural Equation Models*: 136–62. Newbury Park, CA, Sage.

Bumpus, H. C. (1899). The elimination of the unfit as illustrated by the introduced sparrow. *Biological Lectures Delivered at the Marine Biological Laboratory of Woods Hole* 6: 209–26.

Burke, J. (1996). *The Pinball Effect: How Renaissance Water Gardens Made the Carburetor Possible – and Other Journeys through Knowledge*. Boston, Little, Brown.

Cleveland, W. S., and Devlin, S. J. (1988). Locally-weighted regression: an approach to regression analysis by local fitting. *Journal of the American Statistical Association* 83: 596–610.

Cleveland, W. S., Devlin, S. J., and Grosse, E. (1988). Regression by local fitting. *Journal of Econometrics* 37: 87–114.

Cleveland, W. S., Grosse, E., and Shyu, W. M. (1992). Local regression models, in Chambers, J. M., and Hastie, T. J. (eds.), *Statistical Models in S*: 309–76. Pacific Grove, CA, Wadsworth & Brooks.

Conover, W. J., and Iman, R. L. (1981). Rank transformations as a bridge between parametric and nonparametric statistics. *American Statistician* 35: 124–9.

Cowan, I. R., and Farquhar, G. D. (1977). Stomatal function in relation to leaf metabolism environment, in Jennings, D. H. (ed.), *Integration of Activity in the Higher Plant*: 471–505. Cambridge University Press.

Cowles, M., and Davis, C. (1982a). Is the .05 level subjectively reasonable? *Canadian Journal of Behavioural Sciences* 14: 248–52.

  (1982b). On the origins of the .05 level of statistical significance. *American Psychologist* 37: 553–8.

D'Agostino, R. B., Belanger, A., and D'Agostino, R. B. J. (1990). A suggestion for using powerful and informative tests of normality. *American Statistician* 44: 316–21.

Davenport, C. B. (1917). Inheritance of stature. *Genetics* 2: 313–89.

Davis, W. R. (1993). The FC1 rule of identification for confirmatory factor analysis. *Sociological Methods and Research* 21: 403–37.

De Robertis, E. D. P., and De Robertis, E. M. F. (1980). *Cell and Molecular Biology*. Boston, Thomson Learning.

DeCarlo, L. T. (1997). On the meaning and use of kurtosis. *Psychological Methods* 2: 292–307.

Duhem, P. (1914). *La théorie physique: Son objet, sa structure*. Paris, Rivière.

Dunn, G., Everitt, B., and Pickles, A. (1993). *Modelling Covariances and Latent Variables Using EQS*. London, Chapman & Hall.

Eliason, S. R. (1993). *Maximum Likelihood Estimation: Logic and Practice*. Newbury Park, CA, Sage.

Epstein, R. J. (1987). *A History of Econometrics*. New York, Elsevier Science.

Farebrother, R. (1987). Algorithm AS 231: the distribution of a noncentral chi-square variable with nonnegative degrees of freedom. *Applied Statistics* 36: 402–5.

Feiblman, J. K. (1972). *Scientific Method*. The Hague, Martinus Nijhoff.

Felsenstein, J. (1985). Phylogenies and the comparative method. *American Naturalist* 125: 1–15.

Fisher, F. M. (1970). A correspondence principle for simultaneous equation models. *Econometrica* 38: 73–92.

Fisher, R. A. (1925). *Statistical Methods for Research Workers*. Edinburgh, Oliver & Boyd.

  (1926). *The Design of Experiments*. Edinburgh, Oliver & Boyd.

  (1950). *Contributions to Mathematical Statistics*. New York, Wiley.

  (1959). *Smoking: The Cancer Controversy*. Edinburgh, Oliver & Boyd.

  (1970). *The Design of Experiments*, 8th edn. New York, Hafner.

Forrest, D. W. (1974). *Francis Galton: The Life and Work of a Victorian Genius*. New York, Taplinger.

Galton, F. (1869). *Hereditary Genius: An Inquiry into Its Laws and Consequences*. London, Macmillan.

Geiger, D., Verma, T., and Pearl, J. (1990). Identifying independence in Bayesian networks. *Networks* 20: 507–34.

Glymour, G., Scheines, R., Spirtes, R., and Kelly, K. (1987). *Discovering Causal Structure: Artificial Intelligence, Philosophy of Science, and Statistical Modeling*. Orlando, Academic Press.

Goldberger, A. S. (1972). Structural equation methods in the social sciences. *Econometrica* 40: 979–1002.

Good, P. (1993). *Permutation Tests: A Practical Guide to Resampling Methods for Testing Hypotheses*. New York, Springer.

  (1994). *Permutation Tests: A Practical Guide to Resampling Methods for Testing Hypotheses*, 2nd edn. New York, Springer.

Grace, J. B. (2006). *Structural Equation Modeling and Natural Systems*. Cambridge University Press.

Grace, J. B., and Bollen, K. A. (2008). Representing general theoretical concepts in structural equation models: the role of composite variables. *Environmental and Ecological Statistics* 15:191–213.

Griliches, Z. (1974). Errors in variables and other unobservables. *Econometrica* 42: 971–98.

Grime, J. P. (1979). *Plant Strategies and Vegetation Processes*. New York, Wiley.

Haavelmo, T. (1943). The statistical implications of a system of simultaneous equations. *Econometrica* 11: 1–12.

Harvey, P. H., and Pagel, M. D. (1991). *The Comparative Method in Evolutionary Biology*. Oxford University Press.

Hastie, T. J., and Tibshirani, R. (1990). *Generalized Additive Models*. London, Chapman & Hall.

Heise, D. (1975). *Causal Analysis*. New York, Wiley.

Hoogland, J. J., and Boomstra, A. (1998). Robustness studies in covariance structure modelling: an overview and a meta-analysis. *Sociological Methods and Research* 26: 239–367.

Hotelling, H. (1953). New light on the correlation coefficient and its transformations. *Journal of the Royal Statistical Society, Series B* 15: 193–232.

Howson, C., and Urbach, P. (1989). *Scientific Reasoning: The Bayesian Approach*. La Salle, IL, Open Court.

Hox, J. J. (1993). Factor analysis of multilevel data: gauging the Muthén model, in Oud, J. H. L., and van Blokland-Vogelsang, R. A. W. (eds.), *Advances in Longitudinal and Multivariate Analysis in the Behavioural Sciences*: 141–56. Nijmegen, ITS.

Jobson, J. D. (1992). *Applied Multivariate Data Analysis*, vol. I, *Regression and Experimental Design*. New York, Springer.

Jordano, P. (1995). Frugivore-mediated selection on fruit and seed size: birds and St. Lucie's cherry, *Prunus mahaleb*. *Ecology* 76: 2627–39.

Jöreskog, K. G. (1967). Some contributions to maximum likelihood factor analysis. *Psychometrika* 32: 443–82.

(1969). A general approach to confirmatory maximum likelihood factor analysis. *Psychometrika* 34: 183–202.

(1970). A general method for analysis of covariance structures. *Biometrika* 57: 239–51.

(1973). A general method for estimating a linear structural equation system, in Goldberger, A. S., and Duncan, O. D. (eds.), *Structural Equation Models in the Social Sciences*: 85–112. New York, Academic Press.

Keesling, J. W. (1972). Maximum likelihood approaches to causal analysis, PhD thesis. University of Chicago.

Kempthorpe, O. (1979). *The Design and Analysis of Experiments*. Huntington, NY, Robert E. Krieger.

Kendall, M. G., and Gibbons, J. D. (1990). *Rank Correlation Methods*. New York, Oxford University Press.

Kendall, M. G., and Stuart, A. (1983). *The Advanced Theory of Statistics*. London, Charles Griffin.

Kikuzawa, K. (1995). The basis for variation in leaf longevity of plants. *Vegetatio* 121: 89–100.

Korn, E. L. (1984). The ranges of limiting values of some partial correlations under conditional independence. *American Statistician* 38: 61–2.

Lande, R., and Arnold, S. J. (1983). The measurement of selection on correlated characters. *Evolution* 37: 1210–26.

Li, C. C. (1975). *Path Analysis: A Primer*. Pacific Grove, CA, Boxwood Press.

Little, R. J. A., and Rubin, D. B. (2002). *Statistical Analysis with Missing Data*, 2nd edn. Hoboken, NJ, Wiley.

Mach, E. (1883). *The Science of Mechanics: A Critical and Historical Account of Its Development*, 5th edn, with revisions from 9th German edn. La Salle, IL, Open Court.

Manly, B. F. J. (1997). *Randomization, Bootstrap and Monte Carlo Methods in Biology*, 2nd edn. London, Chapman & Hall.

Mardia, K. V. (1970). Measures of multivariate skewness and kurtosis with applications. *Biometrika* 57: 519–30.

——— (1974). Applications of some measures of multivariate skewness and kurtosis in testing normality and robustness studies. *Sankhya, Series B* 36: 115–28.

Mardia, K. V., Kent, J. T., and Bibby, J. M. (1979). *Multivariate Analysis*. London, Academic Press.

Martins, E. P., and Hansen, T. F. (1997). Phylogenies and the comparative method: a general approach to incorporating phylogenetic information into the anlaysis of interspecific data. *American Naturalist* 149: 646–67.

Mayo, D. G. (1996). *Error and the Growth of Experimental Knowledge*. Chicago University Press.

McDonald, R. P. (1994). The bilevel reticular action model for path analysis with latent variables. *Sociological Methods and Research* 22: 399–413.

Meziane, D. (1998). Étude de la variation interspécifique de la vitesse spécifique de croissance et modélisation de l'effet des attributs morphologiques, physiologiques et d'allocation de biomasse, PhD thesis. Université de Sherbrooke.

Mulaik, S. A. (1986). Toward a synthesis of deterministic and probabilistic formulations of causal relations by the functional relation concept. *Philosophy of Science* 53: 313–32.

Muthén, B. O. (1990). *Mean and Covariance Structure Analysis of Hierarchical Data*, Statistical Series paper no. 62. Los Angeles, University of California.

——— (1994). Multilevel covariance structure analysis. *Sociological Methods and Research* 22: 376–98.

——— (1997). Latent variable modeling of longitudinal and multilevel data, in Raftery, A. E. (ed.), *Sociological Methodology 1997*: 453–81. Washington, DC, American Sociological Association.

Muthén, B. O., and Satorra, A. (1995). Complex sample data in structural equation modeling, in Marsden, P. V. (ed.), *Sociological Methodology*: 267–316. Washington, DC, American Sociological Association.

Niles, H. E. (1922). Correlation, causation and Wright's theory of 'path coefficients'. *Genetics* 7: 258–73.

Norton, B. J. (1975). Biology and philosophy: the methodological foundations of biometry. *Journal of the History of Biology* 8: 85–93.

Passmore, J. (1966). *A Hundred Years of Philosophy*. Harmondsworth, Penguin Books.

Pearl, J. (1988). *Probabilistic Reasoning in Intelligent Systems: Networks of Plausible Inference*. San Francisco, Morgan Kaufmann.

——— (1997). The new challenge: from a century of statistics to an age of causation. *Computing Science and Statistics* 29: 415–23.

——— (2000). *Causality*. Cambridge University Press.

Pearl, J., and Dechter, R. (1996). Identifying independencies in causal graphs with feedback, in Horvitz, E., and Jensen, F. V. (eds.), *Proceedings of the Twelfth Conference on Uncertainty in Artificial Intelligence*: 240–6. San Francisco, Morgan Kaufmann.

Pearson, E. S., and Kendall, M. G. (1970). *Studies in the History of Statistics and Probability.* London, Griffin.

Pearson, K. (1892). *The Grammar of Science.* London, Adam & Charles Black.

(1911). *The Grammar of Science,* 3rd edn. London, Adam & Charles Black.

Peters, R. H. (1991). *A Critique for Ecology.* Cambridge University Press.

Pollack, J. L. (1986). *Contemporary Theories of Knowledge.* Totowa, NJ, Rowman & Littlefield.

Popper, K. (1980). *The Logic of Scientific Discovery.* London, Hutchinson.

Press, W. H., Flannery, B. P., Teukolsky, S. A., and Vetterling, W. T. (1986). *Numerical Recipes: The Art of Scientific Computing.* Cambridge University Press.

Provine, W. B. (1986). *Sewall Wright and Evolutionary Biology.* University of Chicago Press.

Pugesek, B. H., and Tomer, A. (1996). The Bumpus house sparrow data: a reanalysis using structural equation models. *Evolutionary Ecology* 10: 387–404.

Rao, M. M. (1984). *Probability Theory with Applications.* Orlando, Academic Press.

Rapport, S., and Wright, T. (1963). *Science: Method and Meaning.* New York University Press.

Richardson, T. (1996a). A discovery algorithm for directed cyclic graphs, in Horvitz, E., and Jensen, F. V. (eds.), *Proceedings of the Twelfth Conference on Uncertainty in Artificial Intelligence*: 454–61. San Francisco, Morgan Kaufmann.

(1996b). Models of feedback: interpretation and discovery. PhD thesis, Pittsburgh, Carnegie Mellon University.

Rigdon, E. E. (1995). A necessary and sufficient identification rule for structural models estimated in practice. *Multivariate Behavioral Research* 30: 359–83.

Rosseel, Y. (2012). lavaan: an R package for structural equation modeling. *Journal of Statistical Software* 48: 1–36.

Rubin, D. B. (1996). Multiple imputation after 18+ years. *Journal of the American Statistical Association* 57: 473–89.

Santos, J. C., and Cannetella, D. C. (2011). Phenotypic integration emerges from aposematism and scale in poison frogs. *Proceedings of the National Association of Science* 108: 6175–80.

Satorra, A., and Bentler, P. M. (1988). Scaling corrections for chi-square statistics in covariance structure analysis, in *Proceedings of the Business and Economic Statistics Section: Papers Presented at the Annual Meeting of the American Statistical Association*: 308–13. Alexandria, VA, American Statistical Association.

Schafer, J. L. (1997). *Analysis of Incomplete Multivariate Data.* London, Chapman & Hall.

Scott, A. J., and Holt, D. (1982). The effect of two-stage sampling on ordinary least squares methods. *Journal of the American Statistical Association* 77: 848–54.

Shipley, B. (1995). Structured interspecific determinants of specific leaf area in 34 species of herbaceous angiosperms. *Functional Ecology* 9: 312–19.

(1997). Exploratory path analysis with applications in ecology and evolution. *American Naturalist* 149: 1113–38.

(1999). Exploring hypothesis space: examples from organismal biology, in Glymour, C., and Cooper, G. F. (eds.), *Computation, Causation, and Discovery*: 441–52. Menlo Park, CA, AAAI Press.

(2000). A new inferential test for path models based on directed acyclic graphs. *Structural Equation Modeling* 7: 206–18.

(2009). Confirmatory path analysis in a generalized multilevel context. *Ecology* 90: 363–8.

Shipley, B., and Hunt, R. (1996). Regression smoothers for estimating parameters of growth analyses. *Annals of Botany* 76: 569–76.

Shipley, B., and Lechowicz, M. J. (2000). The functional coordination of leaf morphology and gas exchange in 40 wetland plant species. *Ecoscience* 7: 183–94.

Shipley, B., Lechowicz, M. J., Wright, I. J., and Reich, P. B. (2006). Fundamental trade-offs generating the worldwide leaf economics spectrum. *Ecology* 87: 535–41.

Shipley, B., and Peters, R. H. (1990). A test of the Tilman model of plant strategies: relative growth rate and biomass partitioning. *American Naturalist* 136: 139–53.

Shirahata, S. (1980). Rank tests of partial correlation. *Bulletin of Mathematical Statistics* 19: 9–18.

Simon, H. (1977). *Models of Discovery*. Dordrecht, D. Reidel.

Sokal, R. R., and Rohlf, F. J. (1981). *Biometry*. New York, Freeman.

Spearman, C. (1904). General intelligence objectively determined and measured. *American Journal of Psychology* 15: 201–93.

Spirtes, P. (1995). Directed cyclic graphical representation of feedback models, in Besnard, P., and Hanks, S. (eds.), *Proceedings of the Eleventh Conference on Uncertainty in Artificial Intelligence*: 491–8. San Francisco, Morgan Kaufmann.

Spirtes, P., Glymour, C., and Scheines, R. (1990). Causality from probability, in McGee, G. (ed.), *Evolving Knowledge in Natural Science and Artificial Intelligence*: 181–99. London, Pitman. (1993). *Causation, Prediction, and Search*. New York, Springer.

Spirtes, P., Richardson, T., Meek, C., and Scheines, R. (1998). Using path diagrams as a structural equation modeling tool. *Sociological Methods and Research* 27: 182–225.

Steiger, J. H. (1989). *EzPATH: A Supplementary Manual for SYSTAT and SYGRAPH*. Evanston, IL, SYSTAT Inc. (1990). Structural model evaluation and modification: an interval estimation approach. *Multivariate Behavioral Research* 25: 173–80.

Tanaka, J. S. (1993). Multifaceted conceptions of fit in structural equation models, in Bollen, K. A., and Long, J. S. (eds.), *Testing Structural Equation Models*: 10–39. Newbury Park, CA, Sage.

Van Buuren, S., and Groothuis-Oudshoorn, K. (2011). Multivariate imputation by chained equations. *Journal of Statistical Software* 45: 1–67.

Van Hulst, R. (1979). On the dynamics of vegetation: Markov chains as models of succession. *Vegetatio* 40: 3–14.

Verma, T., and Pearl, J. (1988). Causal networks: semantics and expressiveness, in Shachter, R., Levitt, T., Kanal, L. N., and Lemmer, J. F. (eds.), *Proceedings of the Fourth Conference on Uncertainty in Artificial Intelligence*: 352–9. New York, Elsevier Science. (1990). Equivalence and synthesis of causal models, in Bonissone, P. P., Henrion, M., Kanal, L. N., and Lemmer, J. F. (eds.), *Proceedings of the Sixth Conference on Uncertainty in Artificial Intelligence*: 255–68. New York, Elsevier Science.

Von Hardenberg, A., and Gonzalez-Voyer, A. (2012). Disentangling evolutionary cause–effect relationships with phylogenetic confirmatory path analysis. *Evolution* 67: 378–87.

Wahba, G. (1991). *Spline Models for Observational Data*. Philadelphia, SIAM Press.

Wishart, J. (1928). Sampling errors in the theory of two factors. *British Journal of Psychology* 19: 180–7.

Wright, I. J., Reich, P. B., Westoby, M., Ackerly, D. D., Baruch, Z., Bongers, F., Cavender-Bares, J., Chapin, T., Cornelissen, J. H. C., Diemer, M., Flexas, J., Garnier, E., Groom, P. K., Gulias, J., Hikosaka, K., Lamont, B. B., Lee, T., Lee, W., Lusk, C., Midgley, J. J., Navas, M.-L., Niinemets, Ü., Oleksyn, J., Osada, N., Poorter, H., Poot, P., Prior, L., Pyankov, V. I., Roumet, C., Thomas, S. C., Tjoelker, M. G., Veneklaas, E. J., and Villar, R. (2004). The worldwide leaf economics spectrum. *Nature* 428: 821–7.

Wright, S. (1918). On the nature of size factors. *Genetics* 3: 367–74.

(1920). The relative importance of heredity and environment in determining the piebald pattern of guinea pigs. *Proceedings of the National Academy of Science* 6: 320–32.

(1921). Correlation and causation. *Journal of Agricultural Research* 10: 557–85.

(1925). *Corn and Hog Correlations*, USDA Bulletin no. 1300. Washington, DC, US Department of Agriculture.

(1984). Diverse uses of path analysis, in Chakravarti, A., *Human Population Genetics*: 1–34. New York, Van Nostrand Reinhold.

Index

Printed in the United States
By Bookmasters